CW00386192

Robot Modeling and Control

Mark W. Spong
Seth Hutchinson
M. Vidyasagar

John Wiley & Sons, Inc.

SENIOR ACQUISITIONS EDITOR	Catherine Fields Shultz
SENIOR EDITORIAL ASSISTANT	Maureen Clendenny
EDITORIAL ASSISTANT	Dana Kellogg
SENIOR PRODUCTION EDITOR	Ken Santor
MARKETING MANAGER	Phyllis Diaz Cerys
COVER DESIGNER	Hope Miller

This book was set in LaTeX by the authors and printed and bound by Courier Digital Solutions.
The cover was printed by Courier Digital Solutions.

This book is printed on acid free paper. ∞

ISBN-10 0-471-64990-2
ISBN-13 978-0-471-64990-8

Printed in the United States of America

30 29 28 27 26 25 24 23 22

To the women in our lives

Lila - MWS
Cynthia - SH
Shakunthala and Aparna - MV

Preface

The field of robotics has changed in numerous and exciting ways since the early 1980's when robot manipulators were touted as the ultimate solution to automated manufacturing. Early predictions were that entire factories of the future would require few, if any, human operators. Some predicted that even electric lighting would be unnecessary as robots would "happily" carry out their work in total darkness. These predictions seem naive today but it is nevertheless interesting to examine some of the reasons why they failed to materialize. The first reason can be stated very simply: robotics is difficult or, somewhat equivalently, humans are very good at what they do. Automated manufacturing is not simply a matter of removing a human worker from an assembly line and installing a robot. Rather, it involves complex systems integration problems. Often, the entire workcell must be redesigned, beginning with an analysis of the assembly process itself and leading to part and fixture redesign, workcell layout, sensor development, control system design, software verification, and a host of other interconnected issues. The result is that any savings in labor costs often did not outweigh the development costs, except for relatively simple tasks like spot welding, spray painting, and palletizing.

As a result, robotics fell out of favor in the late 1980's. We are now witnessing a resurgence of interest in robotics, not only in manufacturing, but in other areas such as medical robotics, search and rescue, entertainment, and service robotics. Recent years have seen robots exploring the surface of Mars, locating sunken ships, searching out land mines, and finding victims in collapsed buildings. Robotics is now seen as part of the larger field of mechatronics, which is defined as the synergistic integration of mechanics, electronics, controls, and computer science. The robot is the ultimate mechatronic system.

The present text began as a second edition of M.W. Spong and M. Vidyasagar, *Robot Dynamics and Control*, John Wiley & Sons, Inc., 1989. Early on it became apparent, due to both the length of elapsed time since the first edition and also to the evolution of the field in the intervening years, that the end product would be an entirely new book. We have retained the philosophy and a good portion of the material from that earlier book but have added much that is new. The material on motion planning, computer vision, and visual servo control is entirely new. All of the control chapters have been rewritten to reflect the maturation of the field of robot control that took place during an intensive period of research at the end of the 1980's and early 1990's. The fundamentals of kinematics and dynamics remain largely the same but has been expanded and improved from a pedagogical standpoint.

Organization of the Text

This text is organized into twelve chapters. The first six chapters are at a relatively elementary level whereas the last six chapters are more advanced. The chapters can be conceptually divided into three groups. After an introductory chapter, Chapters 2 through 5 deal with issues related to the geometry of robot motion. Chapters 6 through 10 deal with dynamics and control. Finally, Chapters 11 and 12 discuss computer vision and how it can be incorporated directly into the robot control loop. A more specific description of the chapters is as follows.

Chapter 1 is an introduction to the terminology and history of robotics and discusses the most common robot design and applications.

Chapter 2 presents the mathematics of rigid motions; rotations, translations, and homogeneous transformations.

Chapter 3 presents solutions to the forward kinematics problem using the Denavit-Hartenberg representation and to the inverse kinematics problem using the geometric approach, which is especially suited for manipulators with spherical wrists.

Chapter 4 is a lengthy chapter on velocity kinematics and the manipulator Jacobian. The geometric Jacobian is derived in the so-called cross product form. We also introduce the so-called analytical Jacobian for later use in task space control. Chapter 4 also discusses the important notion of manipulability.

Chapter 5 is an introduction to the problems of motion planning and trajectory generation. Several of the most popular methods for motion planning and obstacle avoidance are presented, including the method of artificial potential fields, randomized algorithms, and probabilistic roadmap methods. The problem of trajectory generation is presented as essentially a problem of polynomial spline interpolation. Trajectory generation based on cubic and quintic polynomials as well as trapezoidal velocity trajectories are derived for interpolation in joint space.

Chapter 6 is an introduction to independent joint control. Linear control based on PD, PID, and state space methods is presented for the tracking and disturbance rejection problem for linear actuator and drive-train dynamics. The concept of feedforward control is introduced for tracking time-varying reference trajectories.

Chapter 7 is a detailed account of robot dynamics. The Euler-Lagrange equations are derived from first principles and their structural properties are discussed in detail. The recursive Newton-Euler formulation of robot dynamics is also presented.

Chapter 8 discusses multivariable control. This chapter summarizes much of the research in robot control that took place in the late 1980's and early 1990's. Simple derivations of the most common robust and adaptive control algorithms are presented that prepare the reader for the extensive literature in robot control.

Chapter 9 treats the force control problem. Both impedance control and hybrid control are discussed. We also present the lesser known hybrid impedance control method which allows one to control impedance and regulate motion and force at the same time. To our knowledge this is the first textbook that discusses the hybrid impedance approach to robot force control.

Chapter 10 is an introduction to geometric nonlinear control. This chapter is considerably more advanced than the other chapters and can be reserved for graduate level courses in nonlinear control and robotics. However, the material is presented in a readable style that should be accessible for advanced undergraduates. We derive and prove the necessary and sufficient conditions for local feedback linearization of single-input/single-output systems which we then apply to the flexible joint control problem. We also briefly discuss Chow's Theorem for the problem of control of systems subject to nonholonomic constraints.

Chapter 11 is an introduction to computer vision. We present those aspects of vision that are most useful for robotics applications, such as threshholding, image segmentation, and camera calibration.

Chapter 12 discusses the visual servo control problem, which is the problem of controlling robots using feedback from cameras mounted either on the robot or in the workspace.

This text is suitable for several quarter or semester long courses in robotics, either as a sequence or as stand-alone courses. The first six chapters can be used for a junior/senior level introduction to robotics for students with a minimal background in linear control systems. One of the key changes we made in this text from the earlier text was to place the independent joint control chapter before the dynamics chapter. The independent joint control problem largely involves the control of actuator and drive train dynamics; hence most of the subject can be taught without prior knowledge of Euler-Lagrange dynamics.

Below we outline two possible courses that can be taught from this book:

Course 1: Introduction to Robotics
Level: Junior/Senior undergraduate
For a one quarter course (10 weeks):

> **Chapter 1:** Introduction
> **Chapter 2:** Rigid Motions and Homogeneous Transformations
> **Chapter 3:** Forward and Inverse Kinematics
> **Chapter 4:** Jacobians

For a one semester course (16 weeks) add:

> **Chapter 5:** Motion Planning and Trajectory Generation
> **Chapter 6:** Independent Joint Control
> **Chapter 11:** Computer Vision

Course 2: Robot Dynamics and Control
Level: Senior undergraduate/graduate
For a one quarter course (10 weeks):

> **Chapters 1–5:** Rapid Review of Kinematics (selected sections)
> **Chapter 6:** Independent Joint Control
> **Chapter 7:** Dynamics
> **Chapter 8:** Multivariable Control
> **Chapter 9:** Force Control

For a one semester course (16 weeks) add:

> **Chapter 10:** Geometric Nonlinear Control
> **Chapter 11:** Computer Vision
> **Chapter 12:** Visual Servo Control

We have taught both of these one semester courses at the University of Illinois. The students who take the first course typically come from Computer Science, Electrical and Computer Engineering, General Engineering and Mechanical Engineering Departments. For this reason, we have tried to make these chapters accessible to a wide variety of engineering students. The second course has typically been taken by upper level students pursuing graduate studies in robotics or control, and therefore these chapters are written at a more advanced level.

Acknowledgements

We would like to offer a special thanks to Peter Hokayem and Daniel Herring who did an outstanding job of producing most of the figures in the book. In addition, Benjamin Sapp provided most of the figures for Chapter 11 and Nick Gans provided many figures for Chapter 12. We would like to thank Francois Chaumette for discussions regarding the formulation of the interaction matrix in Chapter 12 and to Martin Corless for discussion on the robust control problem in Chapter 8.

We are indebted to several reviewers for their very detailed and thoughtful reviews, especially Brad Bishop, Kevin Lynch, Matt Mason, Eric Westervelt and Ning Xi.

We would like to thank our students, Nikhil Chopra, Chris Graesser, James Davidson, Nick Gans, Jon Holm, Silvia Mastellone, Adrian Lee, Oscar Martinez, Erick Rodriguez and Kunal Srivastava who suffered through many early drafts of the manuscript and provided feedback and suggestions for improvement as well as finding numerous typos.

Mark W. Spong
Seth Hutchinson
M. Vidyasagar

Contents

Chapter 1

INTRODUCTION

Robotics is a relatively young field of modern technology that crosses traditional engineering boundaries. Understanding the complexity of robots and their application requires knowledge of electrical engineering, mechanical engineering, systems and industrial engineering, computer science, economics, and mathematics. New disciplines of engineering, such as manufacturing engineering, applications engineering, and knowledge engineering have emerged to deal with the complexity of the field of robotics and factory automation.

This book is concerned with fundamentals of robotics, including **kinematics**, **dynamics**, **motion planning**, **computer vision**, and **control**. Our goal is to provide an introduction to the most important concepts in these subjects as applied to industrial robot manipulators and other mechanical systems.

The term **robot** was first introduced by the Czech playwright Karel Capek in his 1920 play *Rossum's Universal Robots*, the word *robota* being the Czech word for work. Since then the term has been applied to a great variety of mechanical devices, such as teleoperators, underwater vehicles, autonomous land rovers, etc. Virtually anything that operates with some degree of autonomy, usually under computer control, has at some point been called a robot. In this text the term robot will mean a computer controlled industrial manipulator of the type shown in Figure 1.1.

This type of robot is essentially a mechanical arm operating under computer control. Such devices, though far from the robots of science fiction, are nevertheless extremely complex electromechanical systems whose analytical description requires advanced methods, presenting many challenging and interesting research problems. An official definition of such a robot comes from the **Robot Institute of America** (RIA):

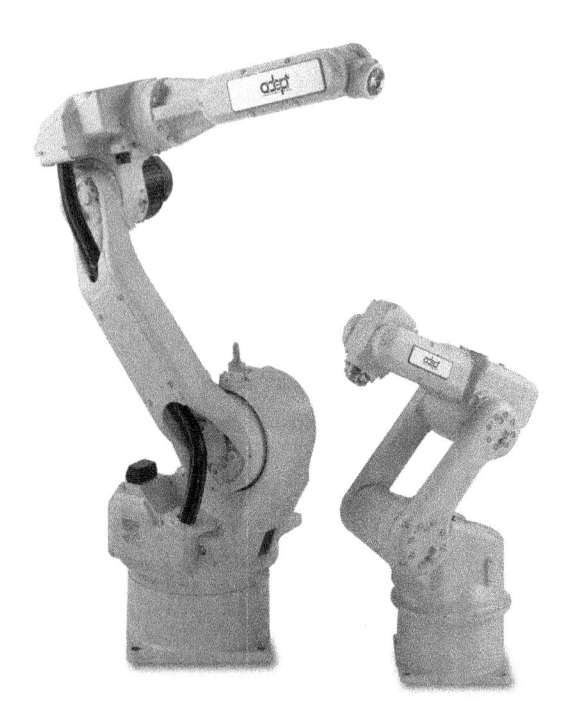

Figure 1.1: Examples of typical industrial manipulators, the AdeptSix 600 robot (left) and the AdeptSix 300 robot (right). Both are six-axis, high performance robots designed for materials handling or assembly applications. (Photo courtesy of Adept Technology, Inc.)

Definition: *A robot is a reprogrammable, multifunctional manipulator designed to move material, parts, tools, or specialized devices through variable programmed motions for the performance of a variety of tasks.*

The key element in the above definition is the reprogrammability, which gives a robot its utility and adaptability. The so-called robotics revolution is, in fact, part of the larger computer revolution.

Even this restricted definition of a robot has several features that make it attractive in an industrial environment. Among the advantages often cited in favor of the introduction of robots are decreased labor costs, increased precision and productivity, increased flexibility compared with specialized machines, and more humane working conditions as dull, repetitive, or hazardous jobs are performed by robots.

The robot, as we have defined it, was born out of the marriage of two earlier technologies: **teleoperators** and **numerically controlled milling machines**. Teleoperators, or master-slave devices, were developed during the second world war to handle radioactive materials. Computer numerical control (CNC) was developed because of the high precision required in the machining of certain items, such as components of high performance aircraft. The first robots essentially combined the mechanical linkages of the teleoperator with the autonomy and programmability of CNC machines.

The first successful applications of robot manipulators generally involved some sort of material transfer, such as injection molding or stamping, in which the robot merely attends a press to unload and either transfer or stack the finished parts. These first robots could be programmed to execute a sequence of movements, such as moving to a location A, closing a gripper, moving to a location B, etc., but had no external sensor capability. More complex applications, such as welding, grinding, deburring, and assembly require not only more complex motion but also some form of external sensing such as vision, tactile, or force sensing, due to the increased interaction of the robot with its environment.

Worldwide there are currently over 800,000 industrial robots in operation, mostly in Japan, the European Union and North America (see Figure 1.2). After a period of stagnation in the late 1980's, the sale of industrial robots began to rise in the 1990's and sales growth is likely to remain strong for the remainder of this decade.

It should be pointed out that the important applications of robots are by no means limited to those industrial jobs where the robot is directly replacing a human worker. In fact, there are over 600,000 household robots currently in use primarily as vacuum cleaning and lawn mowing robots. There are many other applications of robotics in areas where the use of humans is impractical or undesirable. Among these are undersea and planetary exploration, satellite retrieval and repair, the defusing of explosive devices, and work in radioactive environments. Finally, prostheses, such as artificial limbs, are themselves robotic devices requiring methods of analysis and design similar to those of industrial manipulators.

1.1 MATHEMATICAL MODELING OF ROBOTS

In this text we will be primarily concerned with developing and analyzing mathematical models for robots. In particular, we will develop methods to represent basic geometric aspects of robotic manipulation, dynamic aspects of manipulation, and the various sensors available in modern robotic systems. Equipped with these mathematical models, we will develop methods for planning and controlling robot motions to perform specified tasks. We begin here by describing some of the basic notation and terminology

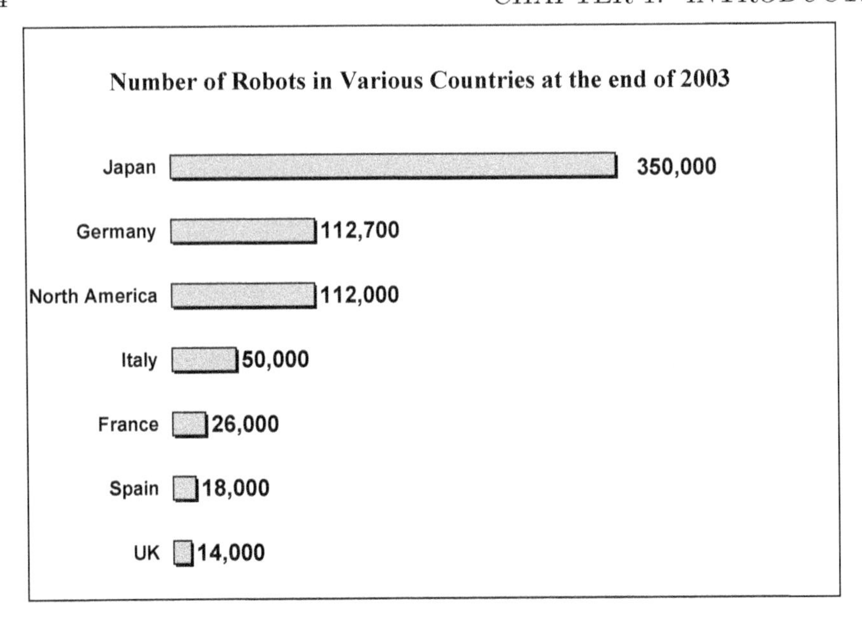

Figure 1.2: Number of robots in use at the end of 2003. Japan has the largest number of industrial robots, followed by the European Union and North America. Source: UNECE - United Nations Economic Commission for Europe, October, 2004.

that we will use in later chapters to develop mathematical models for robot manipulators.

1.1.1 Symbolic Representation of Robots

Robot manipulators are composed of **links** connected by **joints** to form a **kinematic chain**. Joints are typically rotary (revolute) or linear (prismatic). A **revolute** joint is like a hinge and allows relative rotation between two links. A **prismatic** joint allows a linear relative motion between two links. We denote revolute joints by R and prismatic joints by P, and draw them as shown in Figure 1.3. For example, a three-link arm with three revolute joints will be referred to as an RRR arm.

Each joint represents the interconnection between two links. We denote the axis of rotation of a revolute joint, or the axis along which a prismatic joint translates by z_i if the joint is the interconnection of links i and $i+1$. The **joint variables**, denoted by θ for a revolute joint and d for the prismatic joint, represent the relative displacement between adjacent links. We will make this precise in Chapter 3.

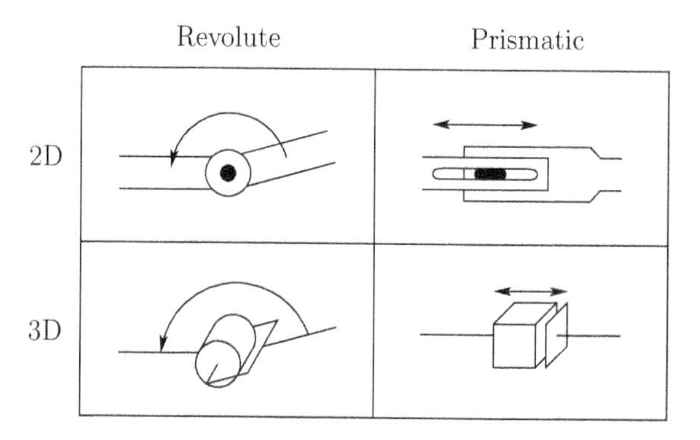

Figure 1.3: Symbolic representation of robot joints. Each joint allows a single degree of freedom of motion between adjacent links of the manipulator. The revolute joint (shown in 2D and 3D on the left) produces a relative rotation between adjacent links. The prismatic joint (shown in 2D and 3D on the right) produces a linear or telescoping motion between adjacent links.

1.1.2 The Configuration Space

A **configuration** of a manipulator is a complete specification of the location of every point on the manipulator. The set of all configurations is called the **configuration space**. In our case, if we know the values for the joint variables (i.e., the joint angle for revolute joints, or the joint offset for prismatic joints), then it is straightforward to infer the position of any point on the manipulator, since the individual links of the manipulator are assumed to be rigid and the base of the manipulator is assumed to be fixed. Therefore, in this text, we will represent a configuration by a set of values for the joint variables. We will denote this vector of values by q, and say that the robot is in configuration q when the joint variables take on the values $q_1, \ldots, q_n$, with $q_i = \theta_i$ for a revolute joint and $q_i = d_i$ for a prismatic joint.

An object is said to have n **degrees of freedom** (DOF) if its configuration can be minimally specified by n parameters. Thus, the number of DOF is equal to the dimension of the configuration space. For a robot manipulator, the number of joints determines the number of DOF. A rigid object in three-dimensional space has six DOF: three for **positioning** and three for **orientation**. Therefore, a manipulator should typically possess at least six independent DOF. With fewer than six DOF the arm cannot reach every point in its work space with arbitrary orientation. Certain applications such as reaching around or behind obstacles may require more than six DOF. A manipulator having more than six DOF is referred to as a **kinematically redundant** manipulator.

1.1.3 The State Space

A configuration provides an instantaneous description of the geometry of a manipulator, but says nothing about its dynamic response. In contrast, the **state** of the manipulator is a set of variables that, together with a description of the manipulator's dynamics and future inputs, is sufficient to determine the future time response of the manipulator. The **state space** is the set of all possible states. In the case of a manipulator arm, the dynamics are Newtonian, and can be specified by generalizing the familiar equation $F = ma$. Thus, a state of the manipulator can be specified by giving the values for the joint variables q and for joint velocities $\dot{q}$ (acceleration is related to the derivative of joint velocities).

1.1.4 The Workspace

The **workspace** of a manipulator is the total volume swept out by the end effector as the manipulator executes all possible motions. The workspace is constrained by the geometry of the manipulator as well as mechanical constraints on the joints. For example, a revolute joint may be limited to less than a full 360° of motion. The workspace is often broken down into a **reachable workspace** and a **dexterous workspace**. The reachable workspace is the entire set of points reachable by the manipulator, whereas the dexterous workspace consists of those points that the manipulator can reach with an arbitrary orientation of the end effector. Obviously the dexterous workspace is a subset of the reachable workspace. The workspaces of several robots are shown later in this chapter.

1.2 ROBOTS AS MECHANICAL DEVICES

There are a number of physical aspects of robotic manipulators that we will not necessarily consider when developing our mathematical models. These include mechanical aspects (e.g., how are the joints actually implemented), accuracy and repeatability, and the tooling attached at the end effector. In this section, we briefly describe some of these.

1.2.1 Classification of Robotic Manipulators

Robot manipulators can be classified by several criteria, such as their **power source**, or the way in which the joints are actuated; their **geometry**, or kinematic structure; their **method of control**; and their intended **application area**. Such classification is useful primarily in order to determine which robot is right for a given task. For example, an hydraulic robot would not be suitable for food handling or clean room applications whereas a SCARA robot would not be suitable for automobile spray painting. We explain this in more detail below.

Power Source

Most robots are either electrically, hydraulically, or pneumatically powered. Hydraulic actuators are unrivaled in their speed of response and torque producing capability. Therefore hydraulic robots are used primarily for lifting heavy loads. The drawbacks of hydraulic robots are that they tend to leak hydraulic fluid, require much more peripheral equipment (such as pumps, which require more maintenance), and they are noisy. Robots driven by DC or AC motors are increasingly popular since they are cheaper, cleaner and quieter. Pneumatic robots are inexpensive and simple but cannot be controlled precisely. As a result, pneumatic robots are limited in their range of applications and popularity.

Method of Control

Robots are classified by control method into **servo** and **nonservo** robots. The earliest robots were nonservo robots. These robots are essentially **open-loop** devices whose movements are limited to predetermined mechanical stops, and they are useful primarily for materials transfer. In fact, according to the definition given above, fixed stop robots hardly qualify as robots. Servo robots use **closed-loop** computer control to determine their motion and are thus capable of being truly multifunctional, reprogrammable devices.

Servo controlled robots are further classified according to the method that the controller uses to guide the end effector. The simplest type of robot in this class is the **point-to-point** robot. A point-to-point robot can be taught a discrete set of points but there is no control of the path of the end effector in between taught points. Such robots are usually taught a series of points with a **teach pendant**. The points are then stored and played back. Point-to-point robots are limited in their range of applications. With **continuous path** robots, on the other hand, the entire path of the end effector can be controlled. For example, the robot end effector can be taught to follow a straight line between two points or even to follow a contour such as a welding seam. In addition, the velocity and/or acceleration of the end effector can often be controlled. These are the most advanced robots and require the most sophisticated computer controllers and software development.

Application Area

Robot manipulators are often classified by application area into **assembly** and **nonassembly robots**. Assembly robots tend to be small, electrically driven and either revolute or SCARA (described below) in design. Typical nonassembly application areas to date have been in welding, spray painting, material handling, and machine loading and unloading.

One of the primary differences between assembly and nonassembly applications is the increased level of precision required in assembly due to significant interaction with objects in the workspace. For example, an assembly task may require part insertion (the so-called **peg-in-hole problem**) or gear meshing. A slight mismatch between the parts can result in wedging and jamming, which can cause large interaction forces and failure of the task. As a result assembly tasks are difficult to accomplish without special fixtures and jigs, or without sensing and controlling the interaction forces.

Geometry

Most industrial manipulators at the present time have six or fewer DOF. These manipulators are usually classified kinematically on the basis of the first three joints of the arm, with the wrist being described separately. The majority of these manipulators fall into one of five geometric types: **articulated (RRR)**, **spherical (RRP)**, **SCARA (RRP)**, **cylindrical (RPP)**, or **Cartesian (PPP)**. We discuss each of these below in Section 1.3.

Each of these five manipulator arms is a **serial link** robot. A sixth distinct class of manipulators consists of the so-called **parallel robot**. In a parallel manipulator the links are arranged in a closed rather than open kinematic chain. Although we include a brief discussion of parallel robots in this chapter, their kinematics and dynamics are more difficult to derive than those of serial link robots and hence are usually treated only in more advanced texts.

1.2.2 Robotic Systems

A robot manipulator should be viewed as more than just a series of mechanical linkages. The mechanical arm is just one component in an overall **robotic system**, illustrated in Figure 1.4, which consists of the **arm**, **external power source**, **end-of-arm tooling**, **external and internal sensors**, **computer interface**, and **control computer**. Even the programmed software should be considered as an integral part of the overall system, since the manner in which the robot is programmed and controlled can have a major impact on its performance and subsequent range of applications.

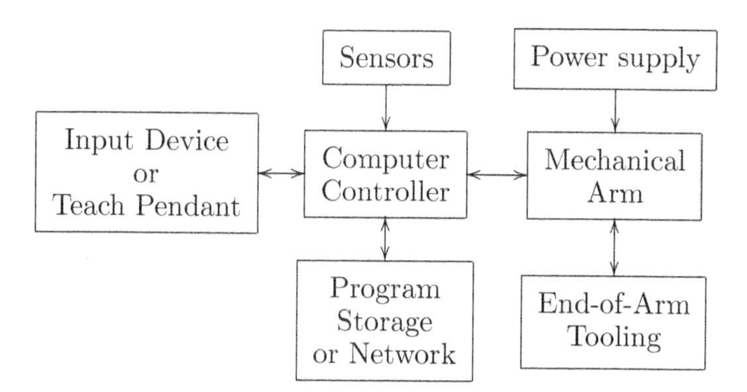

Figure 1.4: The integration of a mechanical arm, sensing, computation, user interface and tooling forms a complex robotic system. Many modern robotic systems have integrated computer vision, force/torque sensing, and advanced programming and user interface features.

1.2.3 Accuracy and Repeatability

The **accuracy** of a manipulator is a measure of how close the manipulator can come to a given point within its workspace. **Repeatability** is a measure of how close a manipulator can return to a previously taught point. The primary method of sensing positioning errors is with position encoders located at the joints, either on the shaft of the motor that actuates the joint or on the joint itself. There is typically no direct measurement of the end-effector position and orientation. One relies instead on the assumed geometry of the manipulator and its rigidity to calculate the end-effector position from the measured joint positions. Accuracy is affected therefore by computational errors, machining accuracy in the construction of the manipulator, flexibility effects such as the bending of the links under gravitational and other loads, gear backlash, and a host of other static and dynamic effects. It is primarily for this reason that robots are designed with extremely high rigidity. Without high rigidity, accuracy can only be improved by some sort of direct sensing of the end-effector position, such as with computer vision.

Once a point is taught to the manipulator, however, say with a teach pendant, the above effects are taken into account and the proper encoder values necessary to return to the given point are stored by the controlling computer. Repeatability therefore is affected primarily by the controller resolution. **Controller resolution** means the smallest increment of motion that the controller can sense. The resolution is computed as the total distance traveled divided by 2^n, where n is the number of bits of encoder accuracy. In this context, linear axes, that is, prismatic joints, typically have higher resolution than revolute joints, since the straight line distance traversed by the tip of a linear axis between two points is less than the corresponding arc length traced by the tip of a rotational link.

In addition, as we will see in later chapters, rotational axes usually result in a large amount of kinematic and dynamic coupling among the links, with a resultant accumulation of errors and a more difficult control problem. One may wonder then what the advantages of revolute joints are in manipulator design. The answer lies primarily in the increased dexterity and compactness of revolute joint designs. For example, Figure 1.5 shows that for the same range of motion, a rotational link can be made much smaller than a link with linear motion.

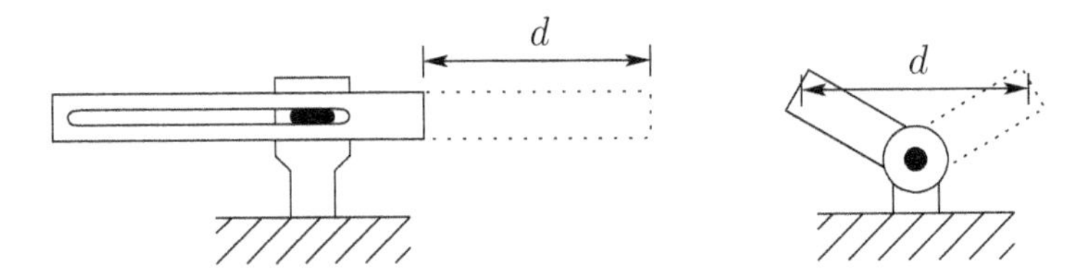

Figure 1.5: Linear vs. rotational link motion showing that a smaller revolute joint can cover the same distance d as a larger prismatic joint. The tip of a prismatic link can cover a distance equal to the length of the link. The tip of a rotational link of length a, by contrast, can cover a distance of $2a$ by rotating 180 degrees.

Thus, manipulators made from revolute joints occupy a smaller working volume than manipulators with linear axes. This increases the ability of the manipulator to work in the same space with other robots, machines, and people. At the same time revolute joint manipulators are better able to maneuver around obstacles and have a wider range of possible applications.

1.2.4 Wrists and End Effectors

The joints in the kinematic chain between the arm and end effector are referred to as the **wrist**. The wrist joints are nearly always all revolute. It is increasingly common to design manipulators with **spherical wrists**, by which we mean wrists whose three joint axes intersect at a common point, known as the **wrist center point**. Such a spherical wrist is shown in Figure 1.6.

The spherical wrist greatly simplifies kinematic analysis, effectively allowing one to decouple the position and orientation of the end effector. Typically the manipulator will possess three DOF for position, which are

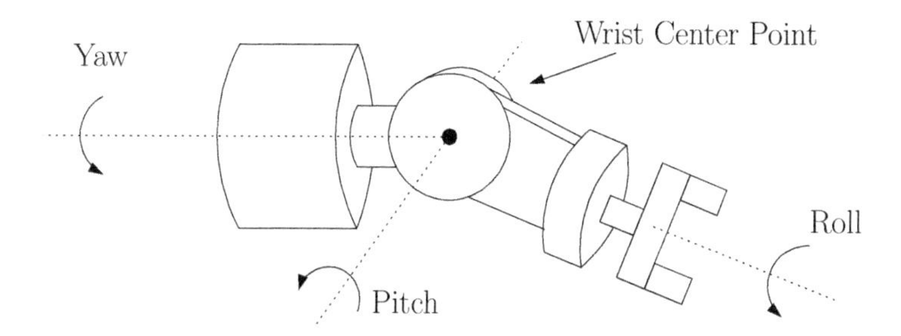

Figure 1.6: The spherical wrist. The axes of rotation of the spherical wrist are typically denoted roll, pitch, and yaw and intersect at a point called the wrist center point.

produced by three or more joints in the arm. The number of DOF for orientation will then depend on the DOF of the wrist. It is common to find wrists having one, two, or three DOF depending on the application. For example, the SCARA robot shown in Figure 1.14 has four DOF: three for the arm, and one for the wrist, which has only a rotation about the final z-axis.

The arm and wrist assemblies of a robot are used primarily for positioning the **hand**, **end effector**, and any **tool** it may carry. It is the end effector or tool that actually performs the task. The simplest type of end effector is a gripper, such as shown in Figure 1.7 which is usually capable of only two actions, **opening** and **closing**. While this is adequate for materials transfer, some parts handling, or gripping simple tools, it is not adequate for other tasks such as welding, assembly, grinding, etc.

Figure 1.7: Examples of robot grippers. Shown here from left to right are a two-fingered parallel jaw gripper, a scissor-type gripper, and a vertical gripper. (Photos courtesy of ASG-Jergen's, Cleveland Ohio.)

A great deal of research is therefore devoted to the design of special purpose end effectors as well as of tools that can be rapidly changed as the task dictates. There is also much research on the development of anthropomorphic hands such as that shown in Figure 1.8. Since we are concerned with the analysis and control of the manipulator itself and not in the particular application or end effector, we will not discuss the design of end effectors or the study of grasping and manipulation.

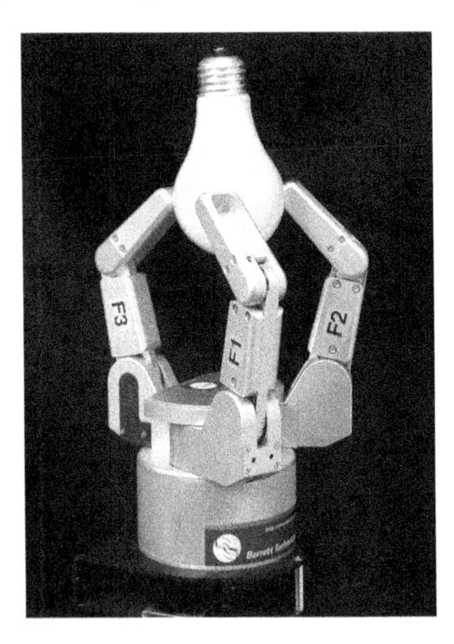

Figure 1.8: A three-fingered anthropomorphic hand developed by Barrett Technologies. Such grippers allow for more dexterity and the ability to manipulate objects of various sizes and geometries. (Photo courtesy of Barrett Technologies.)

1.3 COMMON KINEMATIC ARRANGEMENTS

There are many possible ways to construct kinematic chains using prismatic and revolute joints. However, in practice, only a few kinematic designs are used. Here we briefly describe the most typical arrangements.

1.3.1 Articulated Manipulator (RRR)

The articulated manipulator is also called a **revolute**, **elbow**, or **anthropomorphic** manipulator. The ABB IRB1400 articulated arm is shown in Figure 1.9. In the anthropomorphic design the three links are designated as the body, upper arm, and forearm, respectively, as shown in Figure 1.9. The joint axes are designated as the **waist** (z_0), **shoulder** (z_1), and **elbow** (z_2). Typically, the joint axis z_2 is parallel to z_1 and both z_1 and z_2 are

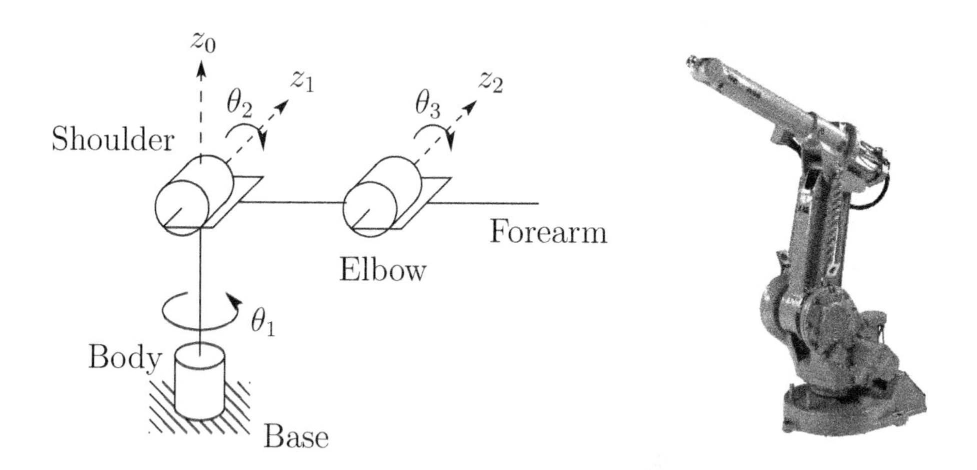

Figure 1.9: The ABB IRB1400 Robot, a six-DOF elbow manipulator (right). The symbolic representation of this manipulator (left) shows why it is referred to as an anthropomorphic robot. The links and joints are analogous to human joints and limbs. (Photo courtesy of ABB.)

perpendicular to z_0. The workspace of the revolute manipulator is shown in Figure 1.10. The revolute manipulator provides for relatively large freedom of movement in a compact space.

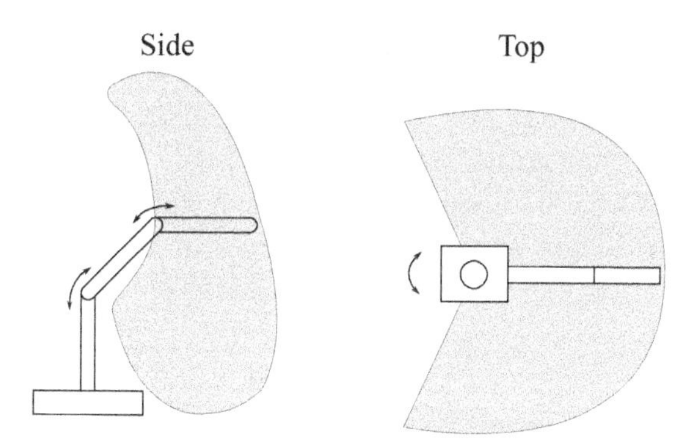

Figure 1.10: Workspace of the elbow manipulator. The elbow manipulator provides a larger workspace than other kinematic designs relative to its size.

An alternate revolute joint design is the **parallelogram linkage** such as the ABB IRB6400, shown in Figure 1.11. The parallelogram linkage is less dexterous than the elbow manipulator but has several advantages that make it an attractive and popular design. The most notable feature of the parallelogram linkage manipulator is that the actuator for joint 3 is located on link 1. Since the weight of the motor is born by link 1, links 2 and 3 can be made more lightweight and the motors themselves can be less powerful. Also, the dynamics of the parallelogram manipulator are simpler than those of the elbow manipulator making it easier to control.

Figure 1.11: The ABB IRB6400 manipulator utilizes a parallelogram linkage design. The motor that actuates the elbow joint is located on the shoulder, which reduces the weight of the upper arm. A general principle in manipulator design is to locate as much of the mass of the robot away from the distal links as possible. (Photo courtesy of ABB.)

1.3.2 Spherical Manipulator (RRP)

By replacing the third or elbow joint in the revolute manipulator by a prismatic joint, one obtains the spherical manipulator shown in Figure 1.12. The term **spherical manipulator** derives from the fact that the joint coordinates coincide with the spherical coordinates of the end effector relative to a coordinate frame located at the shoulder joint. Figure 1.12 shows the Stanford Arm, one of the most well-known spherical robots.

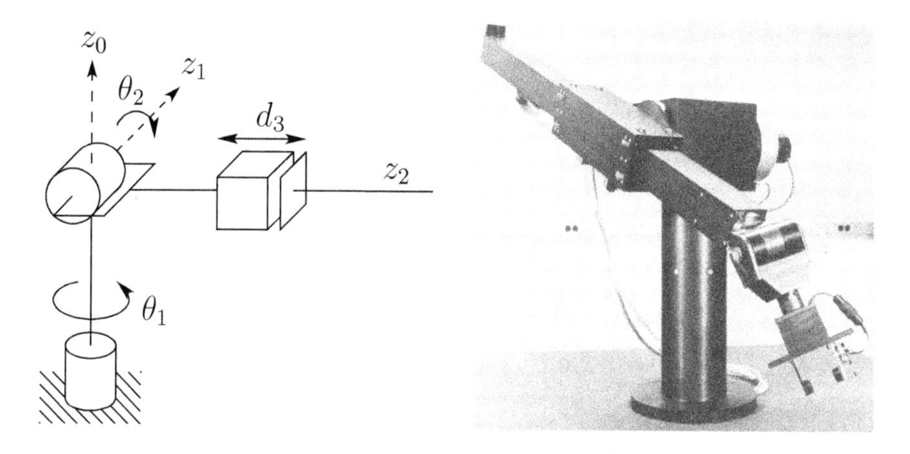

Figure 1.12: The Stanford Arm is an example of a spherical manipulator. The earliest manipulator designs were spherical robots. (Photo courtesy of the Coordinated Science Lab, University of Illinois at Urbana-Champaign.)

1.3.3 SCARA Manipulator (RRP)

The **SCARA** arm (for **S**elective **C**ompliant **A**rticulated **R**obot for **A**ssembly) shown in Figure 1.14 is a popular manipulator, which, as its name suggests, is tailored for assembly operations. Although the SCARA has an RRP structure, it is quite different from the spherical manipulator in both appearance and in its range of applications. Unlike the spherical design, which has z_0 perpendicular to z_1, and z_1 perpendicular to z_2, the SCARA has z_0, z_1, and z_2 mutually parallel. Figure 1.13 shows the symbolic representation of the SCARA arm and Figure 1.14 shows the Adept Cobra Smart600.

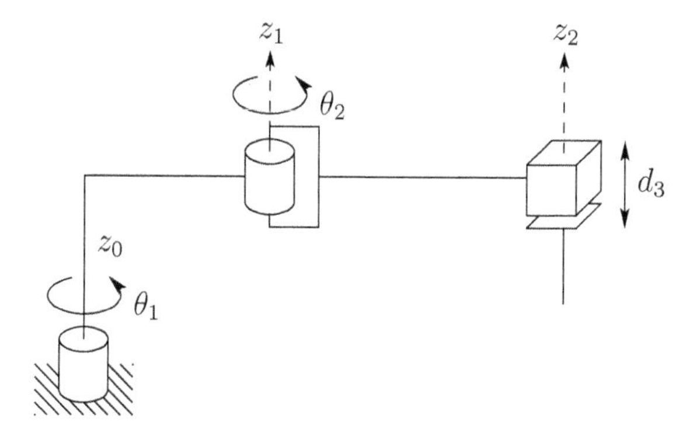

Figure 1.13: Symbolic representation of the SCARA arm.

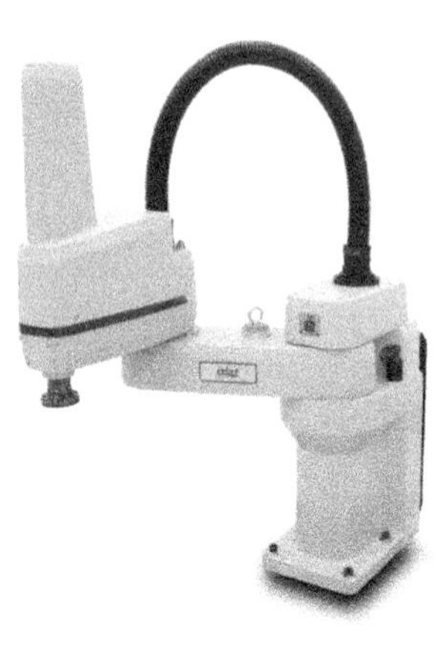

Figure 1.14: The Adept Cobra Smart600 SCARA Robot. The SCARA design is ideal for table top assembly, pick-and-place tasks, and certain types of packaging applications. (Photo Courtesy of Adept Technology, Inc.)

1.3.4 Cylindrical Manipulator (RPP)

The cylindrical manipulator is shown in Figure 1.15. The first joint is revolute and produces a rotation about the base, while the second and third joints are prismatic. As the name suggests, the joint variables are the cylindrical coordinates of the end effector with respect to the base.

1.3.5 Cartesian Manipulator (PPP)

A manipulator whose first three joints are prismatic is known as a Cartesian manipulator. The joint variables of the Cartesian manipulator are the Cartesian coordinates of the end effector with respect to the base. As might be expected, the kinematic description of this manipulator is the simplest of all manipulators. Cartesian manipulators are useful for table-top assembly applications and, as gantry robots, for transfer of material or cargo. An example of a Cartesian robot, from Epson, is shown in Figure 1.16.

The workspaces of the spherical, SCARA, cylindrical, and Cartesian geometries are shown in Figure 1.17

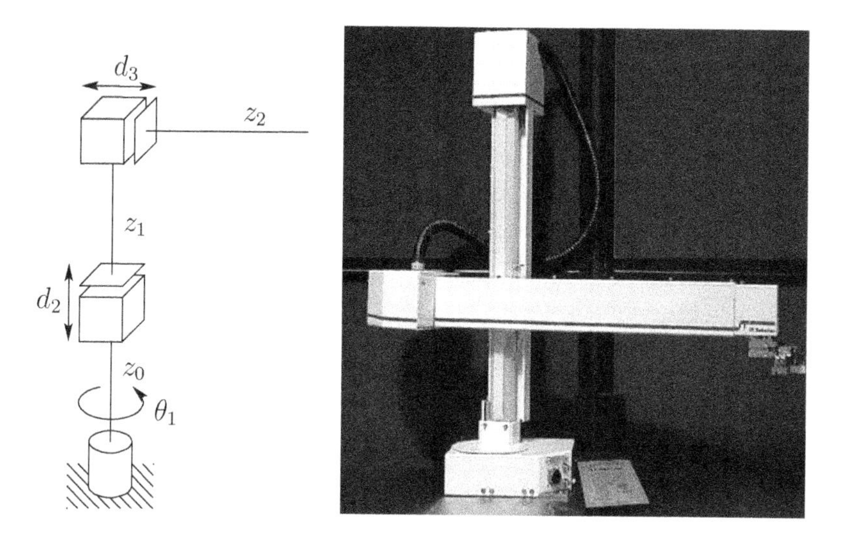

Figure 1.15: The Seiko RT3300 Robot cylindrical robot. Cylindrical robots are often used in materials transfer tasks. (Photo courtesy of Epson Robots.)

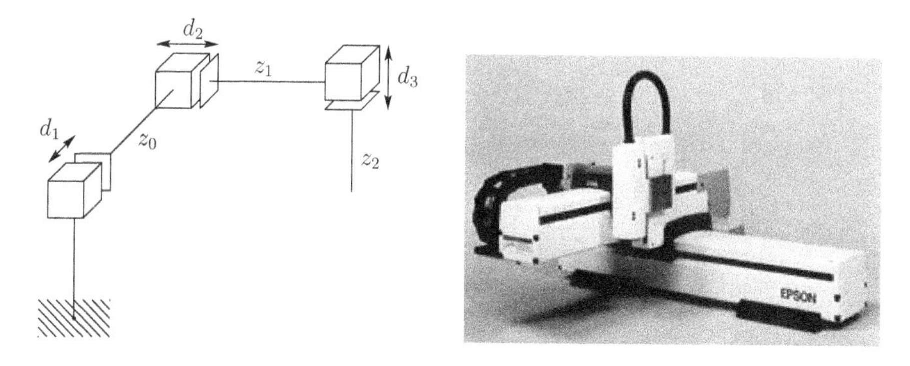

Figure 1.16: The Epson Cartesian Robot. Cartesian robot designs allow increased structural rigidity and hence higher precision. Cartesian robots are often used in pick and place operations. (Photo courtesy of Epson Robots.)

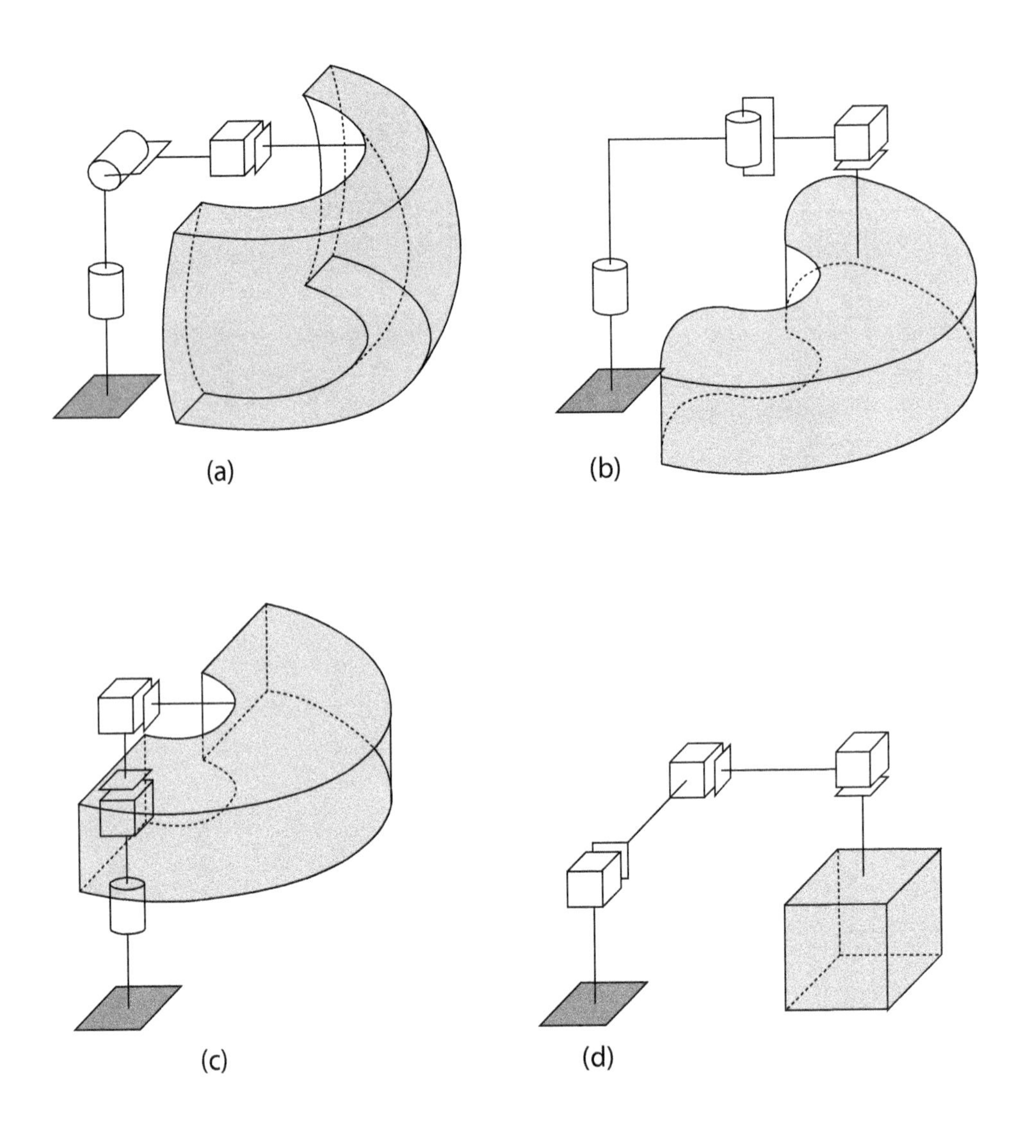

Figure 1.17: Comparison of the workspaces of the (a) spherical, (b) SCARA, (c) cylindrical, and (d) Cartesian robots. The nature of the workspace dictates the types of application for which each design can be used.

1.3.6 Parallel Manipulator

A **parallel manipulator** is one in which some subset of the links form a closed chain. More specifically, a parallel manipulator has two or more kinematic chains connecting the base to the end effector. Figure 1.18 shows the ABB IRB940 Tricept robot, which is a parallel manipulator. The closed-chain kinematics of parallel robots can result in greater structural rigidity, and hence greater accuracy, than open chain robots. The kinematic description of parallel robots is fundamentally different from that of serial link robots and therefore requires different methods of analysis.

Figure 1.18: The ABB IRB940 Tricept parallel robot. Parallel robots generally have much higher structural rigidity than serial link robots. (Photo courtesy of ABB.)

1.4 OUTLINE OF THE TEXT

A typical application involving an industrial manipulator is shown in Figure 1.19. The manipulator is shown with a grinding tool that it must use to remove a certain amount of metal from a surface. In the present text we are concerned with the following question: *What are the basic issues to be resolved and what must we learn in order to be able to program a robot to perform such tasks?* The ability to answer this question for a full six degree-of-freedom manipulator represents the goal of the present text. The answer is too complicated to be presented at this point. We can, however,

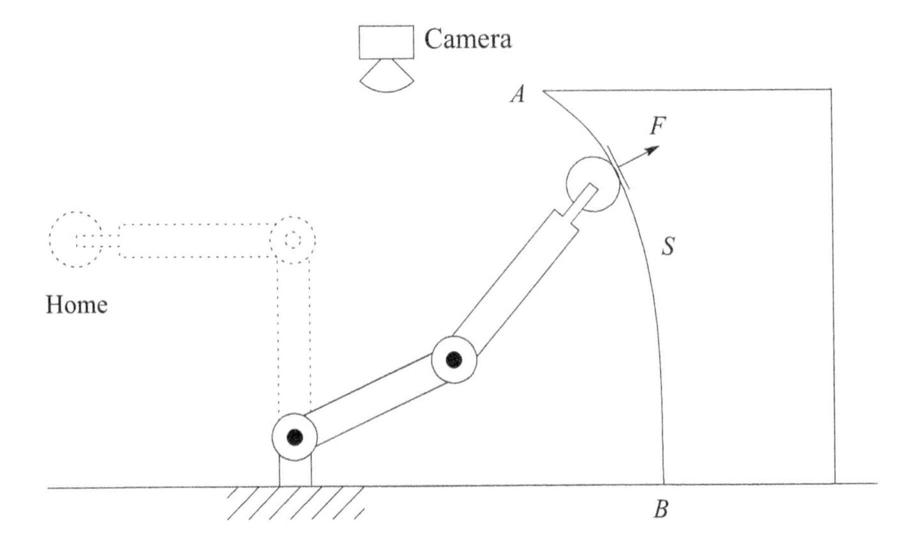

Figure 1.19: Two-link planar robot example. Each chapter of the text discusses a fundamental concept applicable to the task shown.

use the simple two-link planar mechanism to illustrate some of the major issues involved and to preview the topics covered in this text.

Suppose we wish to move the manipulator from its **home** position to position A, from which point the robot is to follow the contour of the surface S to the point B, at constant velocity, while maintaining a prescribed force F normal to the surface. In so doing the robot will cut or grind the surface according to a predetermined specification. To accomplish this and even more general tasks, we must solve a number of problems. Below we give examples of these problems, all of which will be treated in more detail in the remainder of the text.

Forward Kinematics

The first problem encountered is to describe both the position of the tool and the locations A and B (and most likely the entire surface S) with respect to a common coordinate system. In Chapter 2 we describe representations of coordinate systems and transformations among various coordinate systems.

Typically, the manipulator will be able to sense its own position in some manner using internal sensors (position encoders located at joints 1 and 2) that can measure directly the joint angles θ_1 and θ_2. We also need therefore to express the positions A and B in terms of these joint angles. This leads to the **forward kinematics problem** studied in Chapter 3, which is to determine the position and orientation of the end effector or tool in terms of the joint variables.

It is customary to establish a fixed coordinate system, called the **world** or **base** frame to which all objects including the manipulator are referenced. In this case we establish the base coordinate frame $o_0x_0y_0$ at the base of the robot, as shown in Figure 1.20.

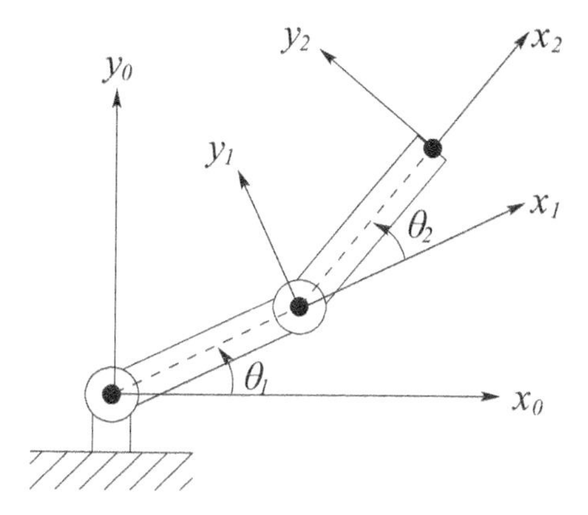

Figure 1.20: Coordinate frames attached to the links of a two-link planar robot. Each coordinate frame moves as the corresponding link moves. The mathematical description of the robot motion is thus reduced to a mathematical description of moving coordinate frames.

The coordinates (x, y) of the tool are expressed in this coordinate frame as

$$x = a_1 \cos\theta_1 + a_2\cos(\theta_1 + \theta_2) \tag{1.1}$$

$$y = a_1 \sin\theta_1 + a_2\sin(\theta_1 + \theta_2) \tag{1.2}$$

in which a_1 and a_2 are the lengths of the two links, respectively. Also the **orientation of the tool frame** relative to the base frame is given by the direction cosines of the x_2 and y_2 axes relative to the x_0 and y_0 axes, that is,

$$\begin{array}{ll} x_2 \cdot x_0 = \cos(\theta_1 + \theta_2)\ ; & y_2 \cdot x_0 = -\sin(\theta_1 + \theta_2) \\ x_2 \cdot y_0 = \sin(\theta_1 + \theta_2)\ ; & y_2 \cdot y_0 = \ \ \cos(\theta_1 + \theta_2) \end{array} \tag{1.3}$$

which we may combine into a **rotation matrix**

$$\left[\begin{array}{cc} x_2 \cdot x_0 & y_2 \cdot x_0 \\ x_2 \cdot y_0 & y_2 \cdot y_0 \end{array}\right] = \left[\begin{array}{cc} \cos(\theta_1 + \theta_2) & -\sin(\theta_1 + \theta_2) \\ \sin(\theta_1 + \theta_2) & \cos(\theta_1 + \theta_2) \end{array}\right] \tag{1.4}$$

Equations (1.1), (1.2), and (1.4) are called the **forward kinematic equations** for this arm. For a six-DOF robot these equations are quite complex and cannot be written down as easily as for the two-link manipulator. The general procedure that we discuss in Chapter 3 establishes coordinate frames at each joint and allows one to transform systematically among these frames using matrix transformations. The procedure that we use is referred to as the **Denavit-Hartenberg** convention. We then use **homogeneous coordinates** and **homogeneous transformations** to simplify the transformation among coordinate frames.

Inverse Kinematics

Now, given the joint angles θ_1, θ_2 we can determine the end-effector coordinates x and y. In order to command the robot to move to location A we need the inverse; that is, we need the joint variables θ_1, θ_2 in terms of the x and y coordinates of A. This is the problem of **inverse kinematics**. In other words, given x and y in Equations (1.1) and (1.2), we wish to solve for the joint angles. Since the forward kinematic equations are nonlinear, a solution may not be easy to find, nor is there a unique solution in general. We can see in the case of a two-link planar mechanism that there may be no solution, for example if the given (x, y) coordinates are out of reach of the manipulator. If the given (x, y) coordinates are within the manipulator's reach there may be two solutions as shown in Figure 1.21, the so-called

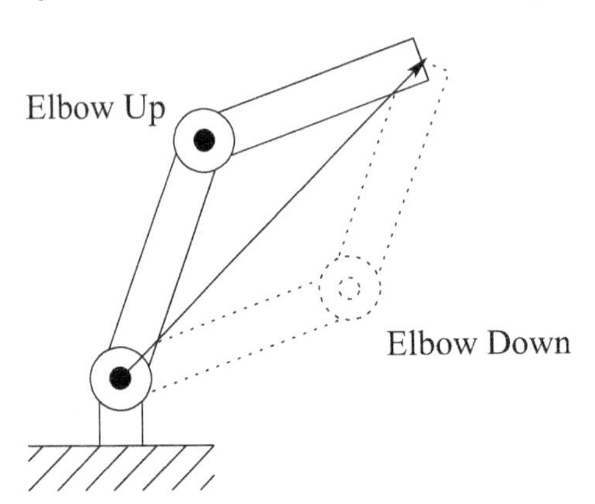

Figure 1.21: The two-link elbow robot has two solutions to the inverse kinematics except at singular configurations, the elbow up solution and the elbow down solution.

elbow up and **elbow down** configurations, or there may be exactly one solution if the manipulator must be fully extended to reach the point. There may even be an infinite number of solutions in some cases (Problem 1-20).

Consider the diagram of Figure 1.22. Using the **law of cosines**[1] we see

[1]See Appendix A

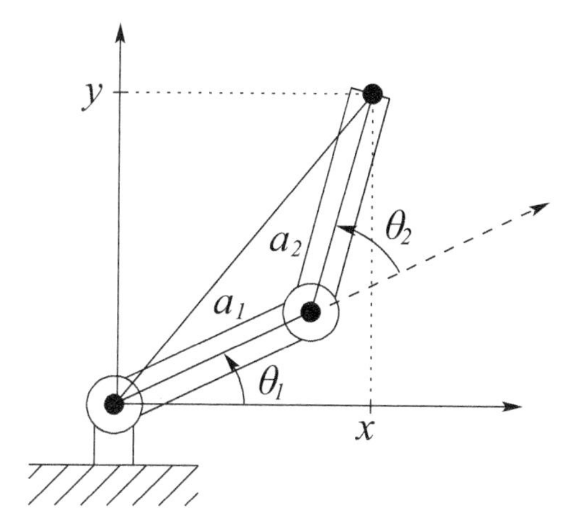

Figure 1.22: Solving for the joint angles of a two-link planar arm.

that the angle θ_2 is given by

$$\cos\theta_2 = \frac{x^2 + y^2 - a_1^2 - a_2^2}{2a_1a_2} := D \tag{1.5}$$

We could now determine θ_2 as $\theta_2 = \cos^{-1}(D)$. However, a better way to find θ_2 is to notice that if $\cos(\theta_2)$ is given by Equation (1.5), then $\sin(\theta_2)$ is given as

$$\sin(\theta_2) = \pm\sqrt{1 - D^2} \tag{1.6}$$

and, hence, θ_2 can be found by

$$\theta_2 = \tan^{-1}\frac{\pm\sqrt{1 - D^2}}{D} \tag{1.7}$$

The advantage of this latter approach is that both the elbow-up and elbow-down solutions are recovered by choosing the negative and positive signs in Equation (1.7), respectively.

It is left as an exercise (Problem 1-18) to show that θ_1 is now given as

$$\theta_1 = \tan^{-1}(y/x) - \tan^{-1}\left(\frac{a_2\sin\theta_2}{a_1 + a_2\cos\theta_2}\right) \tag{1.8}$$

Notice that the angle θ_1 depends on θ_2. This makes sense physically since we would expect to require a different value for θ_1, depending on which solution is chosen for θ_2.

Velocity Kinematics

To follow a contour at constant velocity, or at any prescribed velocity, we must know the relationship between the tool velocity and the joint velocities. In this case we can differentiate Equations (1.1) and (1.2) to obtain

$$\begin{aligned} \dot{x} &= -a_1 \sin\theta_1 \cdot \dot{\theta}_1 - a_2 \sin(\theta_1+\theta_2)(\dot{\theta}_1+\dot{\theta}_2) \\ \dot{y} &= a_1 \cos\theta_1 \cdot \dot{\theta}_1 + a_2 \cos(\theta_1+\theta_2)(\dot{\theta}_1+\dot{\theta}_2) \end{aligned} \tag{1.9}$$

Using the vector notation $x = \begin{bmatrix} x \\ y \end{bmatrix}$ and $\theta = \begin{bmatrix} \theta_1 \\ \theta_2 \end{bmatrix}$, we may write these equations as

$$\begin{aligned} \dot{x} &= \begin{bmatrix} -a_1 \sin\theta_1 - a_2 \sin(\theta_1+\theta_2) & -a_2 \sin(\theta_1+\theta_2) \\ a_1 \cos\theta_1 + a_2 \cos(\theta_1+\theta_2) & a_2 \cos(\theta_1+\theta_2) \end{bmatrix} \dot{\theta} \\ &= J\dot{\theta} \end{aligned} \tag{1.10}$$

The matrix J defined by Equation (1.10) is called the **Jacobian** of the manipulator and is a fundamental object to determine for any manipulator. In Chapter 4 we present a systematic procedure for deriving the manipulator Jacobian.

The determination of the joint velocities from the end-effector velocities is conceptually simple since the velocity relationship is linear. Thus, the joint velocities are found from the end-effector velocities via the inverse Jacobian

$$\dot{\theta} = J^{-1}\dot{x} \tag{1.11}$$

where J^{-1} is given by

$$J^{-1} = \frac{1}{a_1 a_2 \sin\theta_2} \begin{bmatrix} a_2 \cos(\theta_1+\theta_2) & a_2 \sin(\theta_1+\theta_2) \\ -a_1 \cos\theta_1 - a_2 \cos(\theta_1+\theta_2) & -a_1 \sin\theta_1 - a_2 \sin(\theta_1+\theta_2) \end{bmatrix}$$

The determinant of the Jacobian in Equation (1.10) is equal to $a_1 a_2 \sin\theta_2$. Therefore, this Jacobian does not have an inverse when $\theta_2 = 0$ or $\theta_2 = \pi$, in which case the manipulator is said to be in a **singular configuration**, such as shown in Figure 1.23 for $\theta_2 = 0$.

The determination of such singular configurations is important for several reasons. At singular configurations there are infinitesimal motions that are unachievable; that is, the manipulator end effector cannot move in certain directions. In the above example the end effector cannot move in the positive x_2 direction when $\theta_2 = 0$. Singular configurations are also related to the nonuniqueness of solutions of the inverse kinematics. For example, for a given end-effector position of the two-link planar manipulator, there

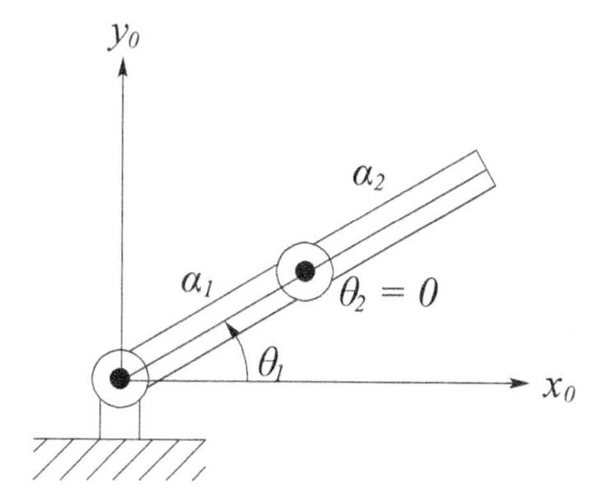

Figure 1.23: A singular configuration results when the elbow is straight. In this configuration the two-link robot has only one DOF.

are in general two possible solutions to the inverse kinematics. Note that a singular configuration separates these two solutions in the sense that the manipulator cannot go from one to the other without passing through a singularity. For many applications it is important to plan manipulator motions in such a way that singular configurations are avoided.

Path Planning and Trajectory Generation

The robot control problem is typically decomposed hierarchically into three tasks: **path planning**, **trajectory generation**, and **trajectory tracking**. The path planning problem, considered in Chapter 5, is to determine a path in task space (or configuration space) to move the robot to a goal position while avoiding collisions with objects in its workspace. These paths encode position and orientation information without timing considerations, that is, without considering velocities and accelerations along the planned paths. The trajectory generation problem, also considered in Chapter 5, is to generate reference trajectories that determine the time history of the manipulator along a given path or between initial and final configurations. These are typically given in joint space as polynomial functions of time. We discuss the most common polynomial interpolation schemes used to generate these trajectories.

Independent Joint Control

Once reference trajectories for the robot are specified, it is the task of the control system to track them. In Chapter 6 we discuss the motion control problem. We treat the **twin problems of tracking and disturbance rejection**, which are to determine the control inputs necessary to follow, or **track**, a reference trajectory, while simultaneously **rejecting** disturbances due to unmodeled dynamic effects such as friction and noise. We first model the actuator and drive-train dynamics and discuss the design of independent joint control algorithms. A block diagram of a single-input/single-output (SISO) feedback control system is shown in Figure 1.24.

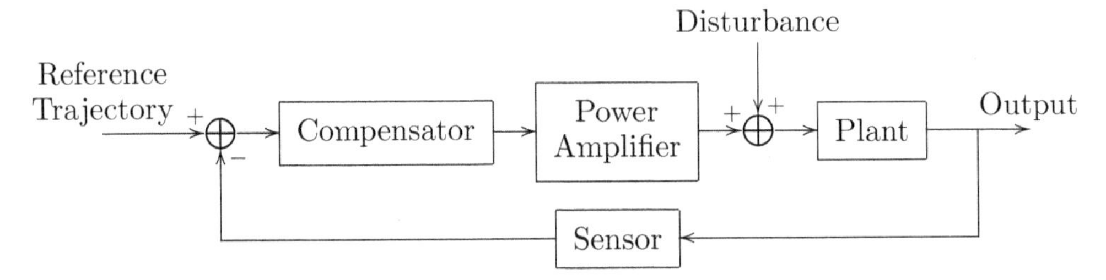

Figure 1.24: Basic structure of a feedback control system. The compensator measures the error between a reference and a measured output and produces a signal to the plant that is designed to drive the error to zero despite the presences of disturbances.

We detail the standard approaches to robot control based on both frequency domain and state space techniques. We also introduce the notion of **feedforward control** for tracking time varying trajectories.

Dynamics

The simple control strategies considered in Chapter 6 are based on the actuator and drive-train dynamics but ignore the coupling effects due to the motion of the links. In Chapter 7 we develop techniques based on **Lagrangian dynamics** for systematically deriving the equations of motion of rigid-link robots. Deriving the dynamic equations of motion for robots is not a simple task due to the large number of degrees of freedom and the nonlinearities present in the system. We also discuss the so-called **recursive Newton-Euler** method for deriving the robot equations of motion. The Newton-Euler formulation is well-suited to real-time computation for both simulation and control.

Multivariable Control

In Chapter 8 we discuss more advanced control techniques based on the Lagrangian dynamic equations of motion derived in Chapter 7. We introduce the fundamental notions of **computed torque** and **inverse dynamics** as a means for compensating the complex nonlinear interaction forces among the links of the manipulator. Robust and adaptive control are also introduced in using the **second method of Lyapunov**. Chapter 10 provides some additional advanced techniques from geometric nonlinear control theory that are useful for controlling high performance robots. We also discuss the control of so-called **nonholonomic systems** such as mobile robots.

Force Control

In the example robot task above, once the manipulator has reached location A, it must follow the contour S maintaining a constant force normal to the surface. Conceivably, knowing the location of the object and the shape of the contour, one could carry out this task using position control alone. This would be quite difficult to accomplish in practice, however. Since the manipulator itself possesses high rigidity, any errors in position due to uncertainty in the exact location of the surface or tool would give rise to extremely large forces at the end effector that could damage the tool, the surface, or the robot. A better approach is to measure the forces of interaction directly and use a **force control** scheme to accomplish the task. In Chapter 9 we discuss force control and compliance, along with common approaches to force control, namely **hybrid control** and **impedance control**.

Computer Vision

Cameras have become reliable and relatively inexpensive sensors in many robotic applications. Unlike joint sensors, which give information about the internal configuration of the robot, cameras can be used not only to measure the position of the robot but also to locate objects in the robot's workspace. In Chapter 11 we discuss the use of computer vision to determine position and orientation of objects.

Vision-Based Control

In some cases, we may wish to control the motion of the manipulator relative to some target as the end effector moves through free space. Here, force control cannot be used. Instead, we can use computer vision to close the control loop around the vision sensor. This is the topic of Chapter 12. There are several approaches to vision-based control, but we will focus on the method of Image-Based Visual Servo (IBVS). With IBVS, an error measured in image coordinates is directly mapped to a control input that governs the motion of the camera. This method has become very popular in recent years, and it relies on mathematical development analogous to that given in Chapter 4.

PROBLEMS

1-1 What are the key features that distinguish robots from other forms of automation such as CNC milling machines?

1-2 Briefly define each of the following terms: forward kinematics, inverse kinematics, trajectory planning, workspace, accuracy, repeatability, resolution, joint variable, spherical wrist, end effector.

1-3 What are the main ways to classify robots?

1-4 Make a list of 10 robot applications. For each application discuss which type of manipulator would be best suited; which least suited. Justify your choices in each case.

1-5 List several applications for nonservo robots; for point-to-point robots; for continuous path robots.

1-6 List five applications that a continuous path robot could do that a point-to-point robot could not do.

1-7 List five applications for which computer vision would be useful in robotics.

1-8 List five applications for which either tactile sensing or force feedback control would be useful in robotics.

1-9 Find out how many industrial robots are currently in operation in Japan. How many are in operation in the United States? What country ranks third in the number of industrial robots in use?

1-10 Suppose we could close every factory today and reopen them tomorrow fully automated with robots. What would be some of the economic and social consequences of such a development?

1-11 Suppose a law were passed banning all future use of industrial robots. What would be some of the economic and social consequences of such an act?

1-12 Discuss applications for which redundant manipulators would be useful.

1-13 Referring to Figure 1.25, suppose that the tip of a single link travels a distance d between two points. A linear axis would travel the distance d while a rotational link would travel through an arc length $\ell\theta$ as shown. Using the law of cosines, show that the distance d is given by

$$d = \ell\sqrt{2(1 - \cos\theta)}$$

which is of course less than $\ell\theta$. With 10-bit accuracy, $\ell = 1$ meter, and $\theta = 90°$, what is the resolution of the linear link? of the rotational link?

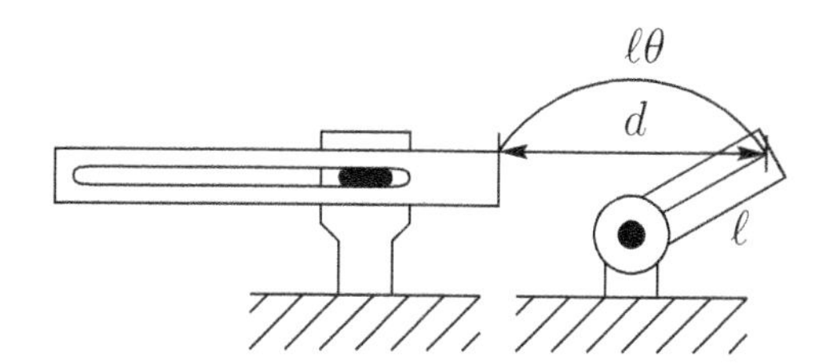

Figure 1.25: Diagram for Problem 1-15.

1-14 For the single-link revolute arm shown in Figure 1.25, if the length of the link is 50 cm and the arm travels 180 degrees, what is the control resolution obtained with an 8-bit encoder?

1-15 Repeat Problem 1.14 assuming that the 8-bit encoder is located on the motor shaft that is connected to the link through a 50:1 gear reduction. Assume perfect gears.

1-16 Why is accuracy generally less than repeatability?

1-17 How could manipulator accuracy be improved using endpoint sensing? What difficulties might endpoint sensing introduce into the control problem?

1-18 Derive Equation (1.8).

1-19 For the two-link manipulator of Figure 1.20 suppose $a_1 = a_2 = 1$.

1. Find the coordinates of the tool when $\theta_1 = \frac{\pi}{6}$ and $\theta_2 = \frac{\pi}{2}$.
2. If the joint velocities are constant at $\dot{\theta}_1 = 1$, $\dot{\theta}_2 = 2$, what is the velocity of the tool? What is the instantaneous tool velocity when $\theta_1 = \theta_2 = \frac{\pi}{4}$?
3. Write a computer program to plot the joint angles as a function of time given the tool locations and velocities as a function of time in Cartesian coordinates.
4. Suppose we desire that the tool follow a straight line between the points (0,2) and (2,0) at constant speed s. Plot the time history of joint angles.

1-20 For the two-link planar manipulator of Figure 1.20 is it possible for there to be an infinite number of solutions to the inverse kinematic equations? If so, explain how this can occur.

1-21 Explain why it might be desirable to reduce the mass of distal links in a manipulator design. List some ways this can be done. Discuss any possible disadvantages of such designs.

NOTES AND REFERENCES

We give below some of the important milestones in the history of modern robotics.

1947 — The first servoed electric powered teleoperator is developed.

1948 — A teleoperator is developed incorporating force feedback.

1949 — Research on numerically controlled milling machine is initiated.

1954 — George Devol designs the first programmable robot

1956 — Joseph Engelberger, a Columbia University physics student, buys the rights to Devol's robot and founds the Unimation Company.

1961 — The first Unimate robot is installed in a Trenton, New Jersey plant of General Motors to tend a die casting machine.

1961 — The first robot incorporating force feedback is developed.

1963 — The first robot vision system is developed.

1971 — The Stanford Arm is developed at Stanford University.

1973 — The first robot programming language (WAVE) is developed at Stanford.

1974 — Cincinnati Milacron introduced the T^3 robot with computer control.

1975 — Unimation Inc. registers its first financial profit.

1976 — The Remote Center Compliance (RCC) device for part insertion in assembly is developed at Draper Labs in Boston.

1976 — Robot arms are used on the Viking I and II space probes and land on Mars.

1978 — Unimation introduces the PUMA robot, based on designs from a General Motors study.

1979 — The SCARA robot design is introduced in Japan.

1981 — The first direct-drive robot is developed at Carnegie-Mellon University.

1982 — Fanuc of Japan and General Motors form GM Fanuc to market robots in North America.

1983 — Adept Technology is founded and successfully markets the direct-drive robot.

1986 — The underwater robot, Jason, of the Woods Hole Oceanographic Institute, explores the wreck of the Titanic, found a year earlier by Dr. Robert Barnard.

1988 — Stäubli Group purchases Unimation from Westinghouse.

1988 — The IEEE Robotics and Automation Society is formed.

1993 — The experimental robot, ROTEX, of the German Aerospace Agency (DLR) was flown aboard the space shuttle Columbia and performed a variety of tasks under both teleoperated and sensor-based offline programmed modes.

1996 — Honda unveils its Humanoid robot; a project begun in secret in 1986.

1997 — The first robot soccer competition, RoboCup-97, is held in Nagoya, Japan and draws 40 teams from around the world.

1997 — The Sojourner mobile robot travels to Mars aboard NASA's Mars PathFinder mission.

2001 — Sony begins to mass produce the first household robot, a robot dog named Aibo.

2001 — The Space Station Remote Manipulation System (SSRMS) is launched in space on board the space shuttle Endeavor to facilitate continued construction of the space station.

2001 — The first telesurgery is performed when surgeons in New York perform a laparoscopic gall bladder removal on a woman in Strasbourg, France.

2001 — Robots are used to search for victims at the World Trade Center site after the September 11th tragedy.

2002 — Honda's Humanoid Robot ASIMO rings the opening bell at the New York Stock Exchange on February 15th.

2005 — ROKVISS (Robotic Component Verification on board the International Space Station), the experimental teleoperated arm built by the German Aerospace Center (DLR), undergoes its first tests in space.

Many books have been written about basic and advanced topics in robotics. Below is an incomplete list of references.

- H. Asada and J-J. Slotine. *Robot Analysis and Control.* Wiley, New York, 1986.
- G. A. Bekey, *Autonomous Robots.* MIT Press, Cambridge, MA, 2005.
- M. Brady et al., editors. *Robot Motion: Planning and Control.* MIT Press, Cambridge, MA, 1983.
- H. Choset, K. M. Lynch, S. Hutchinson, G. Kantor, W. Burgard, L. E. Kavraki, and S. Thrun. *Principles of Robot Motion: Theory, Algorithms, and Implementations.* MIT Press, Cambridge, MA, 2005.
- J. Craig. *Introduction to Robotics: Mechanics and Control.* Addison Wesley, Reading, MA, 1986.
- R. Dorf. *Robotics and Automated Manufacturing.* Reston, VA, 1983.
- J. Engleberger. *Robotics in Practice.* Kogan Page, London, 1980.
- K.S. Fu, R. C. Gonzalez, and C.S.G. Lee. *Robotics: Control Sensing, Vision, and Intelligence.* McGraw-Hill, St Louis, 1987.
- B. K. Ghosh, N. Xi and T. J. Tarn. *Control in Robotics and Automation: Sensor-Based Integration*, Academic Press, San Diego, CA, 1999.
- T. R. Kurfess. *Robotics and Automation Handbook*, CRC Press, Boca Raton, FL, 2005.
- J. C. Latombe. *Robot Motion Planning.* Kluwer Academic Publishers, Boston, 1991.
- M. T. Mason. *Mechanics of Robotic Manipulation*, MIT Press, Cambridge, MA, 2001.
- R.M. Murray, Z. Li, and S.S. Sastry. *A Mathematical Introduction to Robotics.* CRC Press, Boca Raton, FL, 1994.
- S. B. Niku. *Introduction to Robotics: Analysis, Systems, Applications.* Prentice Hall, Upper Saddle River, NJ, 2001.
- R. Paul. *Robot Manipulators: Mathematics, Programming and Control.* MIT Press, Cambridge, MA, 1982.
- L. Sciavicco and B. Siciliano. *Modelling and Control of Robot Manipulators, 2nd Edition*, Springer-Verlag, London, 2000.
- M. Shahinpoor. *Robot Engineering Textbook.* Harper and Row, New York, 1987.
- W. Snyder. *Industrial Robots: Computer Interfacing and Control.* Prentice-Hall, Englewood Cliffs, NJ, 1985.

- M.W. Spong, F.L. Lewis and C.T. Abdallah. *Robot Control: Dynamics, Motion Planning, and Analysis*, IEEE Press, Boca Raton, FL, 1992.
- M. Spong and M. Vidyasagar. *Robot Dynamics and Control.* John Wiley and Sons, NY, NY, 1989.
- W. Wolovich. *Robotics: Basic Analysis and Design.* Holt, Rinehart, and Winston, New York, 1985.

There is a great deal of ongoing research in robotics. Current research results can be found in journals such as

- *IEEE Transactions on Robotics* (previously *IEEE Transactions on Robotics and Automation*)
- *IEEE Robotics and Automation Magazine*
- *International Journal of Robotics Research*
- *Robotics and Autonomous Systems*
- *Journal of Robotic Systems*
- *Robotica*
- *Journal of Intelligent and Robotic Systems*
- *Autonomous Robots*
- *Advanced Robotics*

Chapter 2

RIGID MOTIONS AND HOMOGENEOUS TRANSFORMATIONS

A large part of robot kinematics is concerned with establishing various coordinate frames to represent the positions and orientations of rigid objects, and with transformations among these coordinate frames. Indeed, the geometry of three-dimensional space and of rigid motions plays a central role in all aspects of robotic manipulation. In this chapter we study the operations of rotation and translation, and introduce the notion of homogeneous transformations.[1] Homogeneous transformations combine the operations of rotation and translation into a single matrix multiplication, and are used in Chapter 3 to derive the so-called forward kinematic equations of rigid manipulators.

We begin by examining representations of points and vectors in a Euclidean space equipped with multiple coordinate frames. Following this, we introduce the concept of a rotation matrix to represent relative orientations among coordinate frames. Then we combine these two concepts to build homogeneous transformation matrices, which can be used to simultaneously represent the position and orientation of one coordinate frame relative to another. Furthermore, homogeneous transformation matrices can be used to perform coordinate transformations. Such transformations allow us to represent various quantities in different coordinate frames, a facility that we will often exploit in subsequent chapters.

---

[1]Since we make extensive use of elementary matrix theory, the reader may wish to review Appendix B before beginning this chapter.

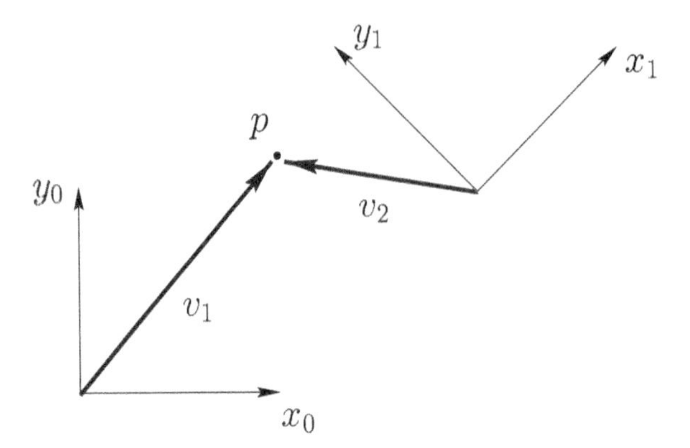

Figure 2.1: Two coordinate frames, a point p, and two vectors v_1 and v_2.

2.1 REPRESENTING POSITIONS

Before developing representation schemes for points and vectors, it is instructive to distinguish between the two fundamental approaches to geometric reasoning: the **synthetic** approach and the **analytic** approach. In the former, one reasons directly about geometric entities (e.g., points or lines), while in the latter, one represents these entities using coordinates or equations, and reasoning is performed via algebraic manipulations. The latter approach requires the choice of a reference coordinate frame. A coordinate frame consists of an origin (a single point in space), and two or three orthogonal coordinate axes, for two- and three-dimensional spaces, respectively.

Consider Figure 2.1, which shows two coordinate frames that differ in orientation by an angle of 45°. Using the synthetic approach, without ever assigning coordinates to points or vectors, one can say that x_0 is perpendicular to y_0, or that $v_1 \times v_2$ defines a vector that is perpendicular to the plane containing v_1 and v_2, in this case pointing out of the page.

In robotics, one typically uses analytic reasoning, since robot tasks are often defined using Cartesian coordinates. Of course, in order to assign coordinates it is necessary to specify a reference coordinate frame. Consider again Figure 2.1. We could specify the coordinates of the point p with respect to either frame $o_0x_0y_0$ or frame $o_1x_1y_1$. In the former case, we might assign to p the coordinate vector $[5, 6]^T$, and in the latter case $[-2.8, 4.2]^T$. So that the reference frame will always be clear, we will adopt a notation in which a superscript is used to denote the reference frame. Thus, we would write

$$p^0 = \begin{bmatrix} 5 \\ 6 \end{bmatrix}, \qquad p^1 = \begin{bmatrix} -2.8 \\ 4.2 \end{bmatrix}$$

Geometrically, a point corresponds to a specific location in space. We stress here that p is a geometric entity, a point in space, while both p^0 and p^1 are coordinate vectors that represent the location of this point in space with respect to coordinate frames $o_0x_0y_0$ and $o_1x_1y_1$, respectively.

Since the origin of a coordinate frame is just a point in space, we can assign coordinates that represent the position of the origin of one coordinate frame with respect to another. In Figure 2.1, for example, we have

$$o_1^0 = \begin{bmatrix} 10 \\ 5 \end{bmatrix}, \qquad o_0^1 = \begin{bmatrix} -10.6 \\ 3.5 \end{bmatrix}$$

In cases where there is only a single coordinate frame, or in which the reference frame is obvious, we will often omit the superscript. This is a slight abuse of notation, and the reader is advised to bear in mind the difference between the geometric entity called p and any particular coordinate vector that is assigned to represent p. The former is independent of the choice of coordinate frames, while the latter obviously depends on the choice of coordinate frames.

While a point corresponds to a specific location in space, a *vector* specifies a direction and a magnitude. Vectors can be used, for example, to represent displacements or forces. Therefore, while the point p is not equivalent to the vector v_1, the displacement from the origin o_0 to the point p is given by the vector v_1. In this text, we will use the term *vector* to refer to what are sometimes called *free vectors*, that is, vectors that are not constrained to be located at a particular point in space. Under this convention, it is clear that points and vectors are not equivalent, since points refer to specific locations in space, but a vector can be moved to any location in space. Under this convention, two vectors are equal if they have the same direction and the same magnitude.

When assigning coordinates to vectors, we use the same notational convention that we used when assigning coordinates to points. Thus, v_1 and v_2 are geometric entities that are invariant with respect to the choice of coordinate frames, but the representation by coordinates of these vectors depends directly on the choice of reference coordinate frame. In the example of Figure 2.1, we would obtain

$$v_1^0 = \begin{bmatrix} 5 \\ 6 \end{bmatrix}, \qquad v_1^1 = \begin{bmatrix} 7.77 \\ 0.8 \end{bmatrix}, \qquad v_2^0 = \begin{bmatrix} -5.1 \\ 1 \end{bmatrix}, \qquad v_2^1 = \begin{bmatrix} -2.89 \\ 4.2 \end{bmatrix}$$

In order to perform algebraic manipulations using coordinates, it is essential that all coordinate vectors be defined with respect to the same coordinate frame. In the case of free vectors, it is enough that they be defined with respect to "parallel" coordinate frames, that is, frames whose respective coordinate axes are parallel, since only their magnitude and direction are specified and not their absolute locations in space.

Using this convention, an expression of the form $v_1^1 + v_2^2$, where v_1^1 and v_2^2 are as in Figure 2.1, is not defined since the frames $o_0x_0y_0$ and $o_1x_1y_1$ are not parallel. Thus, we see a clear need not only for a representation system that allows points to be expressed with respect to various coordinate frames, but also for a mechanism that allows us to transform the coordinates of points from one coordinate frame to another. Such coordinate transformations are the topic for much of the remainder of this chapter.

2.2 REPRESENTING ROTATIONS

In order to represent the relative position and orientation of one rigid body with respect to another, we will attach coordinate frames to each body, and then specify the geometric relationships between these coordinate frames. In Section 2.1 we saw how one can represent the position of the origin of one frame with respect to another frame. In this section, we address the problem of describing the orientation of one coordinate frame relative to another frame. We begin with the case of rotations in the plane, and then generalize our results to the case of orientations in a three-dimensional space.

2.2.1 Rotation in the Plane

Figure 2.2 shows two coordinate frames, with frame $o_1x_1y_1$ being obtained by rotating frame $o_0x_0y_0$ by an angle θ. Perhaps the most obvious way to represent the relative orientation of these two frames is to merely specify the angle of rotation θ. There are two immediate disadvantages to such a representation. First, there is a discontinuity in the mapping from relative orientation to the value of θ in a neighborhood of $\theta = 0$. In particular, for $\theta = 2\pi - \epsilon$, small changes in orientation can produce large changes in the value of θ, for example, a rotation by ϵ causes θ to "wrap around" to zero. Second, this choice of representation does not scale well to the three-dimensional case.

A slightly less obvious way to specify the orientation is to specify the coordinate vectors for the axes of frame $o_1x_1y_1$ with respect to coordinate frame $o_0x_0y_0$:

$$R_1^0 = \left[x_1^0 \mid y_1^0\right]$$

in which x_1^0 and y_1^0 are the coordinates in frame $o_0x_0y_0$ of unit vectors x_1

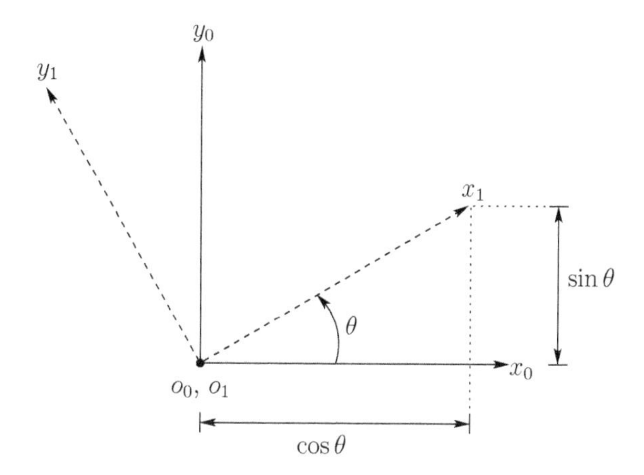

Figure 2.2: Coordinate frame $o_1x_1y_1$ is oriented at an angle θ with respect to $o_0x_0y_0$.

and y_1, respectively.[2] A matrix in this form is called a **rotation matrix**. Rotation matrices have a number of special properties that we will discuss below.

In the two-dimensional case, it is straightforward to compute the entries of this matrix. As illustrated in Figure 2.2,

$$x_1^0 = \begin{bmatrix} \cos\theta \\ \sin\theta \end{bmatrix}, \qquad y_1^0 = \begin{bmatrix} -\sin\theta \\ \cos\theta \end{bmatrix}$$

which gives

$$R_1^0 = \begin{bmatrix} \cos\theta & -\sin\theta \\ \sin\theta & \cos\theta \end{bmatrix} \tag{2.1}$$

Note that we have continued to use the notational convention of allowing the superscript to denote the reference frame. Thus, R_1^0 is a matrix whose column vectors are the coordinates of the unit vectors along the axes of frame $o_1x_1y_1$ expressed relative to frame $o_0x_0y_0$.

Although we have derived the entries for R_1^0 in terms of the angle θ, it is not necessary that we do so. An alternative approach, and one that scales nicely to the three-dimensional case, is to build the rotation matrix by projecting the axes of frame $o_1x_1y_1$ onto the coordinate axes of frame $o_0x_0y_0$. Recalling that the dot product of two unit vectors gives the projection of

[2]We will use x_i, y_i to denote both coordinate axes and unit vectors along the coordinate axes depending on the context.

one onto the other, we obtain

$$x_1^0 = \begin{bmatrix} x_1 \cdot x_0 \\ x_1 \cdot y_0 \end{bmatrix}, \qquad y_1^0 = \begin{bmatrix} y_1 \cdot x_0 \\ y_1 \cdot y_0 \end{bmatrix}$$

which can be combined to obtain the rotation matrix

$$R_1^0 = \begin{bmatrix} x_1 \cdot x_0 & y_1 \cdot x_0 \\ x_1 \cdot y_0 & y_1 \cdot y_0 \end{bmatrix}$$

Thus, the columns of R_1^0 specify the direction cosines of the coordinate axes of $o_1x_1y_1$ relative to the coordinate axes of $o_0x_0y_0$. For example, the first column $[x_1 \cdot x_0, x_1 \cdot y_0]^T$ of R_1^0 specifies the direction of x_1 relative to the frame $o_0x_0y_0$. Note that the right-hand sides of these equations are defined in terms of geometric entities, and not in terms of their coordinates. Examining Figure 2.2 it can be seen that this method of defining the rotation matrix by projection gives the same result as was obtained in Equation (2.1).

If we desired instead to describe the orientation of frame $o_0x_0y_0$ with respect to the frame $o_1x_1y_1$ (that is, if we desired to use the frame $o_1x_1y_1$ as the reference frame), we would construct a rotation matrix of the form

$$R_0^1 = \begin{bmatrix} x_0 \cdot x_1 & y_0 \cdot x_1 \\ x_0 \cdot y_1 & y_0 \cdot y_1 \end{bmatrix}$$

Since the dot product is commutative, (that is, $x_i \cdot y_j = y_j \cdot x_i$), we see that

$$R_0^1 = (R_1^0)^T$$

In a geometric sense, the orientation of $o_0x_0y_0$ with respect to the frame $o_1x_1y_1$ is the inverse of the orientation of $o_1x_1y_1$ with respect to the frame $o_0x_0y_0$. Algebraically, using the fact that coordinate axes are mutually orthogonal, it can readily be seen that

$$(R_1^0)^T = (R_1^0)^{-1}$$

The column vectors of R_1^0 are of unit length and mutually orthogonal (Problem 2-4). Such a matrix is said to be **orthogonal**. It can also be shown (Problem 2-5) that $\det R_1^0 = \pm 1$. If we restrict ourselves to right-handed coordinate frames, as defined in Appendix B, then $\det R_1^0 = +1$ (Problem 2-5). It is customary to refer to the set of all such $n \times n$ matrices by the symbol $SO(n)$, which denotes the **Special Orthogonal** group of order n.

For any $R \in SO(n)$ the following properties hold.

- $R^T = R^{-1} \in SO(n)$
- The columns (and therefore the rows) of R are mutually orthogonal
- Each column (and therefore each row) of R is a unit vector
- $\det R = 1$

To provide further geometric intuition for the notion of the inverse of a rotation matrix, note that in the two-dimensional case, the inverse of the rotation matrix corresponding to a rotation by angle θ can also be easily computed simply by constructing the rotation matrix for a rotation by the angle $-\theta$:

$$\begin{bmatrix} \cos(-\theta) & -\sin(-\theta) \\ \sin(-\theta) & \cos(-\theta) \end{bmatrix} = \begin{bmatrix} \cos\theta & \sin\theta \\ -\sin\theta & \cos\theta \end{bmatrix} = \begin{bmatrix} \cos\theta & -\sin\theta \\ \sin\theta & \cos\theta \end{bmatrix}^T$$

2.2.2 Rotations in Three Dimensions

The projection technique described above scales nicely to the three-dimensional case. In three dimensions, each axis of the frame $o_1x_1y_1z_1$ is projected onto coordinate frame $o_0x_0y_0z_0$. The resulting rotation matrix is given by

$$R_1^0 = \begin{bmatrix} x_1 \cdot x_0 & y_1 \cdot x_0 & z_1 \cdot x_0 \\ x_1 \cdot y_0 & y_1 \cdot y_0 & z_1 \cdot y_0 \\ x_1 \cdot z_0 & y_1 \cdot z_0 & z_1 \cdot z_0 \end{bmatrix}$$

As was the case for rotation matrices in two dimensions, matrices in this form are orthogonal, with determinant equal to 1. In this case, 3×3 rotation matrices belong to the group $SO(3)$.

Example 2.1

Suppose the frame $o_1x_1y_1z_1$ is rotated through an angle θ about the z_0-axis, and we wish to find the resulting transformation matrix R_1^0. By convention, the right hand rule (see Appendix B) defines the positive sense for the angle θ to be such that rotation by θ about the z-axis would advance a right-hand threaded screw along the positive z-axis. From Figure 2.3 we see that

$$x_1 \cdot x_0 = \cos\theta, \qquad y_1 \cdot x_0 = -\sin\theta,$$
$$x_1 \cdot y_0 = \sin\theta, \qquad y_1 \cdot y_0 = \cos\theta$$

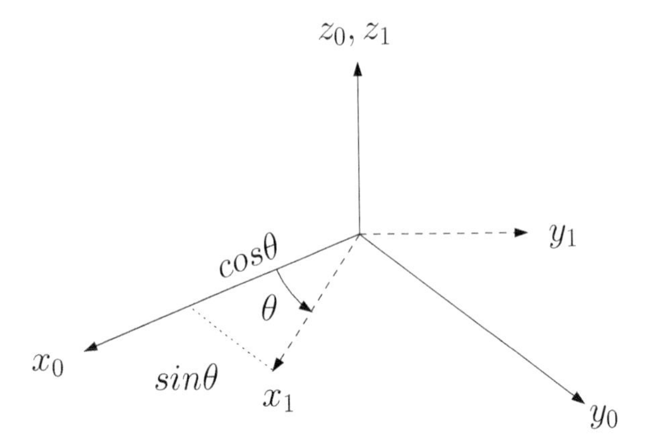

Figure 2.3: Rotation about z_0 by an angle θ.

and

$$z_0 \cdot z_1 \quad = \quad 1$$

while all other dot products are zero. Thus, the rotation matrix R_1^0 has a particularly simple form in this case, namely

$$R_1^0 \quad = \quad \left[\begin{array}{ccc} \cos\theta & -\sin\theta & 0 \\ \sin\theta & \cos\theta & 0 \\ 0 & 0 & 1 \end{array} \right] \tag{2.2}$$

◇

The rotation matrix given in Equation (2.2) is called a **basic rotation matrix** (about the z-axis). In this case we find it useful to use the more descriptive notation $R_{z,\theta}$ instead of R_1^0 to denote the matrix. It is easy to verify that the basic rotation matrix $R_{z,\theta}$ has the properties

$$R_{z,0} \quad = \quad I \tag{2.3}$$

$$R_{z,\theta} R_{z,\phi} \quad = \quad R_{z,\theta+\phi} \tag{2.4}$$

which together imply

$$\left(R_{z,\theta}\right)^{-1} \quad = \quad R_{z,-\theta} \tag{2.5}$$

Similarly, the basic rotation matrices representing rotations about the x

and y-axes are given as (Problem 2-8)

$$R_{x,\theta} = \begin{bmatrix} 1 & 0 & 0 \\ 0 & \cos\theta & -\sin\theta \\ 0 & \sin\theta & \cos\theta \end{bmatrix} \tag{2.6}$$

$$R_{y,\theta} = \begin{bmatrix} \cos\theta & 0 & \sin\theta \\ 0 & 1 & 0 \\ -\sin\theta & 0 & \cos\theta \end{bmatrix} \tag{2.7}$$

which also satisfy properties analogous to Equations (2.3)–(2.5).

Example 2.2

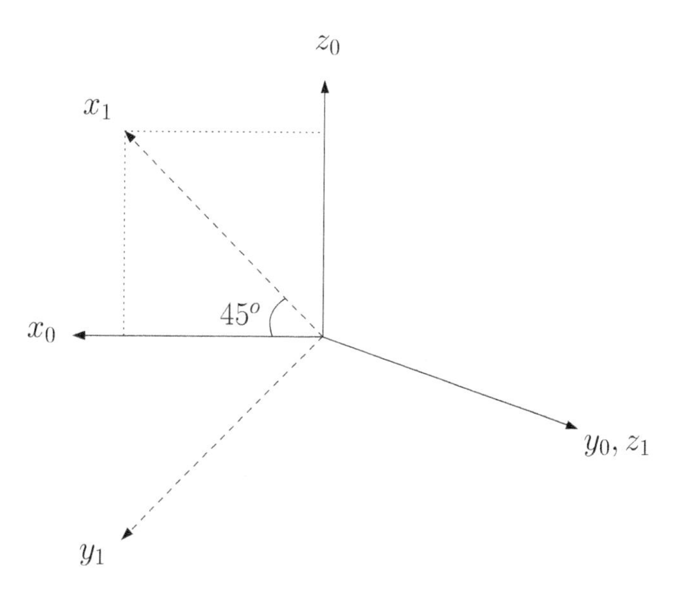

Figure 2.4: Defining the relative orientation of two frames.

Consider the frames $o_0x_0y_0z_0$ *and* $o_1x_1y_1z_1$ *shown in Figure 2.4. Projecting the unit vectors* x_1, y_1, z_1 *onto* x_0, y_0, z_0 *gives the coordinates of* x_1, y_1, z_1 *in the* $o_0x_0y_0z_0$ *frame. We see that the coordinates of* x_1, y_1, *and* z_1 *are given by*

$$x_1 = \begin{bmatrix} \frac{1}{\sqrt{2}} \\ 0 \\ \frac{1}{\sqrt{2}} \end{bmatrix}, \quad y_1 = \begin{bmatrix} \frac{1}{\sqrt{2}} \\ 0 \\ \frac{-1}{\sqrt{2}} \end{bmatrix}, \quad z_1 = \begin{bmatrix} 0 \\ 1 \\ 0 \end{bmatrix}$$

The rotation matrix R_1^0 *specifying the orientation of* $o_1x_1y_1z_1$ *relative to*

$o_0x_0y_0z_0$ has these as its column vectors, that is,

$$R_1^0 = \begin{bmatrix} \frac{1}{\sqrt{2}} & \frac{1}{\sqrt{2}} & 0 \\ 0 & 0 & 1 \\ \frac{1}{\sqrt{2}} & \frac{-1}{\sqrt{2}} & 0 \end{bmatrix}$$

◇

2.3 ROTATIONAL TRANSFORMATIONS

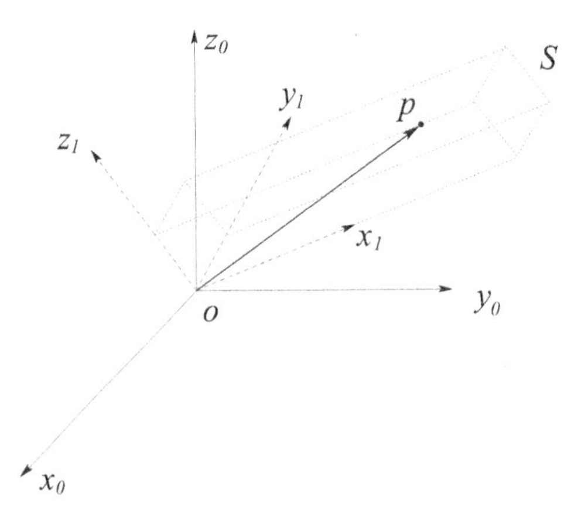

Figure 2.5: Coordinate frame attached to a rigid body.

Figure 2.5 shows a rigid object S to which a coordinate frame $o_1x_1y_1z_1$ is attached. Given the coordinates p^1 of the point p (in other words, given the coordinates of p with respect to the frame $o_1x_1y_1z_1$), we wish to determine the coordinates of p relative to a fixed reference frame $o_0x_0y_0z_0$. The coordinates $p^1 = [u, v, w]^T$ satisfy the equation

$$p = ux_1 + vy_1 + wz_1$$

In a similar way, we can obtain an expression for the coordinates p^0 by projecting the point p onto the coordinate axes of the frame $o_0x_0y_0z_0$, giving

$$p^0 = \begin{bmatrix} p \cdot x_0 \\ p \cdot y_0 \\ p \cdot z_0 \end{bmatrix}$$

Combining these two equations we obtain

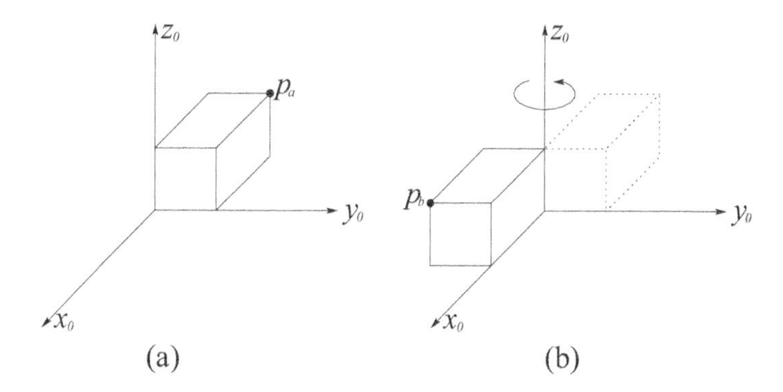

Figure 2.6: The block in (b) is obtained by rotating the block in (a) by π about z_0.

$$
\begin{aligned}
p^0 &= \begin{bmatrix} (ux_1 + vy_1 + wz_1) \cdot x_0 \\ (ux_1 + vy_1 + wz_1) \cdot y_0 \\ (ux_1 + vy_1 + wz_1) \cdot z_0 \end{bmatrix} \\
&= \begin{bmatrix} ux_1 \cdot x_0 + vy_1 \cdot x_0 + wz_1 \cdot x_0 \\ ux_1 \cdot y_0 + vy_1 \cdot y_0 + wz_1 \cdot y_0 \\ ux_1 \cdot z_0 + vy_1 \cdot z_0 + wz_1 \cdot z_0 \end{bmatrix} \\
&= \begin{bmatrix} x_1 \cdot x_0 & y_1 \cdot x_0 & z_1 \cdot x_0 \\ x_1 \cdot y_0 & y_1 \cdot y_0 & z_1 \cdot y_0 \\ x_1 \cdot z_0 & y_1 \cdot z_0 & z_1 \cdot z_0 \end{bmatrix} \begin{bmatrix} u \\ v \\ w \end{bmatrix}
\end{aligned}
$$

But the matrix in this final equation is merely the rotation matrix R_1^0, which leads to

$$
p^0 = R_1^0 p^1 \tag{2.8}
$$

Thus, the rotation matrix R_1^0 can be used not only to represent the orientation of coordinate frame $o_1x_1y_1z_1$ with respect to frame $o_0x_0y_0z_0$, but also to transform the coordinates of a point from one frame to another. If a given point is expressed relative to $o_1x_1y_1z_1$ by coordinates p^1, then $R_1^0p^1$ represents the **same point** expressed relative to the frame $o_0x_0y_0z_0$.

We can also use rotation matrices to represent rigid motions that correspond to pure rotation. Consider Figure 2.6. One corner of the block in Figure 2.6(a) is located at the point p_a in space. Figure 2.6(b) shows the same block after it has been rotated about z_0 by the angle π. In Figure 2.6(b), the same corner of the block is now located at point p_b in space. It is possible to derive the coordinates for p_b given only the coordinates for p_a

and the rotation matrix that corresponds to the rotation about z_0. To see how this can be accomplished, imagine that a coordinate frame is rigidly attached to the block in Figure 2.6(a), such that it is coincident with the frame $o_0x_0y_0z_0$. After the rotation by π, the block's coordinate frame, which is rigidly attached to the block, is also rotated by π. If we denote this rotated frame by $o_1x_1y_1z_1$, we obtain

$$R_1^0 = R_{z,\pi} = \begin{bmatrix} -1 & 0 & 0 \\ 0 & -1 & 0 \\ 0 & 0 & 1 \end{bmatrix}$$

In the local coordinate frame $o_1x_1y_1z_1$, the point p_b has the coordinate representation p_b^1. To obtain its coordinates with respect to frame $o_0x_0y_0z_0$, we merely apply the coordinate transformation Equation (2.8), giving

$$p_b^0 = R_{z,\pi} p_b^1$$

It is important to notice that the local coordinates p_b^1 of the corner of the block do not change as the block rotates, since they are defined in terms of the block's own coordinate frame. Therefore, when the block's frame is aligned with the reference frame $o_0x_0y_0z_0$ (that is, before the rotation is performed), the coordinates p_b^1 equals p_a^0, since before the rotation is performed, the point p_a is coincident with the corner of the block. Therefore, we can substitute p_a^0 into the previous equation to obtain

$$p_b^0 = R_{z,\pi} p_a^0$$

This equation shows how to use a rotation matrix to represent a rotational motion. In particular, if the point p_b is obtained by rotating the point p_a as defined by the rotation matrix R, then the coordinates of p_b with respect to the reference frame are given by

$$p_b^0 = R p_a^0$$

This same approach can be used to rotate vectors with respect to a coordinate frame, as the following example illustrates.

Example 2.3

The vector v with coordinates $v^0 = [0, 1, 1]^T$ is rotated about y_0 by $\frac{\pi}{2}$ as shown in Figure 2.7. The resulting vector v_1 has coordinates given by

$$v_1^0 = R_{y,\frac{\pi}{2}} v^0 \tag{2.9}$$

$$= \begin{bmatrix} 0 & 0 & 1 \\ 0 & 1 & 0 \\ -1 & 0 & 0 \end{bmatrix} \begin{bmatrix} 0 \\ 1 \\ 1 \end{bmatrix} = \begin{bmatrix} 1 \\ 1 \\ 0 \end{bmatrix} \tag{2.10}$$

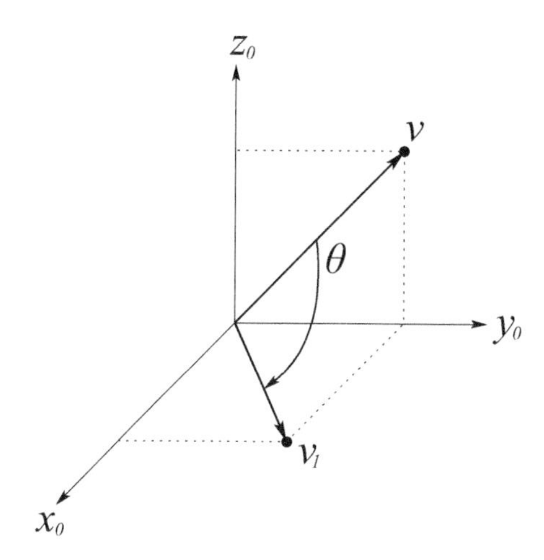

Figure 2.7: Rotating a vector about axis y_0.

Thus, a third interpretation of a rotation matrix R is as an operator acting on vectors in a fixed frame. In other words, instead of relating the coordinates of a fixed vector with respect to two different coordinate frames, Equation (2.9) can represent the coordinates in $o_0x_0y_0z_0$ of a vector v_1 that is obtained from a vector v by a given rotation.

◇

As we have seen, rotation matrices can serve several roles. A rotation matrix, either $R \in SO(3)$ or $R \in SO(2)$, can be interpreted in three distinct ways:

1. It represents a coordinate transformation relating the coordinates of a point p in two different frames.

2. It gives the orientation of a transformed coordinate frame with respect to a fixed coordinate frame.

3. It is an operator taking a vector and rotating it to give a new vector in the same coordinate frame.

The particular interpretation of a given rotation matrix R will be made clear by the context.

2.3.1 Similarity Transformations

A coordinate frame is defined by a set of **basis vectors**, for example, unit vectors along the three coordinate axes. This means that a rotation matrix, as a coordinate transformation, can also be viewed as defining a change of basis from one frame to another. The matrix representation of a general linear transformation is transformed from one frame to another using a so-called **similarity transformation.**[3] For example, if A is the matrix representation of a given linear transformation in $o_0x_0y_0z_0$ and B is the representation of the same linear transformation in $o_1x_1y_1z_1$ then A and B are related as

$$B = (R_1^0)^{-1} A R_1^0 \tag{2.11}$$

where R_1^0 is the coordinate transformation between frames $o_1x_1y_1z_1$ and $o_0x_0y_0z_0$. In particular, if A itself is a rotation, then so is B, and thus the use of similarity transformations allows us to express the same rotation easily with respect to different frames.

Example 2.4

Henceforth, whenever convenient we use the shorthand notation $c_\theta = \cos\theta$, $s_\theta = \sin\theta$ *for trigonometric functions. Suppose frames* $o_0x_0y_0z_0$ *and* $o_1x_1y_1z_1$ *are related by the rotation*

$$R_1^0 = \begin{bmatrix} 0 & 0 & 1 \\ 0 & 1 & 0 \\ -1 & 0 & 0 \end{bmatrix}$$

If $A = R_{z,\theta}$ *relative to the frame* $o_0x_0y_0z_0$*, then, relative to frame* $o_1x_1y_1z_1$ *we have*

$$B = (R_1^0)^{-1} A R_1^0 = \begin{bmatrix} 1 & 0 & 0 \\ 0 & c_\theta & s_\theta \\ 0 & -s_\theta & c_\theta \end{bmatrix}$$

In other words, B *is a rotation about the* z_0*-axis but expressed relative to the frame* $o_1x_1y_1z_1$*. This notion will be useful below and in later sections.*

◇

2.4 COMPOSITION OF ROTATIONS

In this section we discuss the composition of rotations. It is important for subsequent chapters that the reader understand the material in this section thoroughly before moving on.

---

[3] See Appendix B.

2.4.1 Rotation with Respect to the Current Frame

Recall that the matrix R_1^0 in Equation (2.8) represents a rotational transformation between the frames $o_0x_0y_0z_0$ and $o_1x_1y_1z_1$. Suppose we now add a third coordinate frame $o_2x_2y_2z_2$ related to the frames $o_0x_0y_0z_0$ and $o_1x_1y_1z_1$ by rotational transformations. A given point p can then be represented by coordinates specified with respect to any of these three frames: p^0, p^1, and p^2. The relationship among these representations of p is

$$p^0 = R_1^0 p^1 \tag{2.12}$$
$$p^1 = R_2^1 p^2 \tag{2.13}$$
$$p^0 = R_2^0 p^2 \tag{2.14}$$

where each R_j^i is a rotation matrix. Substituting Equation (2.13) into Equation (2.12) gives

$$p^0 = R_1^0 R_2^1 p^2 \tag{2.15}$$

Note that R_1^0 and R_2^0 represent rotations relative to the frame $o_0x_0y_0z_0$ while R_2^1 represents a rotation relative to the frame $o_1x_1y_1z_1$. Comparing Equations (2.14) and (2.15) we can immediately infer

$$R_2^0 = R_1^0 R_2^1 \tag{2.16}$$

Equation (2.16) is the composition law for rotational transformations. It states that, in order to transform the coordinates of a point p from its representation p^2 in the frame $o_2x_2y_2z_2$ to its representation p^0 in the frame $o_0x_0y_0z_0$, we may first transform to its coordinates p^1 in the frame $o_1x_1y_1z_1$ using R_2^1 and then transform p^1 to p^0 using R_1^0.

We may also interpret Equation (2.16) as follows. Suppose that initially all three of the coordinate frames coincide. We first rotate the frame $o_2x_2y_2z_2$ relative to $o_0x_0y_0z_0$ according to the transformation R_1^0. Then, with the frames $o_1x_1y_1z_1$ and $o_2x_2y_2z_2$ coincident, we rotate $o_2x_2y_2z_2$ relative to $o_1x_1y_1z_1$ according to the transformation R_2^1. The resulting frame, $o_2x_2y_2z_2$ has orientation with respect to $o_0x_0y_0z_0$ given by $R_1^0 R_2^1$. We call the frame relative to which the rotation occurs the **current frame**.

Example 2.5

Suppose a rotation matrix R represents a rotation of angle ϕ about the current y-axis followed by a rotation of angle θ about the current z-axis as

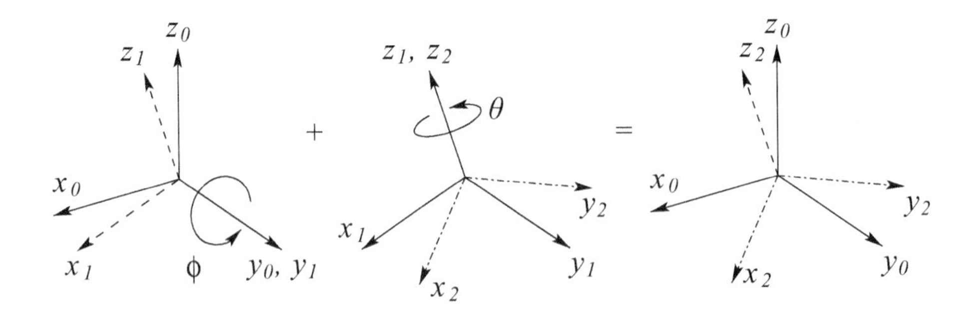

Figure 2.8: Composition of rotations about current axes.

shown in Figure 2.8. Then the matrix R is given by

$$\begin{aligned} R &= R_{y,\phi}R_{z,\theta} \\ &= \begin{bmatrix} c_\phi & 0 & s_\phi \\ 0 & 1 & 0 \\ -s_\phi & 0 & c_\phi \end{bmatrix} \begin{bmatrix} c_\theta & -s_\theta & 0 \\ s_\theta & c_\theta & 0 \\ 0 & 0 & 1 \end{bmatrix} \\ &= \begin{bmatrix} c_\phi c_\theta & -c_\phi s_\theta & s_\phi \\ s_\theta & c_\theta & 0 \\ -s_\phi c_\theta & s_\phi s_\theta & c_\phi \end{bmatrix} \end{aligned} \tag{2.17}$$

◇

It is important to remember that the order in which a sequence of rotations is performed, and consequently the order in which the rotation matrices are multiplied together, is crucial. The reason is that rotation, unlike position, is not a vector quantity and so rotational transformations do not commute in general.

Example 2.6

Suppose that the above rotations are performed in the reverse order, that is, first a rotation about the current z-axis followed by a rotation about the current y-axis. Then the resulting rotation matrix is given by

$$\begin{aligned} R' &= R_{z,\theta}R_{y,\phi} \\ &= \begin{bmatrix} c_\theta & -s_\phi & 0 \\ s_\theta & c_\theta & 0 \\ 0 & 0 & 1 \end{bmatrix} \begin{bmatrix} c_\phi & 0 & s_\phi \\ 0 & 1 & 0 \\ -s_\phi & 0 & c_\phi \end{bmatrix} \\ &= \begin{bmatrix} c_\theta c_\phi & -s_\theta & c_\theta s_\phi \\ s_\theta c_\phi & c_\theta & s_\theta s_\phi \\ -s_\phi & 0 & c_\phi \end{bmatrix} \end{aligned} \tag{2.18}$$

Comparing Equations (2.17) and (2.18) we see that $R \neq R'$.
◇

2.4.2 Rotation with Respect to the Fixed Frame

Many times it is desired to perform a sequence of rotations, each about a given fixed coordinate frame, rather than about successive current frames. For example we may wish to perform a rotation about x_0 followed by a rotation about y_0 (and not y_1!). We will refer to $o_0x_0y_0z_0$ as the **fixed frame**. In this case the composition law given by Equation (2.16) is not valid. It turns out that the correct composition law in this case is simply to multiply the successive rotation matrices *in the reverse order* from that given by Equation (2.16). Note that the rotations themselves are not performed in reverse order. Rather they are performed about the fixed frame instead of about the current frame.

To see this, suppose we have two frames $o_0x_0y_0z_0$ and $o_1x_1y_1z_1$ related by the rotational transformation R_1^0. If $R \in SO(3)$ represents a rotation relative to $o_0x_0y_0z_0$, we know from Section 2.3.1 that the representation for R in the **current** frame $o_1x_1y_1z_1$ is given by $(R_1^0)^{-1}RR_1^0$. Therefore, applying the composition law for rotations about the current axis yields

$$R_2^0 = R_1^0\left[(R_1^0)^{-1}RR_1^0\right] = RR_1^0 \tag{2.19}$$

Thus, when a rotation R is performed with respect to the world coordinate frame, the current rotation matrix is *premultiplied* by R to obtain the desired rotation matrix.

Example 2.7 Rotations about Fixed Axes

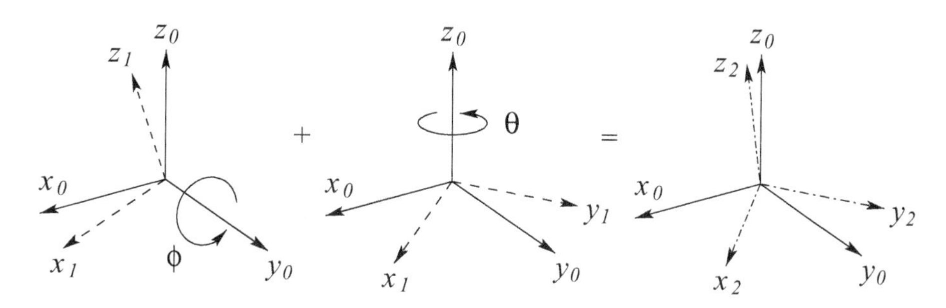

Figure 2.9: Composition of rotations about fixed axes.

Referring to Figure 2.9, suppose that a rotation matrix R represents a rotation of angle ϕ about y_0 followed by a rotation of angle θ about the *fixed* z_0. *The second rotation about the fixed axis is given by* $R_{y,-\phi}R_{z,\theta}R_{y,\phi}$, *which is the basic rotation about the* z*-axis expressed relative to the frame* $o_1x_1y_1z_1$ *using a similarity transformation. Therefore, the composition rule for rotational transformations gives us*

$$R = R_{y,\phi}\left[R_{y,-\phi}R_{z,\theta}R_{y,\phi}\right] = R_{z,\theta}R_{y,\phi} \tag{2.20}$$

It is not necessary to remember the above derivation, only to note by comparing Equation (2.20) with Equation (2.17) that we obtain the same basic rotation matrices, but in the reverse order.

◇

2.4.3 Rules for Composition of Rotational Transformations

We can summarize the rule of composition of rotational transformations by the following recipe. Given a fixed frame $o_0x_0y_0z_0$ and a current frame $o_1x_1y_1z_1$, together with rotation matrix R_1^0 relating them, if a third frame $o_2x_2y_2z_2$ is obtained by a rotation R performed relative to the **current frame** then **postmultiply** R_1^0 by $R = R_2^1$ to obtain

$$R_2^0 \quad = \quad R_1^0 R_2^1 \tag{2.21}$$

If the second rotation is to be performed relative to the **fixed frame** then it is both confusing and inappropriate to use the notation R_2^1 to represent this rotation. Therefore, if we represent the rotation by R, we **premultiply** R_1^0 by R to obtain

$$R_2^0 \quad = \quad R R_1^0 \tag{2.22}$$

In each case R_2^0 represents the transformation between the frames $o_0x_0y_0z_0$ and $o_2x_2y_2z_2$. The frame $o_2x_2y_2z_2$ that results from Equation (2.21) will be different from that resulting from Equation (2.22).

Using the above rule for composition of rotations, it is an easy matter to determine the result of multiple sequential rotational transformations.

Example 2.8

Suppose R *is defined by the following sequence of basic rotations in the order specified:*

1. *A rotation of* θ *about the current* x*-axis*
2. *A rotation of* ϕ *about the current* z*-axis*
3. *A rotation of* α *about the fixed* z*-axis*

4. A rotation of β about the current y-axis

5. A rotation of δ about the fixed x-axis

In order to determine the cumulative effect of these rotations we simply begin with the first rotation $R_{x,\theta}$ and pre- or postmultiply as the case may be to obtain

$$R = R_{x,\delta} R_{z,\alpha} R_{x,\theta} R_{z,\phi} R_{y,\beta} \tag{2.23}$$

◇

2.5 PARAMETERIZATIONS OF ROTATIONS

The nine elements r_{ij} in a general rotational transformation $R \in SO(3)$ are not independent quantities. Indeed a rigid body possesses at most three rotational degrees of freedom, and thus at most three quantities are required to specify its orientation. This can be easily seen by examining the constraints that govern the matrices in $SO(3)$:

$$\sum_i r_{ij}^2 = 1, \quad j \in \{1, 2, 3\} \tag{2.24}$$

$$r_{1i}r_{1j} + r_{2i}r_{2j} + r_{3i}r_{3j} = 0, \quad i \neq j \tag{2.25}$$

Equation (2.24) follows from the fact that the columns of a rotation matrix are unit vectors, and Equation (2.25) follows from the fact that columns of a rotation matrix are mutually orthogonal. Together, these constraints define six independent equations with nine unknowns, which implies that there are three free variables.

In this section we derive three ways in which an arbitrary rotation can be represented using only three independent quantities: the **Euler-angle** representation, the **roll-pitch-yaw** representation, and the **axis/angle** representation.

2.5.1 Euler Angles

A common method of specifying a rotation matrix in terms of three independent quantities is to use the so-called **Euler angles**. Consider the fixed coordinate frame $o_0x_0y_0z_0$ and the rotated frame $o_1x_1y_1z_1$ shown in Figure 2.10. We can specify the orientation of the frame $o_1x_1y_1z_1$ relative to the frame $o_0x_0y_0z_0$ by three angles (ϕ, θ, ψ), known as Euler angles, and obtained by three successive rotations as follows. First rotate about the

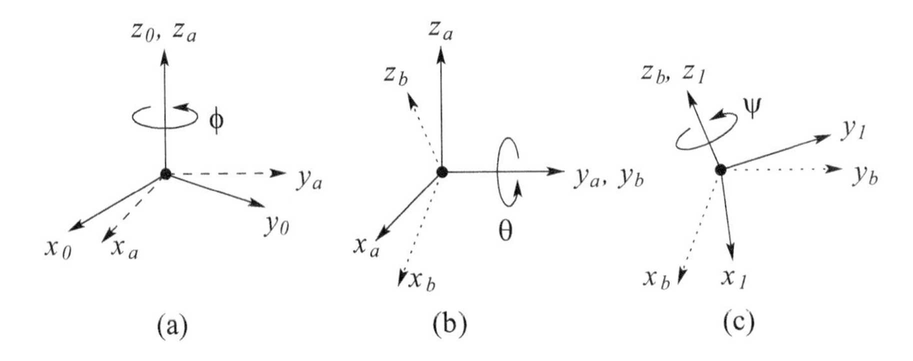

Figure 2.10: Euler angle representation.

z-axis by the angle ϕ. Next rotate about the current y-axis by the angle θ. Finally rotate about the current z-axis by the angle ψ. In Figure 2.10, frame $o_a x_a y_a z_a$ represents the new coordinate frame after the rotation by ϕ, frame $o_b x_b y_b z_b$ represents the new coordinate frame after the rotation by θ, and frame $o_1 x_1 y_1 z_1$ represents the final frame, after the rotation by ψ. Frames $o_a x_a y_a z_a$ and $o_b x_b y_b z_b$ are shown in the figure only to help visualize the rotations.

In terms of the basic rotation matrices the resulting rotational transformation can be generated as the product

$$\begin{aligned} R_{ZYZ} &= R_{z,\phi} R_{y,\theta} R_{z,\psi} \\ &= \begin{bmatrix} c_\phi & -s_\phi & 0 \\ s_\phi & c_\phi & 0 \\ 0 & 0 & 1 \end{bmatrix} \begin{bmatrix} c_\theta & 0 & s_\theta \\ 0 & 1 & 0 \\ -s_\theta & 0 & c_\theta \end{bmatrix} \begin{bmatrix} c_\psi & -s_\psi & 0 \\ s_\psi & c_\psi & 0 \\ 0 & 0 & 1 \end{bmatrix} \\ &= \begin{bmatrix} c_\phi c_\theta c_\psi - s_\phi s_\psi & -c_\phi c_\theta s_\psi - s_\phi c_\psi & c_\phi s_\theta \\ s_\phi c_\theta c_\psi + c_\phi s_\psi & -s_\phi c_\theta s_\psi + c_\phi c_\psi & s_\phi s_\theta \\ -s_\theta c_\psi & s_\theta s_\psi & c_\theta \end{bmatrix} \end{aligned} \tag{2.26}$$

The matrix R_{ZYZ} in Equation (2.26) is called the ZYZ**-Euler angle transformation**.

The more important and more difficult problem is to determine for a particular $R = (r_{ij})$ the set of Euler angles ϕ, θ, and ψ, that satisfy

$$R = \begin{bmatrix} c_\phi c_\theta c_\psi - s_\phi s_\psi & -c_\phi c_\theta s_\psi - s_\phi c_\psi & c_\phi s_\theta \\ s_\phi c_\theta c_\psi + c_\phi s_\psi & -s_\phi c_\theta s_\psi + c_\phi c_\psi & s_\phi s_\theta \\ -s_\theta c_\psi & s_\theta s_\psi & c_\theta \end{bmatrix} \tag{2.27}$$

for a matrix $R \in SO(3)$. This problem will be important later when we address the inverse kinematics problem for manipulators in Section 3.3.

To find a solution for this problem we break it down into two cases. First, suppose that not both of r_{13}, r_{23} are zero. Then from Equation (2.26) we deduce that $s_\theta \neq 0$, and hence that not both of r_{31}, r_{32} are zero. If not both r_{13} and r_{23} are zero, then $r_{33} \neq \pm 1$, and we have $c_\theta = r_{33}$, $s_\theta = \pm\sqrt{1 - r_{33}^2}$ so

$$\theta = \text{Atan2}\left(r_{33}, \sqrt{1 - r_{33}^2}\right) \tag{2.28}$$

or

$$\theta = \text{Atan2}\left(r_{33}, -\sqrt{1 - r_{33}^2}\right) \tag{2.29}$$

where the function Atan2 is the **two-argument arctangent function** defined in Appendix A.

If we choose the value for θ given by Equation (2.28), then $s_\theta > 0$, and

$$\phi = \text{Atan2}(r_{13}, r_{23}) \tag{2.30}$$
$$\psi = \text{Atan2}(-r_{31}, r_{32}) \tag{2.31}$$

If we choose the value for θ given by Equation (2.29), then $s_\theta < 0$, and

$$\phi = \text{Atan2}(-r_{13}, -r_{23}) \tag{2.32}$$
$$\psi = \text{Atan2}(r_{31}, -r_{32}) \tag{2.33}$$

Thus, there are two solutions depending on the sign chosen for θ.

If $r_{13} = r_{23} = 0$, then the fact that R is orthogonal implies that $r_{33} = \pm 1$, and that $r_{31} = r_{32} = 0$. Thus, R has the form

$$R = \begin{bmatrix} r_{11} & r_{12} & 0 \\ r_{21} & r_{22} & 0 \\ 0 & 0 & \pm 1 \end{bmatrix} \tag{2.34}$$

If $r_{33} = 1$, then $c_\theta = 1$ and $s_\theta = 0$, so that $\theta = 0$. In this case, Equation (2.26) becomes

$$\begin{bmatrix} c_\phi c_\psi - s_\phi s_\psi & -c_\phi s_\psi - s_\phi c_\psi & 0 \\ s_\phi c_\psi + c_\phi s_\psi & -s_\phi s_\psi + c_\phi c_\psi & 0 \\ 0 & 0 & 1 \end{bmatrix} = \begin{bmatrix} c_{\phi+\psi} & -s_{\phi+\psi} & 0 \\ s_{\phi+\psi} & c_{\phi+\psi} & 0 \\ 0 & 0 & 1 \end{bmatrix}$$

Thus, the sum $\phi + \psi$ can be determined as

$$\phi + \psi = \text{Atan2}(r_{11}, r_{21}) = \text{Atan2}(r_{11}, -r_{12}) \tag{2.35}$$

Since only the sum $\phi + \psi$ can be determined in this case, there are infinitely many solutions. In this case, we may take $\phi = 0$ by convention. If $r_{33} = -1$, then $c_\theta = -1$ and $s_\theta = 0$, so that $\theta = \pi$. In this case Equation (2.26) becomes

$$\begin{bmatrix} -c_{\phi-\psi} & -s_{\phi-\psi} & 0 \\ s_{\phi-\psi} & c_{\phi-\psi} & 0 \\ 0 & 0 & -1 \end{bmatrix} = \begin{bmatrix} r_{11} & r_{12} & 0 \\ r_{21} & r_{22} & 0 \\ 0 & 0 & -1 \end{bmatrix} \tag{2.36}$$

The solution is thus

$$\phi - \psi = \text{Atan2}(-r_{11}, -r_{12}) \tag{2.37}$$

As before there are infinitely many solutions.

2.5.2 Roll, Pitch, Yaw Angles

A rotation matrix R can also be described as a product of successive rotations about the principal coordinate axes x_0, y_0, and z_0 taken in a specific order. These rotations define the **roll**, **pitch**, and **yaw** angles, which we shall also denote ϕ, θ, ψ, and which are shown in Figure 2.11.

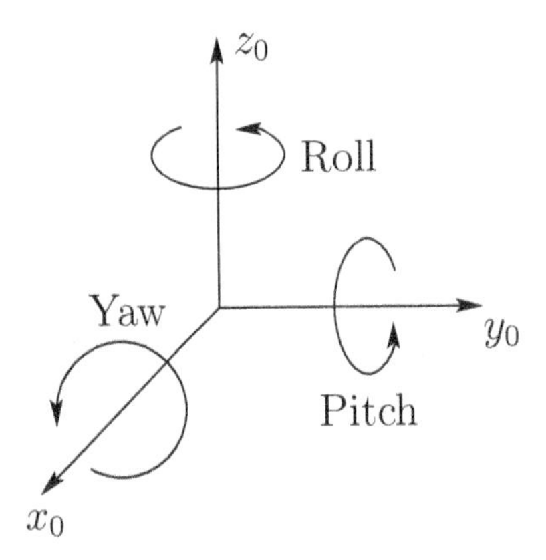

Figure 2.11: Roll, pitch, and yaw angles.

We specify the order of rotation as $x - y - z$, in other words, first a yaw about x_0 through an angle ψ, then pitch about the y_0 by an angle θ, and finally roll about the z_0 by an angle ϕ.[4] Since the successive rotations are

[4]It should be noted that other conventions exist for naming the roll, pitch, and yaw angles.

relative to the fixed frame, the resulting transformation matrix is given by

$$\begin{aligned} R &= R_{z,\phi}R_{y,\theta}R_{x,\psi} \\ &= \begin{bmatrix} c_\phi & -s_\phi & 0 \\ s_\phi & c_\phi & 0 \\ 0 & 0 & 1 \end{bmatrix} \begin{bmatrix} c_\theta & 0 & s_\theta \\ 0 & 1 & 0 \\ -s_\theta & 0 & c_\theta \end{bmatrix} \begin{bmatrix} 1 & 0 & 0 \\ 0 & c_\psi & -s_\psi \\ 0 & s_\psi & c_\psi \end{bmatrix} \\ &= \begin{bmatrix} c_\phi c_\theta & -s_\phi c_\psi + c_\phi s_\theta s_\psi & s_\phi s_\psi + c_\phi s_\theta c_\psi \\ s_\phi c_\theta & c_\phi c_\psi + s_\phi s_\theta s_\psi & -c_\phi s_\psi + s_\phi s_\theta c_\psi \\ -s_\theta & c_\theta s_\psi & c_\theta c_\psi \end{bmatrix} \end{aligned} \tag{2.38}$$

Of course, instead of yaw-pitch-roll relative to the fixed frames we could also interpret the above transformation as roll-pitch-yaw, in that order, each taken with respect to the current frame. The end result is the same matrix as in Equation (2.38).

The three angles ϕ, θ, and ψ can be obtained for a given rotation matrix using a method that is similar to that used to derive the Euler angles above.

2.5.3 Axis/Angle Representation

Rotations are not always performed about the principal coordinate axes. We are often interested in a rotation about an arbitrary axis in space. This provides both a convenient way to describe rotations, and an alternative parameterization for rotation matrices. Let $k = [k_x, k_y, k_z]^T$, expressed in the frame $o_0x_0y_0z_0$, be a unit vector defining an axis. We wish to derive the rotation matrix $R_{k,\theta}$ representing a rotation of θ about this axis.

There are several ways in which the matrix $R_{k,\theta}$ can be derived. One approach is to note that the rotational transformation $R = R_{z,\alpha}R_{y,\beta}$ will bring the world z-axis into alignment with the vector k. Therefore, a rotation about the axis k can be computed using a similarity transformation as

$$R_{k,\theta} = RR_{z,\theta}R^{-1} \tag{2.39}$$

$$= R_{z,\alpha}R_{y,\beta}R_{z,\theta}R_{y,-\beta}R_{z,-\alpha} \tag{2.40}$$

From Figure 2.12 we see that

$$\sin\alpha = \frac{k_y}{\sqrt{k_x^2 + k_y^2}} \qquad \cos\alpha = \frac{k_x}{\sqrt{k_x^2 + k_y^2}} \tag{2.41}$$

$$\sin\beta = \sqrt{k_x^2 + k_y^2} \qquad \cos\beta = k_z \tag{2.42}$$

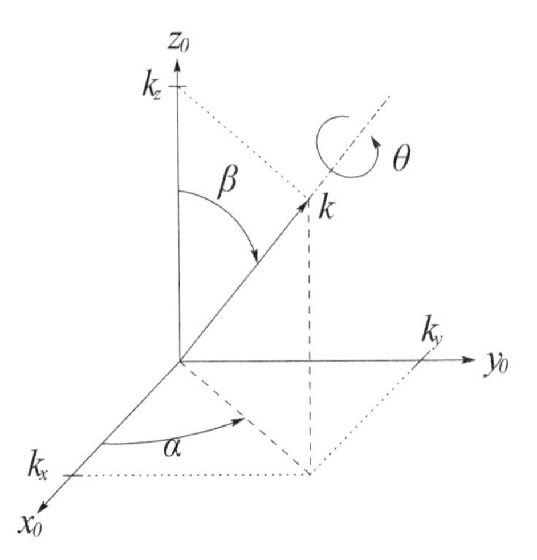

Figure 2.12: Rotation about an arbitrary axis.

Note that the final two equations follow from the fact that k is a unit vector. Substituting Equations (2.41) and (2.42) into Equation (2.40), we obtain after some lengthy calculation (Problem 2-17)

$$R_{k,\theta} = \left[\begin{array}{c|c|c} k_x^2 v_\theta + c_\theta & k_x k_y v_\theta - k_z s_\theta & k_x k_z v_\theta + k_y s_\theta \\ k_x k_y v_\theta + k_z s_\theta & k_y^2 v_\theta + c_\theta & k_y k_z v_\theta - k_x s_\theta \\ k_x k_z v_\theta - k_y s_\theta & k_y k_z v_\theta + k_x s_\theta & k_z^2 v_\theta + c_\theta \end{array}\right] \tag{2.43}$$

where $v_\theta = \text{ vers } \theta = 1 - c_\theta$.

In fact, any rotation matrix $R \in S0(3)$ can be represented by a single rotation about a suitable axis in space by a suitable angle,

$$R = R_{k,\theta} \tag{2.44}$$

where k is a unit vector defining the axis of rotation, and θ is the angle of rotation about k. The pair (k, θ) is called the **axis/angle representation** of R. Given an arbitrary rotation matrix R with components r_{ij}, the equivalent angle θ and equivalent axis k are given by the expressions

$$\theta = \cos^{-1}\left(\frac{r_{11} + r_{22} + r_{33} - 1}{2}\right)$$

and

$$k = \frac{1}{2\sin\theta}\left[\begin{array}{c} r_{32} - r_{23} \\ r_{13} - r_{31} \\ r_{21} - r_{12} \end{array}\right] \tag{2.45}$$

These equations can be obtained by direct manipulation of the entries of the matrix given in Equation (2.43). The axis/angle representation is not unique since a rotation of $-\theta$ about $-k$ is the same as a rotation of θ about k, that is,

$$R_{k,\theta} = R_{-k,-\theta} \tag{2.46}$$

If $\theta = 0$ then R is the identity matrix and the axis of rotation is undefined.

Example 2.9

Suppose R is generated by a rotation of $90°$ about z_0 followed by a rotation of $30°$ about y_0 followed by a rotation of $60°$ about x_0. Then

$$\begin{aligned} R &= R_{x,60}R_{y,30}R_{z,90} \\ &= \begin{bmatrix} 0 & -\frac{\sqrt{3}}{2} & \frac{1}{2} \\ \frac{1}{2} & -\frac{\sqrt{3}}{4} & -\frac{3}{4} \\ \frac{\sqrt{3}}{2} & \frac{1}{4} & \frac{\sqrt{3}}{4} \end{bmatrix} \end{aligned} \tag{2.47}$$

We see that $Tr(R) = 0$ and hence the equivalent angle is given by Equation (2.45) as

$$\theta = \cos^{-1}\left(-\frac{1}{2}\right) = 120° \tag{2.48}$$

The equivalent axis is given from Equation (2.45) as

$$k = \left[\frac{1}{\sqrt{3}}, \frac{1}{2\sqrt{3}} - \frac{1}{2}, \frac{1}{2\sqrt{3}} + \frac{1}{2}\right]^T \tag{2.49}$$

◇

The above axis/angle representation characterizes a given rotation by four quantities, namely the three components of the equivalent axis k and the equivalent angle θ. However, since the equivalent axis k is given as a unit vector only two of its components are independent. The third is constrained by the condition that k is of unit length. Therefore, only three independent quantities are required in this representation of a rotation R. We can represent the equivalent axis/angle by a single vector r as

$$r = [r_x, r_y, r_z]^T = [\theta k_x, \theta k_y, \theta k_z]^T \tag{2.50}$$

Note, since k is a unit vector, that the length of the vector r is the equivalent angle θ and the direction of r is the equivalent axis k.

One should be careful to note that the representation in Equation (2.50) does not mean that two axis/angle representations may be combined using standard rules of vector algebra, as doing so would imply that rotations commute which, as we have seen, is not true in general.

2.6 RIGID MOTIONS

We have now seen how to represent both positions and orientations. We combine these two concepts in this section to define a **rigid motion** and, in the next section, we derive an efficient matrix representation for rigid motions using the notion of homogeneous transformation.

Definition 2.1 *A rigid motion is an ordered pair* (d, R) *where* $d \in \mathbb{R}^3$ *and* $R \in SO(3)$. *The group of all rigid motions is known as the* **Special Euclidean Group** *and is denoted by* $SE(3)$. *We see then that* $SE(3) = \mathbb{R}^3 \times SO(3)$.

A rigid motion is a pure translation together with a pure rotation.[5] Let R_1^0 be the rotation matrix that specifies the orientation of frame $o_1x_1y_1z_1$ with respect to $o_0x_0y_0z_0$, and d be the vector from the origin of frame $o_0x_0y_0z_0$ to the origin of frame $o_1x_1y_1z_1$. Suppose the point p is rigidly attached to coordinate frame $o_1x_1y_1z_1$, with local coordinates p^1. We can express the coordinates of p with respect to frame $o_0x_0y_0z_0$ using

$$p^0 = R_1^0 p^1 + d^0 \tag{2.51}$$

Now consider three coordinate frames $o_0x_0y_0z_0$, $o_1x_1y_1z_1$, and $o_2x_2y_2z_2$. Let d_1 be the vector from the origin of $o_0x_0y_0z_0$ to the origin of $o_1x_1y_1z_1$ and d_2 be the vector from the origin of $o_1x_1y_1z_1$ to the origin of $o_2x_2y_2z_2$. If the point p is attached to frame $o_2x_2y_2z_2$ with local coordinates p^2, we can compute its coordinates relative to frame $o_0x_0y_0z_0$ using

$$p^1 = R_2^1 p^2 + d_2^1 \tag{2.52}$$

and

$$p^0 = R_1^0 p^1 + d_1^0 \tag{2.53}$$

The composition of these two equations defines a third rigid motion, which we can describe by substituting the expression for p^1 from Equation (2.52) into Equation (2.53)

$$p^0 = R_1^0 R_2^1 p^2 + R_1^0 d_2^1 + d_1^0 \tag{2.54}$$

[5]The definition of rigid motion is sometimes broadened to include **reflections**, which correspond to $\det R = -1$. We will always assume in this text that $\det R = +1$ so that $R \in SO(3)$.

Since the relationship between p^0 and p^2 is also a rigid motion, we can equally describe it as

$$p^0 = R_2^0 p^2 + d_2^0 \tag{2.55}$$

Comparing Equations (2.54) and (2.55) we have the relationships

$$R_2^0 = R_1^0 R_2^1 \tag{2.56}$$
$$d_2^0 = d_1^0 + R_1^0 d_2^1 \tag{2.57}$$

Equation (2.56) shows that the orientation transformations can simply be multiplied together and Equation (2.57) shows that the vector from the origin o_0 to the origin o_2 has coordinates given by the sum of d_1^0 (the vector from o_0 to o_1 expressed with respect to $o_0x_0y_0z_0$) and $R_1^0 d_2^1$ (the vector from o_1 to o_2, expressed in the orientation of the coordinate frame $o_0x_0y_0z_0$).

2.7 HOMOGENEOUS TRANSFORMATIONS

One can easily see that the calculation leading to Equation (2.54) would quickly become intractable if a long sequence of rigid motions were considered. In this section we show how rigid motions can be represented in matrix form so that composition of rigid motions can be reduced to matrix multiplication as was the case for composition of rotations.

In fact, a comparison of Equations (2.56) and (2.57) with the matrix identity

$$\begin{bmatrix} R_1^0 & d_1^0 \\ 0 & 1 \end{bmatrix} \begin{bmatrix} R_2^1 & d_1^2 \\ 0 & 1 \end{bmatrix} = \begin{bmatrix} R_1^0 R_2^1 & R_1^0 d_1^2 + d_1^0 \\ 0 & 1 \end{bmatrix} \tag{2.58}$$

where 0 denotes the row vector $(0, 0, 0)$, shows that the rigid motions can be represented by the set of matrices of the form

$$H = \begin{bmatrix} R & d \\ 0 & 1 \end{bmatrix}, \quad R \in SO(3), \quad d \in \mathbb{R}^3 \tag{2.59}$$

Transformation matrices of the form given in Equation (2.59) are called **homogeneous transformations.** A homogeneous transformation is therefore nothing more than a matrix representation of a rigid motion and we will use $SE(3)$ interchangeably to represent both the set of rigid motions and the set of all 4×4 matrices H of the form given in Equation (2.59).

Using the fact that R is orthogonal it is an easy exercise to show that the inverse transformation H^{-1} is given by

$$H^{-1} = \begin{bmatrix} R^T & -R^T d \\ 0 & 1 \end{bmatrix} \tag{2.60}$$

In order to represent the transformation given in Equation (2.51) by a matrix multiplication, we must augment the vectors p^0 and p^1 by the addition of a fourth component of 1 as follows,

$$P^0 = \begin{bmatrix} p^0 \\ 1 \end{bmatrix} \tag{2.61}$$

$$P^1 = \begin{bmatrix} p^1 \\ 1 \end{bmatrix} \tag{2.62}$$

The vectors P^0 and P^1 are known as **homogeneous representations** of the vectors p^0 and p^1, respectively. It can now be seen directly that the transformation given in Equation (2.51) is equivalent to the (homogeneous) matrix equation

$$P^0 = H_1^0 P^1 \tag{2.63}$$

A set of **basic homogeneous transformations** generating $SE(3)$ is given by

$$\text{Trans}_{x,a} = \begin{bmatrix} 1 & 0 & 0 & a \\ 0 & 1 & 0 & 0 \\ 0 & 0 & 1 & 0 \\ 0 & 0 & 0 & 1 \end{bmatrix}, \quad \text{Rot}_{x,\alpha} = \begin{bmatrix} 1 & 0 & 0 & 0 \\ 0 & c_\alpha & -s_\alpha & 0 \\ 0 & s_\alpha & c_\alpha & 0 \\ 0 & 0 & 0 & 1 \end{bmatrix} \tag{2.64}$$

$$\text{Trans}_{y,b} = \begin{bmatrix} 1 & 0 & 0 & 0 \\ 0 & 1 & 0 & b \\ 0 & 0 & 1 & 0 \\ 0 & 0 & 0 & 1 \end{bmatrix}, \quad \text{Rot}_{y,\beta} = \begin{bmatrix} c_\beta & 0 & s_\beta & 0 \\ 0 & 1 & 0 & 0 \\ -s_\beta & 0 & c_\beta & 0 \\ 0 & 0 & 0 & 1 \end{bmatrix} \tag{2.65}$$

$$\text{Trans}_{z,c} = \begin{bmatrix} 1 & 0 & 0 & 0 \\ 0 & 1 & 0 & 0 \\ 0 & 0 & 1 & c \\ 0 & 0 & 0 & 1 \end{bmatrix}, \quad \text{Rot}_{z,\gamma} = \begin{bmatrix} c_\gamma & -s_\gamma & 0 & 0 \\ s_\gamma & c_\gamma & 0 & 0 \\ 0 & 0 & 1 & 0 \\ 0 & 0 & 0 & 1 \end{bmatrix} \tag{2.66}$$

for translation and rotation about the x, y, z-axes, respectively.

The most general homogeneous transformation that we will consider may be written now as

$$H_1^0 = \begin{bmatrix} n_x & s_x & a_x & d_x \\ n_y & s_y & a_y & d_y \\ n_z & s_x & a_z & d_z \\ 0 & 0 & 0 & 1 \end{bmatrix} = \begin{bmatrix} n & s & a & d \\ 0 & 0 & 0 & 1 \end{bmatrix} \tag{2.67}$$

In the above equation $n = [n_x, n_y, n_z]^T$ is a vector representing the direction of x_1 in the $o_0x_0y_0z_0$ frame, $s = [s_x, s_y, s_z]^T$ represents the direction of y_1, and $a = [a_x, a_y, a_z]^T$ represents the direction of z_1. The vector $d = [d_x, d_y, d_z]^T$ represents the vector from the origin o_0 to the origin o_1 expressed in the frame $o_0x_0y_0z_0$. The rationale behind the choice of letters n, s, and a is explained in Chapter 3.

The same interpretation regarding composition and ordering of transformations holds for 4×4 homogeneous transformations as for 3×3 rotations. Given a homogeneous transformation H_1^0 relating two frames, if a second rigid motion, represented by $H \in SE(3)$ is performed relative to the current frame, then

$$H_2^0 = H_1^0 H$$

whereas if the second rigid motion is performed relative to the fixed frame, then

$$H_2^0 = H H_1^0$$

Example 2.10

The homogeneous transformation matrix H that represents a rotation by angle α about the current x-axis followed by a translation of b units along the current x-axis, followed by a translation of d units along the current z-axis, followed by a rotation by angle θ about the current z-axis, is given by

$$\begin{aligned} H &= Rot_{x,\alpha} Trans_{x,b} Trans_{z,d} Rot_{z,\theta} \\ &= \begin{bmatrix} c_\theta & -s_\theta & 0 & b \\ c_\alpha s_\theta & c_\alpha c_\theta & -s_\alpha & -ds_\alpha \\ s_\alpha s_\theta & s_\alpha c_\theta & c_\alpha & dc_\alpha \\ 0 & 0 & 0 & 1 \end{bmatrix} \end{aligned}$$

◇

2.8 SUMMARY

In this chapter, we have seen how matrices in $SE(n)$ can be used to represent the relative position and orientation of two coordinate frames for $n = 2, 3$. We have adopted a notional convention in which a superscript is used to indicate a reference frame. Thus, the notation p^0 represents the coordinates of the point p relative to frame 0.

The relative orientation of two coordinate frames can be specified by a rotation matrix, $R \in SO(n)$, with $n = 2, 3$. In two dimensions, the orientation of frame 1 with respect to frame 0 is given by

$$R_1^0 = \begin{bmatrix} x_1 \cdot x_0 & y_1 \cdot x_0 \\ x_1 \cdot y_0 & y_1 \cdot y_0 \end{bmatrix} = \begin{bmatrix} \cos\theta & -\sin\theta \\ \sin\theta & \cos\theta \end{bmatrix}$$

in which θ is the angle between the two coordinate frames. In the three-dimensional case, the rotation matrix is given by

$$R_1^0 = \begin{bmatrix} x_1 \cdot x_0 & y_1 \cdot x_0 & z_1 \cdot x_0 \\ x_1 \cdot y_0 & y_1 \cdot y_0 & z_1 \cdot y_0 \\ x_1 \cdot z_0 & y_1 \cdot z_0 & z_1 \cdot z_0 \end{bmatrix}$$

In each case, the columns of the rotation matrix are obtained by projecting an axis of the target frame (in this case, frame 1) onto the coordinate axes of the reference frame (in this case, frame 0).

The set of $n \times n$ rotation matrices is known as the special orthogonal group of order n, and is denoted by $SO(n)$. An important property of these matrices is that $R^{-1} = R^T$ for any $R \in SO(n)$.

Rotation matrices can be used to perform coordinate transformations between frames that differ only in orientation. We derived rules for the composition of rotational transformations as

$$R_2^0 = R_1^0 R$$

for the case where the second transformation, R, is performed relative to the current frame and

$$R_2^0 = R R_1^0$$

for the case where the second transformation, R, is performed relative to the fixed frame.

In the three-dimensional case, a rotation matrix can be parameterized using three angles. A common convention is to use the Euler angles (ϕ, θ, ψ), which correspond to successive rotations about the z, y, and z axes. The corresponding rotation matrix is given by

$$R(\phi, \theta, \psi) = R_{z,\phi} R_{y,\theta} R_{z,\psi}$$

Roll, pitch, and yaw angles are similar, except that the successive rotations are performed with respect to the fixed, world frame instead of being performed with respect to the current frame.

Homogeneous transformations combine rotation and translation. In the three-dimensional case, a homogeneous transformation has the form

$$H = \begin{bmatrix} R & d \\ 0 & 1 \end{bmatrix}, R \in SO(3), \ d \in \mathbb{R}^3$$

The set of all such matrices comprises the set $SE(3)$, and these matrices can be used to perform coordinate transformations, analogous to rotational transformations using rotation matrices.

Homogeneous transformation matrices can be used to perform coordinate transformations between frames that differ in orientation and translation. We derived rules for the composition of rotational transformations as

$$H_2^0 = H_1^0 H$$

for the case where the second transformation, H, is performed relative to the current frame and

$$H_2^0 = H H_1^0$$

for the case where the second transformation, H, is performed relative to the fixed frame.

PROBLEMS

2-1 Using the fact that $v_1 \cdot v_2 = v_1^T v_2$, show that the dot product of two free vectors does not depend on the choice of frames in which their coordinates are defined.

2-2 Show that the length of a free vector is not changed by rotation, that is, that $\|v\| = \|Rv\|$.

2-3 Show that the distance between points is not changed by rotation, that is, $\|p_1 - p_2\| = \|Rp_1 - Rp_2\|$.

2-4 If a matrix R satisfies $R^T R = I$, show that the column vectors of R are of unit length and mutually perpendicular.

2-5 If a matrix R satisfies $R^T R = I$, then
a) Show that $\det R = \pm 1$
b) Show that $\det R = +1$ if we restrict ourselves to right-handed coordinate frames.

2-6 Verify Equations (2.3)–(2.5).

2-7 A **group** is a set X together with an operation $*$ defined on that set such that

- $x_1 * x_2 \in X$ for all $x_1, x_2 \in X$
- $(x_1 * x_2) * x_3 = x_1 * (x_2 * x_3)$
- There exists an element $I \in X$ such that $I * x = x * I = x$ for all $x \in X$
- For every $x \in X$, there exists some element $y \in X$ such that $x * y = y * x = I$

Show that SO(n) with the operation of matrix multiplication is a group.

2-8 Derive Equations (2.6) and (2.7).

2-9 Suppose A is a 2×2 rotation matrix. In other words $A^T A = I$ and $\det A = 1$. Show that there exists a unique θ such that A is of the form

$$A = \begin{bmatrix} \cos\theta & -\sin\theta \\ \sin\theta & \cos\theta \end{bmatrix}$$

2-10 Consider the following sequence of rotations:

1. Rotate by ϕ about the world x-axis.
2. Rotate by θ about the current z-axis.
3. Rotate by ψ about the world y-axis.

Write the matrix product that will give the resulting rotation matrix (do not perform the matrix multiplication).

2-11 Consider the following sequence of rotations:

1. Rotate by ϕ about the world x-axis.
2. Rotate by θ about the world z-axis.
3. Rotate by ψ about the current x-axis.

Write the matrix product that will give the resulting rotation matrix (do not perform the matrix multiplication).

2-12 Consider the following sequence of rotations:

1. Rotate by ϕ about the world x-axis.

2. Rotate by θ about the current z-axis.
3. Rotate by ψ about the current x-axis.
4. Rotate by α about the world z-axis.

Write the matrix product that will give the resulting rotation matrix (do not perform the matrix multiplication).

2-13 Consider the following sequence of rotations:

1. Rotate by ϕ about the world x-axis.
2. Rotate by θ about the world z-axis.
3. Rotate by ψ about the current x-axis.
4. Rotate by α about the world z-axis.

Write the matrix product that will give the resulting rotation matrix (do not perform the matrix multiplication).

2-14 If the coordinate frame $o_1x_1y_1z_1$ is obtained from the coordinate frame $o_0x_0y_0z_0$ by a rotation of $\frac{\pi}{2}$ about the x-axis followed by a rotation of $\frac{\pi}{2}$ about the fixed y-axis, find the rotation matrix R representing the composite transformation. Sketch the initial and final frames.

2-15 Suppose that three coordinate frames $o_1x_1y_1z_1$, $o_2x_2y_2z_2$, and $o_3x_3y_3z_3$ are given, and suppose

$$R_2^1 = \begin{bmatrix} 1 & 0 & 0 \\ 0 & \frac{1}{2} & -\frac{\sqrt{3}}{2} \\ 0 & \frac{\sqrt{3}}{2} & \frac{1}{2} \end{bmatrix}, \quad R_3^1 = \begin{bmatrix} 0 & 0 & -1 \\ 0 & 1 & 0 \\ 1 & 0 & 0 \end{bmatrix}$$

Find the matrix R_3^2.

2-16 Derive equations for the roll, pitch, and yaw angles corresponding to the rotation matrix $R = (r_{ij})$.

2-17 Verify Equation (2.43).

2-18 Verify Equation (2.45).

2-19 If R is a rotation matrix show that $+1$ is an eigenvalue of R. Let k be a unit eigenvector corresponding to the eigenvalue $+1$. Give a physical interpretation of k.

2-20 Let $k = \frac{1}{\sqrt{3}}[1, 1, 1]^T$, $\theta = 90°$. Find $R_{k,\theta}$.

2-21 Show by direct calculation that $R_{k,\theta}$ given by Equation (2.43) is equal to R given by Equation (2.47) if θ and k are given by Equations (2.48) and (2.49), respectively.

2-22 Compute the rotation matrix given by the product

$$R_{x,\theta}R_{y,\phi}R_{z,\pi}R_{y,-\phi}R_{x,-\theta}$$

2-23 Suppose R represents a rotation of $90°$ about y_0 followed by a rotation of $45°$ about z_1. Find the equivalent axis/angle to represent R. Sketch the initial and final frames and the equivalent axis vector k.

2-24 Find the rotation matrix corresponding to the Euler angles $\phi = \frac{\pi}{2}$, $\theta = 0$, and $\psi = \frac{\pi}{4}$. What is the direction of the x_1 axis relative to the base frame?

2-25 Section 2.5.1 described only the Z-Y-Z Euler angles. List all possible sets of Euler angles. Is it possible to have Z-Z-Y Euler angles? Why or why not?

2-26 Unit magnitude complex numbers $a + ib$ with $a^2 + b^2 = 1$ can be used to represent orientation in the plane. In particular, for the complex number $a + ib$, we can define the angle $\theta = \text{Atan2}(a, b)$. Show that multiplication of two complex numbers corresponds to addition of the corresponding angles.

2-27 Show that complex numbers together with the operation of complex multiplication define a group. What is the identity for the group? What is the inverse for $a + ib$?

2-28 Complex numbers can be generalized by defining three independent square roots for -1 that obey the multiplication rules

$$\begin{aligned} -1 &= i^2 = j^2 = k^2, \\ i &= jk = -kj, \\ j &= ki = -ik, \\ k &= ij = -ji \end{aligned}$$

Using these, we define a **quaternion** by $Q = q_0 + iq_1 + jq_2 + kq_3$, which is typically represented by the 4-tuple (q_0, q_1, q_2, q_3). A rotation by θ about the unit vector $n = [n_x, n_y, n_z]^T$ can be represented by the unit quaternion $Q = \left(\cos\frac{\theta}{2}, n_x \sin\frac{\theta}{2}, n_y \sin\frac{\theta}{2}, n_z \sin\frac{\theta}{2}\right)$. Show that such a quaternion has unit norm, that is, $q_0^2 + q_1^2 + q_2^2 + q_3^2 = 1$.

2-29 Using $Q = (\cos\frac{\theta}{2}, n_x \sin\frac{\theta}{2}, n_y \sin\frac{\theta}{2}, n_z \sin\frac{\theta}{2})$, and the results from Section 2.5.3, determine the rotation matrix R that corresponds to the rotation represented by the quaternion (q_0, q_1, q_2, q_3).

2-30 Determine the quaternion Q that represents the same rotation as given by the rotation matrix R.

2-31 The quaternion $Q = (q_0, q_1, q_2, q_3)$ can be thought of as having a scalar component q_0 and a vector component $q = [q_1, q_2, q_3]^T$. Show that the product of two quaternions, $Z = XY$ is given by

$$\begin{aligned} z_0 &= x_0 y_0 - x^T y \\ z &= x_0 y + y_0 x + x \times y, \end{aligned}$$

Hint: Perform the multiplication $(x_0 + ix_1 + jx_2 + kx_3)(y_0 + iy_1 + jy_2 + ky_3)$ and simplify the result.

2-32 Show that $Q_I = (1, 0, 0, 0)$ is the identity element for unit quaternion multiplication, that is, $QQ_I = Q_I Q = Q$ for any unit quaternion Q.

2-33 The conjugate Q^* of the quaternion Q is defined as

$$Q^* = (q_0, -q_1, -q_2, -q_3)$$

Show that Q^* is the inverse of Q, that is, $Q^*Q = QQ^* = (1, 0, 0, 0)$.

2-34 Let v be a vector whose coordinates are given by $[v_x, v_y, v_z]^T$. If the quaternion Q represents a rotation, show that the new, rotated coordinates of v are given by $Q(0, v_x, v_y, v_z)Q^*$, in which $(0, v_x, v_y, v_z)$ is a quaternion with zero as its real component.

2-35 Let the point p be rigidly attached to the end effector coordinate frame with local coordinates (x, y, z). If Q specifies the orientation of the end effector frame with respect to the base frame, and T is the vector from the base frame to the origin of the end effector frame, show that the coordinates of p with respect to the base frame are given by

$$Q(0, x, y, z)Q^* + T \tag{2.68}$$

in which $(0, x, y, z)$ is a quaternion with zero as its real component.

2-36 Verify Equation (2.60).

2-37 Compute the homogeneous transformation representing a translation of 3 units along the x-axis followed by a rotation of $\frac{\pi}{2}$ about the current z-axis followed by a translation of 1 unit along the fixed y-axis. Sketch the frame. What are the coordinates of the origin o_1 with respect to the original frame in each case?

2-38 Consider the diagram of Figure 2.13. Find the homogeneous transfor-

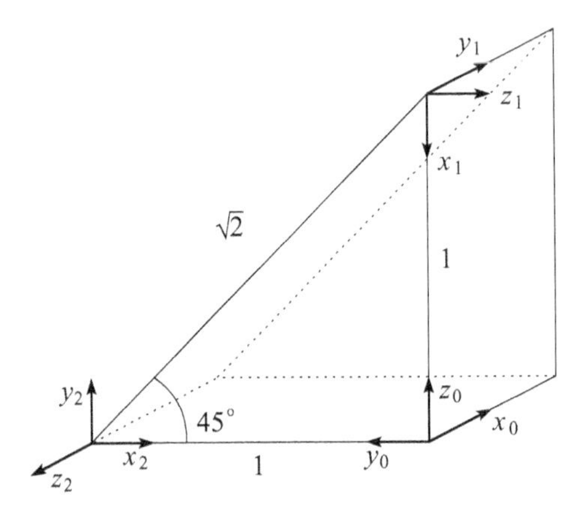

Figure 2.13: Diagram for Problem 2-38.

mations H_1^0, H_2^0, H_2^1 representing the transformations among the three frames shown. Show that $H_2^0 = H_1^0, H_2^1$.

2-39 Consider the diagram of Figure 2.14. A robot is set up 1 meter from a table. The table top is 1 meter high and 1 meter square. A frame $o_1x_1y_1z_1$ is fixed to the edge of the table as shown. A cube measuring 20 cm on a side is placed in the center of the table with frame $o_2x_2y_2z_2$ established at the center of the cube as shown. A camera is situated directly above the center of the block 2 meters above the table top with frame $o_3x_3y_3z_3$ attached as shown. Find the homogeneous transformations relating each of these frames to the base frame $o_0x_0y_0z_0$. Find the homogeneous transformation relating the frame $o_2x_2y_2z_2$ to the camera frame $o_3x_3y_3z_3$.

2-40 In Problem 2-39, suppose that, after the camera is calibrated, it is rotated 90° about z_3. Recompute the above coordinate transformations.

2-41 If the block on the table is rotated 90° about z_2 and moved so that its center has coordinates $[0, .8, .1]^T$ relative to the frame $o_1x_1y_1z_1$, compute the homogeneous transformation relating the block frame to the camera frame; the block frame to the base frame.

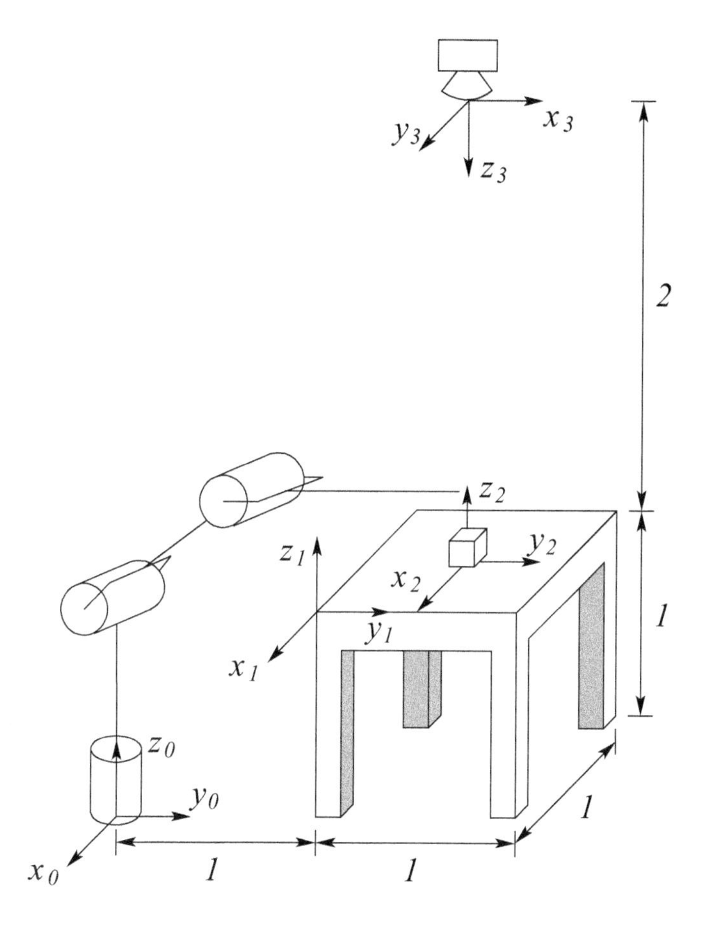

Figure 2.14: Diagram for Problem 2-39.

2-42 Consult an astronomy book to learn the basic details of the Earth's rotation about the sun and about its own axis. Define for the Earth a local coordinate frame whose z-axis is the Earth's axis of rotation. Define $t = 0$ to be the exact moment of the summer solstice, and the global reference frame to be coincident with the Earth's frame at time $t = 0$. Give an expression $R(t)$ for the rotation matrix that represents the instantaneous orientation of the earth at time t. Determine as a function of time the homogeneous transformation that specifies the Earth's frame with respect to the global reference frame.

2-43 In general, multiplication of homogeneous transformation matrices is not commutative. Consider the matrix product

$$H = \text{Rot}_{x,\alpha}\text{Trans}_{x,b}\text{Trans}_{z,d}\text{Rot}_{z,\theta}$$

Determine which pairs of the four matrices on the right hand side commute. Explain why these pairs commute. Find all permutations of these four matrices that yield the same homogeneous transformation matrix, H.

NOTES AND REFERENCES

Rigid body motions and the groups $SO(n)$ and $SE(n)$ are often addressed in mathematics books on the topic of linear algebra. Standard texts for this material include [8], [23], and [40]. These topics are also often covered in applied mathematics texts for physics and engineering, such as [108], [119], and [139]. In addition to these, a detailed treatment of rigid body motion developed with the aid of exponential coordinates and Lie groups is given in [93].

Chapter 3

FORWARD AND INVERSE KINEMATICS

In this chapter we consider the forward and inverse kinematics for serial link manipulators. The problem of kinematics is to describe the motion of the manipulator without consideration of the forces and torques causing the motion. The kinematic description is therefore a geometric one. We first consider the problem of **forward kinematics**, which is to determine the position and orientation of the end effector given the values for the joint variables of the robot. The problem of **inverse kinematics** is to determine the values of the joint variables given the end effector's position and orientation.

3.1 KINEMATIC CHAINS

As described in Chapter 1, a robot manipulator is composed of a set of links connected together by joints. The joints can either be very simple, such as a revolute joint or a prismatic joint, or they can be more complex, such as a ball and socket joint (recall that a revolute joint is like a hinge that allows a relative rotation about a single axis, and a prismatic joint permits a linear motion along a single axis, namely an extension or retraction). The difference between the two situations is that in the first instance the joint has only a single degree-of-freedom of motion: the angle of rotation in the case of a revolute joint, and the amount of linear displacement in the case of a prismatic joint. In contrast, a ball and socket joint has two degrees of freedom. In this book it is assumed throughout that all joints have only a single degree of freedom. This assumption does not involve any real loss of generality, since joints such as a ball and socket joint (two degrees of freedom) or a spherical wrist (three degrees of freedom) can always be thought of as a succession of single degree-of-freedom joints with links of length zero in between.

With the assumption that each joint has a single degree-of-freedom, the action of each joint can be described by a single real number: the angle of rotation in the case of a revolute joint or the displacement in the case of a prismatic joint.

A robot manipulator with n joints will have $n+1$ links, since each joint connects two links. We number the joints from 1 to n, and we number the links from 0 to n, starting from the base. By this convention, joint i connects link $i-1$ to link i. We will consider the location of joint i to be fixed with respect to link $i-1$. *When joint i is actuated, link i moves.* Therefore, link 0 (the first link) is fixed, and does not move when the joints are actuated. Of course the robot manipulator could itself be mobile (e.g., it could be mounted on a mobile platform or on an autonomous vehicle), but we will not consider this case in the present chapter, since it can be handled easily by slightly extending the techniques presented here.

With the ith joint, we associate a *joint variable*, denoted by q_i. In the case of a revolute joint, q_i is the angle of rotation, and in the case of a prismatic joint, q_i is the joint displacement:

$$q_i = \begin{cases} \theta_i & \text{if joint } i \text{ is revolute} \\ d_i & \text{if joint } i \text{ is prismatic} \end{cases} \tag{3.1}$$

To perform the kinematic analysis, we attach a coordinate frame rigidly to each link. In particular, we attach $o_i x_i y_i z_i$ to link i. This means that, whatever motion the robot executes, the coordinates of each point on link i are constant when expressed in the ith coordinate frame. Furthermore, when joint i is actuated, link i and its attached frame, $o_i x_i y_i z_i$, experience a resulting motion. The frame $o_0 x_0 y_0 z_0$, which is attached to the robot base, is referred to as the inertial frame. Figure 3.1 illustrates the idea of attaching frames rigidly to links in the case of an elbow manipulator.

Now, suppose A_i is the homogeneous transformation matrix that gives the position and orientation of $o_i x_i y_i z_i$ with respect to $o_{i-1} x_{i-1} y_{i-1} z_{i-1}$. The matrix A_i is not constant, but varies as the configuration of the robot is changed. However, the assumption that all joints are either revolute or prismatic means that A_i is a function of only a single joint variable, namely q_i. In other words,

$$A_i = A_i(q_i) \tag{3.2}$$

The homogeneous transformation matrix that expresses the position and orientation of $o_j x_j y_j z_j$ with respect to $o_i x_i y_i z_i$ is called a **transformation**

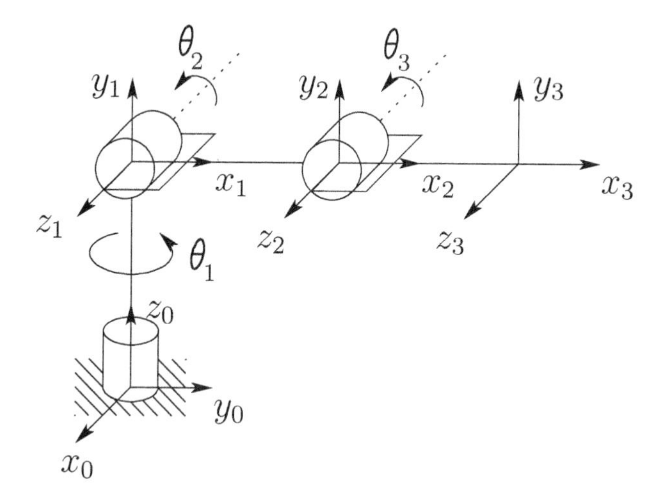

Figure 3.1: Coordinate frames attached to elbow manipulator.

matrix, and is denoted by T_j^i. From Chapter 2 we see that

$$T_j^i = \begin{cases} A_{i+1}A_{i+2}\dots A_{j-1}A_j & \text{if } i < j \\ I & \text{if } i = j \\ (T_i^j)^{-1} & \text{if } j > i \end{cases} \tag{3.3}$$

By the manner in which we have rigidly attached the various frames to the corresponding links, it follows that the position of any point on the end effector when expressed in frame n is a constant independent of the configuration of the robot. We denote the position and orientation of the end effector with respect to the inertial or base frame by a three-vector o_n^0 (which gives the coordinates of the origin of the end-effector frame with respect to the base frame) and the 3×3 rotation matrix R_n^0, and define the homogeneous transformation matrix

$$H = \begin{bmatrix} R_n^0 & o_n^0 \\ 0 & 1 \end{bmatrix} \tag{3.4}$$

Then the position and orientation of the end effector in the inertial frame are given by

$$H = T_n^0 = A_1(q_1)\cdots A_n(q_n) \tag{3.5}$$

Each homogeneous transformation A_i is of the form

$$A_i = \begin{bmatrix} R_i^{i-1} & o_i^{i-1} \\ 0 & 1 \end{bmatrix} \tag{3.6}$$

Hence, for $i < j$

$$T_j^i = A_{i+1} \cdots A_j = \begin{bmatrix} R_j^i & o_j^i \\ 0 & 1 \end{bmatrix} \tag{3.7}$$

The matrix R_j^i expresses the orientation of $o_j x_j y_j z_j$ relative to $o_i x_i y_i z_i$ and is given by the rotational parts of the A-matrices as

$$R_j^i = R_{i+1}^i \cdots R_j^{j-1} \tag{3.8}$$

The coordinate vectors o_j^i are given recursively by the formula

$$o_j^i = o_{j-1}^i + R_{j-1}^i o_j^{j-1} \tag{3.9}$$

These expressions will be useful in Chapter 4 when we study Jacobian matrices.

In principle, that is all there is to forward kinematics; determine the functions $A_i(q_i)$, and multiply them together as needed. However, it is possible to achieve a considerable amount of streamlining and simplification by introducing further conventions, such as the Denavit-Hartenberg representation of a joint, and this is the objective of the next section.

3.2 THE DENAVIT-HARTENBERG CONVENTION

In this section we develop the **forward** or **configuration kinematics** for rigid robots. The forward kinematics problem is concerned with the relationship between the individual joints of the robot manipulator and the position and orientation of the tool or end effector. The joint variables are the angles between the links in the case of revolute or rotational joints, and the link extension in the case of prismatic or sliding joints.

We will develop a set of conventions that provide a systematic procedure for performing this analysis. It is, of course, possible to carry out forward kinematics analysis even without respecting these conventions, as we did for the two-link planar manipulator example in Chapter 1. However, the kinematic analysis of an n-link manipulator can be extremely complex and the conventions introduced below simplify the analysis considerably. Moreover, they give rise to a universal language with which engineers can communicate.

A commonly used convention for selecting frames of reference in robotic applications is the Denavit-Hartenberg, or DH convention. In this convention, each homogeneous transformation A_i is represented as a product of four basic transformations

$$\begin{aligned} A_i &= \text{Rot}_{z,\theta_i}\text{Trans}_{z,d_i}\text{Trans}_{x,a_i}\text{Rot}_{x,\alpha_i} \qquad (3.10)\\ &= \begin{bmatrix} c_{\theta_i} & -s_{\theta_i} & 0 & 0 \\ s_{\theta_i} & c_{\theta_i} & 0 & 0 \\ 0 & 0 & 1 & 0 \\ 0 & 0 & 0 & 1 \end{bmatrix} \begin{bmatrix} 1 & 0 & 0 & 0 \\ 0 & 1 & 0 & 0 \\ 0 & 0 & 1 & d_i \\ 0 & 0 & 0 & 1 \end{bmatrix} \\ &\quad \times \begin{bmatrix} 1 & 0 & 0 & a_i \\ 0 & 1 & 0 & 0 \\ 0 & 0 & 1 & 0 \\ 0 & 0 & 0 & 1 \end{bmatrix} \begin{bmatrix} 1 & 0 & 0 & 0 \\ 0 & c_{\alpha_i} & -s_{\alpha_i} & 0 \\ 0 & s_{\alpha_i} & c_{\alpha_i} & 0 \\ 0 & 0 & 0 & 1 \end{bmatrix} \\ &= \begin{bmatrix} c_{\theta_i} & -s_{\theta_i}c_{\alpha_i} & s_{\theta_i}s_{\alpha_i} & a_i c_{\theta_i} \\ s_{\theta_i} & c_{\theta_i}c_{\alpha_i} & -c_{\theta_i}s_{\alpha_i} & a_i s_{\theta_i} \\ 0 & s_{\alpha_i} & c_{\alpha_i} & d_i \\ 0 & 0 & 0 & 1 \end{bmatrix} \end{aligned}$$

where the four quantities θ_i, a_i, d_i, α_i are parameters associated with link i and joint i. The four parameters a_i, α_i, d_i, and θ_i in Equation (3.10) are generally given the names **link length**, **link twist**, **link offset**, and **joint angle**, respectively. These names derive from specific aspects of the geometric relationship between two coordinate frames, as will become apparent below. Since the matrix A_i is a function of a single variable, three of the above four quantities are constant for a given link, while the fourth parameter, θ_i for a revolute joint and d_i for a prismatic joint, is the joint variable.

From Chapter 2 one can see that an arbitrary homogeneous transformation matrix can be characterized by six numbers, such as, for example, three numbers to specify the fourth column of the matrix and three Euler angles to specify the upper left 3×3 rotation matrix. In the DH representation, in contrast, there are only *four* parameters. How is this possible? The answer is that, while frame i is required to be rigidly attached to link i, we have considerable freedom in choosing the origin and the coordinate axes of the frame. For example, it is not necessary that the origin, o_i, of frame i be placed at the physical end of link i. In fact, it is not even necessary that frame i be placed within the physical link. Frame i could lie in free space so long as frame i is rigidly attached to link i. By a clever choice of the origin and the coordinate axes, it is possible to cut down the number of parameters needed from six to four (or even fewer in some cases). In Section 3.2.1 we will show why, and under what conditions, this can be done, and in Section 3.2.2 we will show exactly how to make the coordinate frame assignments.

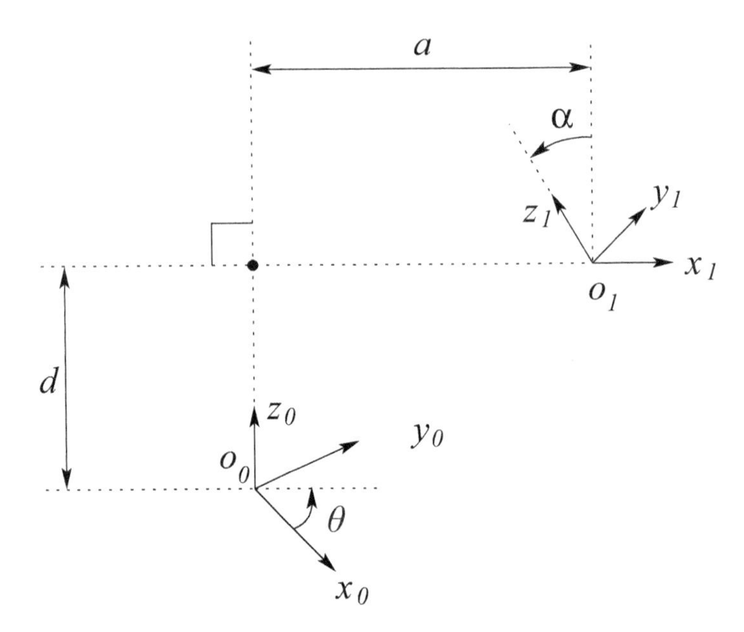

Figure 3.2: Coordinate frames satisfying assumptions DH1 and DH2.

3.2.1 Existence and Uniqueness Issues

Clearly it is not possible to represent any arbitrary homogeneous transformation using only four parameters. Therefore, we begin by determining just which homogeneous transformations can be expressed in the form given by Equation (3.10). Suppose we are given two frames, denoted by frames 0 and 1, respectively. Then there exists a unique homogeneous transformation matrix A that takes the coordinates from frame 1 into those of frame 0. Now, suppose the two frames have the following two additional features.

(DH1) The axis x_1 is perpendicular to the axis z_0.

(DH2) The axis x_1 intersects the axis z_0.

These two properties are illustrated in Figure 3.2. Under these conditions, we claim that there exist unique numbers a, d, θ, α such that

$$A = \text{Rot}_{z,\theta}\text{Trans}_{z,d}\text{Trans}_{x,a}\text{Rot}_{x,\alpha} \tag{3.11}$$

Of course, since θ and α are angles, we really mean that they are unique to within a multiple of 2π. To show that the matrix A can be written in this form, write A as

$$A = \begin{bmatrix} R_1^0 & o_1^0 \\ 0 & 1 \end{bmatrix} \tag{3.12}$$

If (DH1) is satisfied, then x_1 is perpendicular to z_0 and we have $x_1 \cdot z_0 = 0$. Expressing this constraint with respect to $o_0x_0y_0z_0$, using the fact that the first column of R_1^0 is the representation of the unit vector x_1 with respect to frame 0, we obtain

$$\begin{aligned} 0 &= x_1^0 \cdot z_0^0 \\ &= [r_{11}, r_{21}, r_{31}] \begin{bmatrix} 0 \\ 0 \\ 1 \end{bmatrix} = r_{31} \end{aligned}$$

Since $r_{31} = 0$, we now need only show that there exist *unique* angles θ and α such that

$$R_1^0 = R_{x,\theta} R_{x,\alpha} = \begin{bmatrix} c_\theta & -s_\theta c_\alpha & s_\theta s_\alpha \\ s_\theta & c_\theta c_\alpha & -c_\theta s_\alpha \\ 0 & s_\alpha & c_\alpha \end{bmatrix} \tag{3.13}$$

The only information we have is that $r_{31} = 0$, but this is enough. First, since each row and column of R_1^0 must have unit length, $r_{31} = 0$ implies that

$$\begin{aligned} r_{11}^2 + r_{21}^2 &= 1, \\ r_{32}^2 + r_{33}^2 &= 1 \end{aligned}$$

Hence, there exist unique θ and α such that

$$(r_{11}, r_{21}) = (c_\theta, s_\theta), \qquad (r_{33}, r_{32}) = (c_\alpha, s_\alpha)$$

Once θ and α are found, it is routine to show that the remaining elements of R_1^0 must have the form shown in Equation (3.13), using the fact that R_1^0 is a rotation matrix.

Next, assumption (DH2) means that the displacement between o_0 and o_1 can be expressed as a linear combination of the vectors z_0 and x_1. This can be written as $o_1 = o_0 + dz_0 + ax_1$. Again, we can express this relationship in the coordinates of $o_0x_0y_0z_0$, and we obtain

$$\begin{aligned} o_1^0 &= o_0^0 + dz_0^0 + ax_1^0 \\ &= \begin{bmatrix} 0 \\ 0 \\ 0 \end{bmatrix} + d \begin{bmatrix} 0 \\ 0 \\ 1 \end{bmatrix} + a \begin{bmatrix} c_\theta \\ s_\theta \\ 0 \end{bmatrix} \\ &= \begin{bmatrix} ac_\theta \\ as_\theta \\ d \end{bmatrix} \end{aligned}$$

Combining the above results, we obtain Equation (3.10) as claimed. Thus, we see that four parameters are sufficient to specify any homogeneous transformation that satisfies the constraints (DH1) and (DH2).

Now that we have established that each homogeneous transformation matrix satisfying conditions (DH1) and (DH2) above can be expressed as in Equation (3.10), we can give a physical interpretation to each of these four quantities. The parameter a is the distance between the axes z_0 and z_1, and is measured along the axis x_1. The angle α is the angle between the axes z_0 and z_1, measured in a plane normal to x_1. The positive sense for α is determined from z_0 to z_1 by the right hand rule as shown in Figure 3.3. The parameter d is the distance from the origin o_0 to the intersection of the

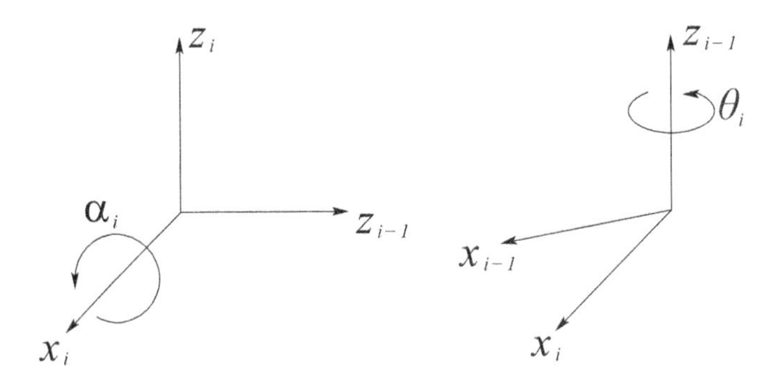

Figure 3.3: Positive sense for α_i and θ_i.

x_1 axis with z_0 measured along the z_0 axis. Finally, θ is the angle from x_0 to x_1 measured in a plane normal to z_0. These physical interpretations will prove useful in developing a procedure for assigning coordinate frames that satisfy the constraints (DH1) and (DH2), and we now turn our attention to developing such a procedure.

3.2.2 Assigning the Coordinate Frames

For a given robot manipulator, one can always choose the frames $0, \ldots, n$ in such a way that the above two conditions are satisfied. In certain circumstances, this will require placing the origin o_i of frame i in a location that may not be intuitively satisfying, but typically this will not be the case. In reading the material below, it is important to keep in mind that the choices of the various coordinate frames are not unique, even when constrained by the requirements above. Thus, it is possible that different engineers will derive differing, but equally correct, coordinate frame assignments for the links of the robot. It is very important to note, however, that the end result (i.e., the matrix T_n^0) will be the same, regardless of the assignment of intermediate DH frames (assuming that the coordinate frames for link n coincide). We will begin by deriving the general procedure. We will then discuss various common special cases for which it is possible to further simplify the homogeneous transformation matrix.

To start, note that the choice of z_i is arbitrary. In particular, from Equation (3.13), we see that by choosing α_i and θ_i appropriately, we can obtain any arbitrary direction for z_i. Thus, for our first step, we assign the axes $z_0, \ldots, z_{n-1}$ in an intuitively pleasing fashion. Specifically, we assign z_i to be the axis of actuation for joint $i+1$. Thus, z_0 is the axis of actuation for joint 1, z_1 is the axis of actuation for joint 2, etc. There are two cases to consider: (i) if joint $i+1$ is revolute, z_i is the axis of revolution of joint $i+1$; (ii) if joint $i+1$ is prismatic, z_i is the axis of translation of joint $i+1$. At first it may seem a bit confusing to associate z_i with joint $i+1$, but recall that this satisfies the convention that we established above, namely that when joint i is actuated, link i and its attached frame, $o_i x_i y_i z_i$, experience a resulting motion.

Once we have established the z-axes for the links, we establish the base frame. The choice of a base frame is nearly arbitrary. We may choose the origin o_0 of the base frame to be any point on z_0. We then choose x_0, y_0 in any convenient manner so long as the resulting frame is right-handed. This sets up frame 0.

Once frame 0 has been established, we begin an iterative process in which we define frame i using frame $i-1$, beginning with frame 1. Figure 3.4 will be useful for understanding the process that we now describe.

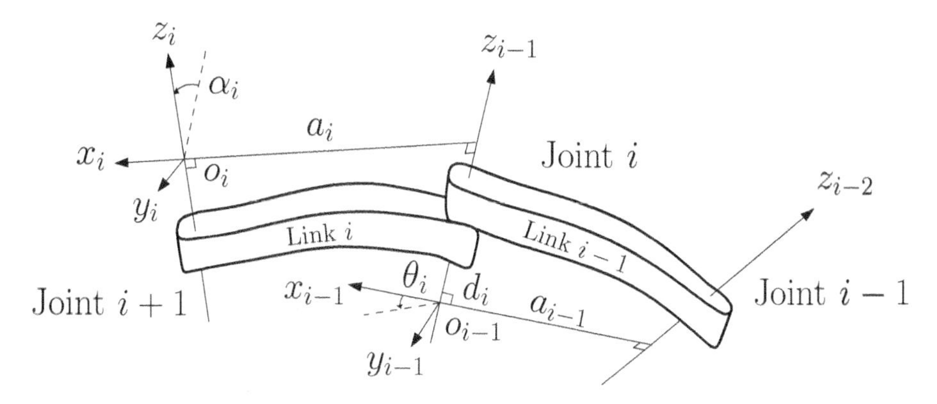

Figure 3.4: Denavit-Hartenberg frame assignment.

In order to set up frame i it is convenient to consider three cases: (i) the axes z_{i-1}, z_i are not coplanar, (ii) the axes z_{i-1}, z_i intersect, (iii) the axes

z_{i-1}, z_i are parallel. Note that in both cases (ii) and (iii) the axes z_{i-1} and z_i are coplanar. This situation is in fact quite common, as we will see in Section 3.2.3. We now consider each of these three cases.

(i) z_{i-1} and z_i are not coplanar: If z_{i-l} and z_i are not coplanar, then there exists a *unique* shortest line segment from z_{i-1} to z_i, perpendicular to both z_{i-1} to z_i. This line segment defines x_i, and the point where it intersects z_i is the origin o_i. By construction, both conditions (DH1) and (DH2) are satisfied and the vector from o_{i-1} to o_i is a linear combination of z_{i-1} and x_i. The specification of frame i is completed by choosing the axis y_i to form a right-handed frame. Since assumptions (DH1) and (DH2) are satisfied, the homogeneous transformation matrix A_i is of the form given in Equation (3.10).

(ii) z_{i-1} is parallel to z_i: If the axes z_{i-1} and z_i are parallel, then there are infinitely many common normals between them and condition (DH1) does not specify x_i completely. In this case we are free to choose the origin o_i anywhere along z_i. One often chooses o_i to simplify the resulting equations. The axis x_i is then chosen either to be directed from o_i toward z_{i-1}, along the common normal, or as the opposite of this vector. A common method for choosing o_i is to choose the normal that passes through o_{i-1} as the x_i axis; o_i is then the point at which this normal intersects z_i. In this case, d_i would be equal to zero. Once x_i is fixed, y_i is determined, as usual by the right hand rule. Since the axes z_{i-1} and z_i are parallel, α_i will be zero in this case.

(iii) z_{i-1} intersects z_i: In this case x_i is chosen normal to the plane formed by z_i and z_{i-1}. The positive direction of x_i is arbitrary. The most natural choice for the origin o_i in this case is at the point of intersection of z_i and z_{i-1}. However, any convenient point along the axis z_i suffices. Note that in this case the parameter a_i will be zero.

This constructive procedure works for frames $0, \ldots, n-1$ in an n-link robot. To complete the construction it is necessary to specify frame n. The final coordinate system $o_n x_n y_n z_n$ is commonly referred to as the **end effector** or **tool frame** (see Figure 3.5). The origin o_n is most often placed symmetrically between the fingers of the gripper. The unit vectors along the x_n, y_n, and z_n axes are labeled as n, s, and a, respectively. The terminology arises from the fact that the direction a is the **approach** direction, in the sense that the gripper typically approaches an object along the a direction. Similarly the s direction is the **sliding** direction, the direction along which the fingers of the gripper slide to open and close, and n is the direction **normal** to the plane formed by a and s.

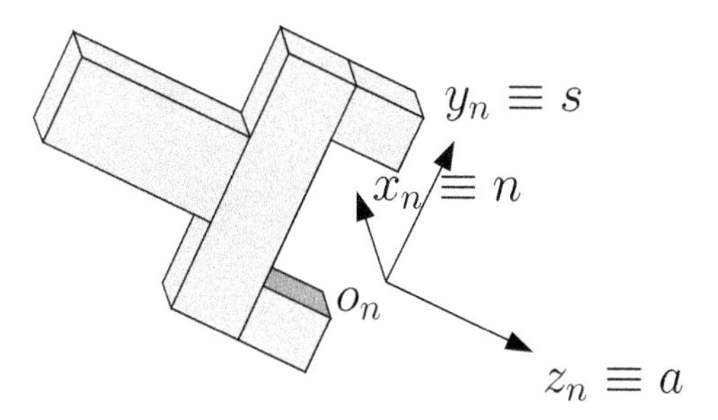

Figure 3.5: Tool frame assignment.

In most contemporary robots the final joint motion is a rotation of the end effector by θ_n and the final two joint axes, z_{n-1} and z_n, coincide. In this case, the transformation between the final two coordinate frames is a translation along z_{n-1} by a distance d_n followed (or preceded) by a rotation of θ_n about z_{n-1}. This is an important observation that will simplify the computation of the inverse kinematics in the next section.

Finally, note the following important fact. In all cases, whether the joint in question is revolute or prismatic, the quantities a_i and α_i are always constant for all i and are characteristic of the manipulator. If joint i is prismatic, then θ_i is also a constant, while d_i is the ith joint variable. Similarly, if joint i is revolute, then d_i is constant and θ_i is the ith joint variable.

3.2.3 Examples

In the DH convention the only variable angle is θ, so we simplify notation by writing c_i for $\cos\theta_i$, etc. We also denote $\theta_1 + \theta_2$ by θ_{12}, and $\cos(\theta_1 + \theta_2)$ by c_{12}, and so on. In the following examples it is important to remember that the DH convention, while systematic, still allows considerable freedom in the choice of some of the manipulator parameters. This is particularly true in the case of parallel joint axes or when prismatic joints are involved.

Example 3.1 Planar Elbow Manipulator

Consider the two-link planar arm of Figure 3.6. The joint axes z_0 and z_1 are normal to the page. We establish the base frame $o_0x_0y_0z_0$ as shown. The origin is chosen at the point of intersection of the z_0 axis with the page and the direction of the x_0 axis is completely arbitrary. Once the base frame is established, the $o_1x_1y_1z_1$ frame is fixed as shown by the DH convention, where the origin o_1 has been located at the intersection of z_1 and the page.

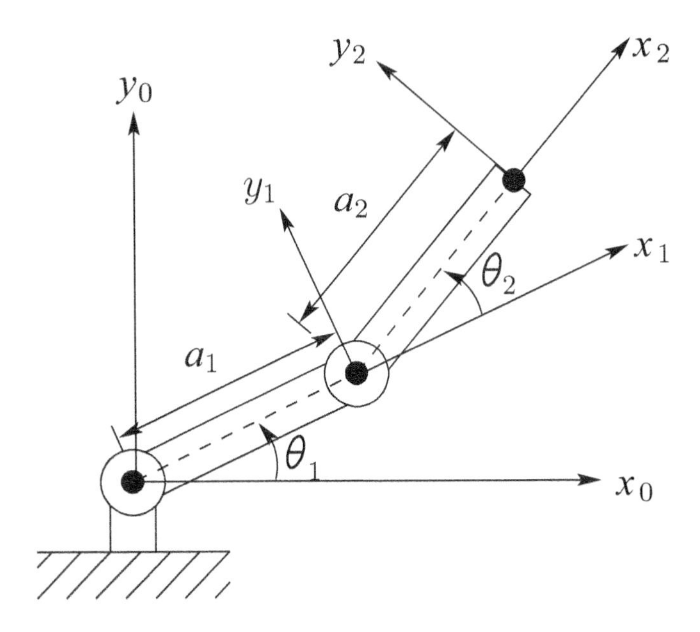

Figure 3.6: Two-link planar manipulator. The z-axes all point out of the page, and are not shown in the figure.

Table 3.1: DH parameters for 2-link planar manipulator.

Link	a_i	α_i	d_i	θ_i
1	a_1	0	0	θ_1^*
2	a_2	0	0	θ_2^*

* variable

The final frame $o_2x_2y_2z_2$ is fixed by choosing the origin o_2 at the end of link 2 as shown. The DH parameters are shown in Table 3.1.

The A-matrices are determined from Equation (3.10) as

$$A_1 = \begin{bmatrix} c_1 & -s_1 & 0 & a_1c_1 \\ s_1 & c_1 & 0 & a_1s_1 \\ 0 & 0 & 1 & 0 \\ 0 & 0 & 0 & 1 \end{bmatrix}, \quad A_2 = \begin{bmatrix} c_2 & -s_2 & 0 & a_2c_2 \\ s_2 & c_2 & 0 & a_2s_2 \\ 0 & 0 & 1 & 0 \\ 0 & 0 & 0 & 1 \end{bmatrix}$$

The T-matrices are thus given by

$$
\begin{aligned}
T_1^0 &= A_1 \\
T_2^0 &= A_1 A_2 = \begin{bmatrix} c_{12} & -s_{12} & 0 & a_1 c_1 + a_2 c_{12} \\ s_{12} & c_{12} & 0 & a_1 s_1 + a_2 s_{12} \\ 0 & 0 & 1 & 0 \\ 0 & 0 & 0 & 1 \end{bmatrix}
\end{aligned}
$$

Notice that the first two entries of the last column of T_2^0 are the x and y components of the origin o_2 in the base frame; that is,

$$
\begin{aligned}
x &= a_1 c_1 + a_2 c_{12} \\
y &= a_1 s_1 + a_2 s_{12}
\end{aligned}
$$

are the coordinates of the end effector in the base frame. The rotational part of T_2^0 gives the orientation of the frame $o_2 x_2 y_2 z_2$ relative to the base frame. ◇

Example 3.2 Three-Link Cylindrical Robot

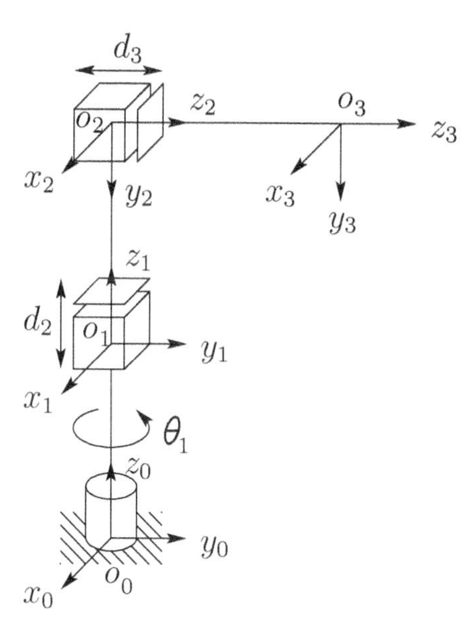

Figure 3.7: Three-link cylindrical manipulator.

Consider now the three-link cylindrical robot represented symbolically by Figure 3.7. We establish o_0 as shown at joint 1. Note that the placement of the origin o_0 along z_0 and the direction of the x_0 axis are arbitrary. Our

Table 3.2: DH parameters for 3-link cylindrical manipulator.

Link	a_i	α_i	d_i	θ_i
1	0	0	d_1	θ_1^*
2	0	-90	d_2^*	0
3	0	0	d_3^*	0

* variable

choice of o_0 is the most natural, but o_0 could just as well be placed at joint 2. Next, since z_0 and z_1 coincide, the origin o_1 is chosen at joint 1 as shown. The x_1 axis is parallel to x_0 when $\theta_1 = 0$ but, of course its direction will change since θ_1 is variable. Since z_2 and z_1 intersect, the origin o_2 is placed at this intersection. The direction of x_2 is chosen parallel to x_1 so that θ_2 is zero. Finally, the third frame is chosen at the end of link 3 as shown.

The DH parameters are shown in Table 3.2. The corresponding A and T matrices are

$$A_1 = \begin{bmatrix} c_1 & -s_1 & 0 & 0 \\ s_1 & c_1 & 0 & 0 \\ 0 & 0 & 1 & d_1 \\ 0 & 0 & 0 & 1 \end{bmatrix}, \quad A_2 = \begin{bmatrix} 1 & 0 & 0 & 0 \\ 0 & 0 & 1 & 0 \\ 0 & -1 & 0 & d_2 \\ 0 & 0 & 0 & 1 \end{bmatrix}$$

$$A_3 = \begin{bmatrix} 1 & 0 & 0 & 0 \\ 0 & 1 & 0 & 0 \\ 0 & 0 & 1 & d_3 \\ 0 & 0 & 0 & 1 \end{bmatrix}$$

$$T_3^0 = A_1 A_2 A_3 = \begin{bmatrix} c_1 & 0 & -s_1 & -s_1 d_3 \\ s_1 & 0 & c_1 & c_1 d_3 \\ 0 & -1 & 0 & d_1 + d_2 \\ 0 & 0 & 0 & 1 \end{bmatrix} \tag{3.14}$$

◇

Example 3.3 Spherical Wrist

Figure 3.8 shows the spherical wrist, a three-link wrist mechanism for which the joint axes z_3, z_4, z_5 intersect at o. The point o is called the **wrist center**. *The DH parameters are shown in Table 3.3. The Stanford manipulator is an example of a manipulator that possesses a wrist of this type.*

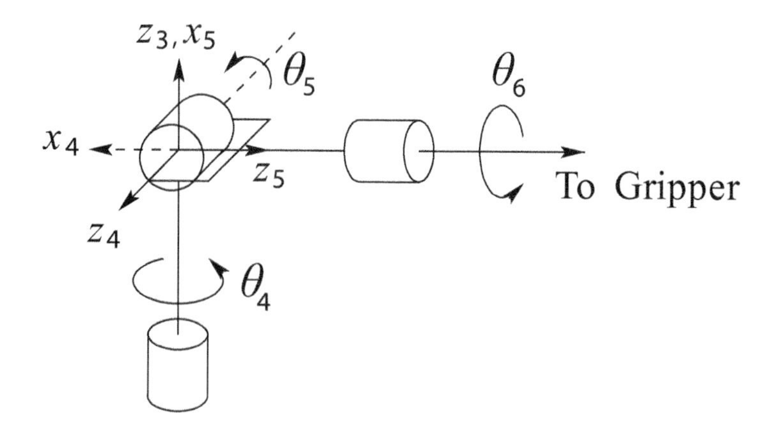

Figure 3.8: The spherical wrist frame assignment.

Table 3.3: DH parameters for spherical wrist.

Link	a_i	α_i	d_i	θ_i
4	0	−90	0	θ_4^*
5	0	90	0	θ_5^*
6	0	0	d_6	θ_6^*

* variable

We show now that the final three joint variables, θ_4, θ_5, θ_6 are the Euler angles ϕ, θ, and ψ, respectively, with respect to the coordinate frame $o_3x_3y_3z_3$. To see this we need only compute the matrices A_4, A_5, and A_6 using Table 3.3 and Equation (3.10). This gives

$$A_4 = \begin{bmatrix} c_4 & 0 & -s_4 & 0 \\ s_4 & 0 & c_4 & 0 \\ 0 & -1 & 0 & 0 \\ 0 & 0 & 0 & 1 \end{bmatrix}, \quad A_5 = \begin{bmatrix} c_5 & 0 & s_5 & 0 \\ s_5 & 0 & -c_5 & 0 \\ 0 & -1 & 0 & 0 \\ 0 & 0 & 0 & 1 \end{bmatrix}$$

$$A_6 = \begin{bmatrix} c_6 & -s_6 & 0 & 0 \\ s_6 & c_6 & 0 & 0 \\ 0 & 0 & 1 & d_6 \\ 0 & 0 & 0 & 1 \end{bmatrix}$$

Multiplying these together yields

$$
\begin{aligned}
T_6^3 &= A_4 A_5 A_6 \\
&= \begin{bmatrix} R_6^3 & o_6^3 \\ 0 & 1 \end{bmatrix} \\
&= \begin{bmatrix} c_4c_5c_6 - s_4s_6 & -c_4c_5s_6 - s_4c_6 & c_4s_5 & c_4s_5d_6 \\ s_4c_5c_6 + c_4s_6 & -s_4c_5s_6 + c_4c_6 & s_4s_5 & s_4s_5d_6 \\ -s_5c_6 & s_5s_6 & c_5 & c_5d_6 \\ 0 & 0 & 0 & 1 \end{bmatrix}
\end{aligned}
\tag{3.15}
$$

Comparing the rotational part R_6^3 of T_6^3 with the Euler angle transformation in Equation (2.26) shows that $\theta_4, \theta_5, \theta_6$ can indeed be identified as the Euler angles ϕ, θ, and ψ with respect to the coordinate frame $o_3x_3y_3z_3$. ◇

Example 3.4 Cylindrical Manipulator with Spherical Wrist

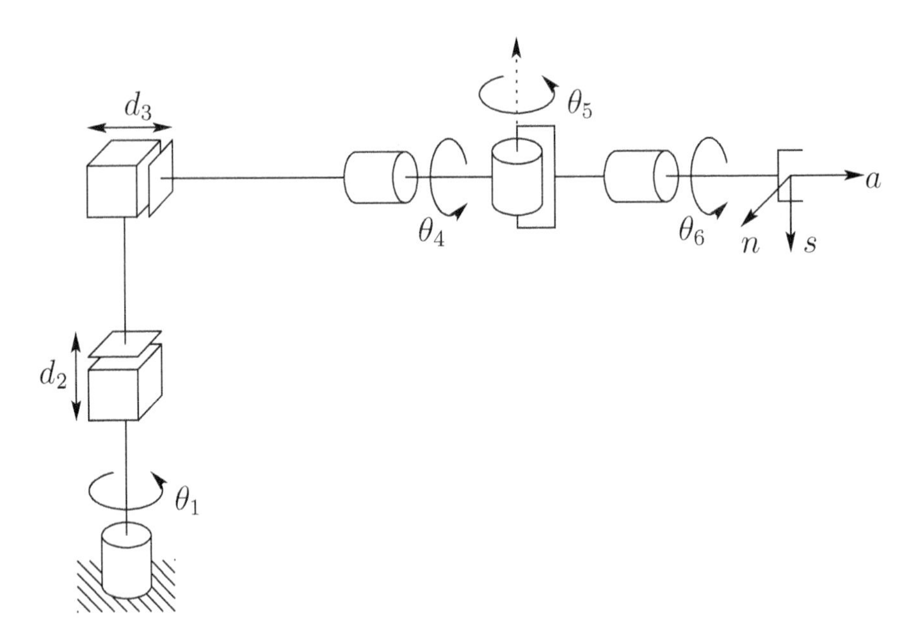

Figure 3.9: Cylindrical robot with spherical wrist.

Suppose that we now attach a spherical wrist to the cylindrical manipulator of Example 3.2 as shown in Figure 3.9. Note that the axis of rotation of joint 4 is parallel to z_2 and thus coincides with the axis z_3 of Example 3.2. The implication of this is that we can immediately combine Equations (3.14) *and (3.15) to derive the forward kinematic equations as*

$$T_6^0 = T_3^0 T_6^3 \tag{3.16}$$

with T_3^0 given by Equation (3.14) and T_6^3 given by Equation (3.15). Therefore the forward kinematic equations of this manipulator are given by

$$T_6^0 = \begin{bmatrix} r_{11} & r_{12} & r_{13} & d_x \\ r_{21} & r_{22} & r_{23} & d_y \\ r_{31} & r_{32} & r_{33} & d_z \\ 0 & 0 & 0 & 1 \end{bmatrix} \tag{3.17}$$

in which

$$\begin{aligned}
r_{11} &= c_1c_4c_5c_6 - c_1s_4s_6 + s_1s_5c_6 \\
r_{21} &= s_1c_4c_5c_6 - s_1s_4s_6 - c_1s_5c_6 \\
r_{31} &= -s_4c_5c_6 - c_4s_6 \\
r_{12} &= -c_1c_4c_5s_6 - c_1s_4c_6 - s_1s_5c_6 \\
r_{22} &= -s_1c_4c_5s_6 - s_1s_4s_6 + c_1s_5c_6 \\
r_{32} &= s_4c_5c_6 - c_4c_6 \\
r_{13} &= c_1c_4s_5 - s_1c_5 \\
r_{23} &= s_1c_4s_5 + c_1c_5 \\
r_{33} &= -s_4s_5 \\
d_x &= c_1c_4s_5d_6 - s_1c_5d_6 - s_1d_3 \\
d_y &= s_1c_4s_5d_6 + c_1c_5d_6 + c_1d_3 \\
d_z &= -s_4s_5d_6 + d_1 + d_2
\end{aligned}$$

Notice how most of the complexity of the forward kinematics for this manipulator results from the orientation of the end effector while the expression for the arm position from Equation (3.14) is fairly simple. The spherical wrist assumption not only simplifies the derivation of the forward kinematics here, but will also greatly simplify the inverse kinematics problem in Section 3.3.

◇

Example 3.5 Stanford Manipulator

Consider now the Stanford Manipulator shown in Figure 3.10. This manipulator is an example of a spherical (RRP) manipulator with a spherical wrist. This manipulator has an offset in the shoulder joint that slightly complicates both the forward and inverse kinematics problems.

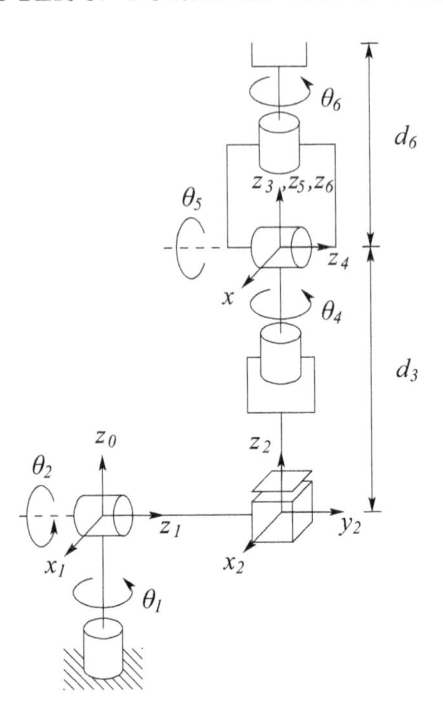

Figure 3.10: DH coordinate frame assignment for the Stanford manipulator.

We first establish the joint coordinate frames using the DH convention. The DH parameters are shown in Table 3.4.

It is straightforward to compute the matrices A_i as

$$A_1 = \begin{bmatrix} c_1 & 0 & -s_1 & 0 \\ s_1 & 0 & c_1 & 0 \\ 0 & -1 & 0 & 0 \\ 0 & 0 & 0 & 1 \end{bmatrix} \quad A_2 = \begin{bmatrix} c_2 & 0 & s_2 & 0 \\ s_2 & 0 & -c_2 & 0 \\ 0 & 1 & 0 & d_2 \\ 0 & 0 & 0 & 1 \end{bmatrix} \tag{3.18}$$

$$A_3 = \begin{bmatrix} 1 & 0 & 0 & 0 \\ 0 & 1 & 0 & 0 \\ 0 & 0 & 1 & d_3 \\ 0 & 0 & 0 & 1 \end{bmatrix} \quad A_4 = \begin{bmatrix} c_4 & 0 & -s_4 & 0 \\ s_4 & 0 & c_4 & 0 \\ 0 & -1 & 0 & 0 \\ 0 & 0 & 0 & 1 \end{bmatrix} \tag{3.19}$$

$$A_5 = \begin{bmatrix} c_5 & 0 & s_5 & 0 \\ s_5 & 0 & -c_5 & 0 \\ 0 & -1 & 0 & 0 \\ 0 & 0 & 0 & 1 \end{bmatrix} \quad A_6 = \begin{bmatrix} c_6 & -s_6 & 0 & 0 \\ s_6 & c_6 & 0 & 0 \\ 0 & 0 & 1 & d_6 \\ 0 & 0 & 0 & 1 \end{bmatrix} \tag{3.20}$$

Table 3.4: DH parameters for the Stanford manipulator.

Link	d_i	a_i	α_i	θ_i
1	0	0	−90	$\theta_1^\star$
2	d_2	0	+90	$\theta_2^\star$
3	$d_3^\star$	0	0	0
4	0	0	−90	$\theta_4^\star$
5	0	0	+90	$\theta_5^\star$
6	d_6	0	0	$\theta_6^\star$

* joint variable

T_6^0 is then given as

$$T_6^0 = A_1 \cdots A_6 = \begin{bmatrix} r_{11} & r_{12} & r_{13} & d_x \\ r_{21} & r_{22} & r_{23} & d_y \\ r_{31} & r_{32} & r_{33} & d_z \\ 0 & 0 & 0 & 1 \end{bmatrix} \tag{3.21}$$

in which

$$\begin{aligned}
r_{11} &= c_1[c_2(c_4c_5c_6 - s_4s_6) - s_2s_5c_6] - d_2(s_4c_5c_6 + c_4s_6) \\
r_{21} &= s_1[c_2(c_4c_5c_6 - s_4s_6) - s_2s_5c_6] + c_1(s_4c_5c_6 + c_4s_6) \\
r_{31} &= -s_2(c_4c_5c_6 - s_4s_6) - c_2s_5c_6 \\
r_{12} &= c_1[-c_2(c_4c_5s_6 + s_4c_6) + s_2s_5s_6] - s_1(-s_4c_5s_6 + c_4c_6) \\
r_{22} &= -s_1[-c_2(c_4c_5s_6 + s_4c_6) + s_2s_5s_6] + c_1(-s_4c_5s_6 + c_4c_6) \\
r_{32} &= s_2(c_4c_5s_6 + s_4c_6) + c_2s_5s_6 \\
r_{13} &= c_1(c_2c_4s_5 + s_2c_5) - s_1s_4s_5 \\
r_{23} &= s_1(c_2c_4s_5 + s_2c_5) + c_1s_4s_5 \\
r_{33} &= -s_2c_4s_5 + c_2c_5 \\
d_x &= c_1s_2d_3 - s_1d_2 + +d_6(c_1c_2c_4s_5 + c_1c_5s_2 - s_1s_4s_5) \\
d_y &= s_1s_2d_3 + c_1d_2 + d_6(c_1s_4s_5 + c_2c_4s_1s_5 + c_5s_1s_2) \\
d_z &= c_2d_3 + d_6(c_2c_5 - c_4s_2s_5)
\end{aligned}$$

◇

Example 3.6 SCARA Manipulator

As another example of the general procedure, consider the SCARA manipulator of Figure 3.11. This manipulator, which is an abstraction of the

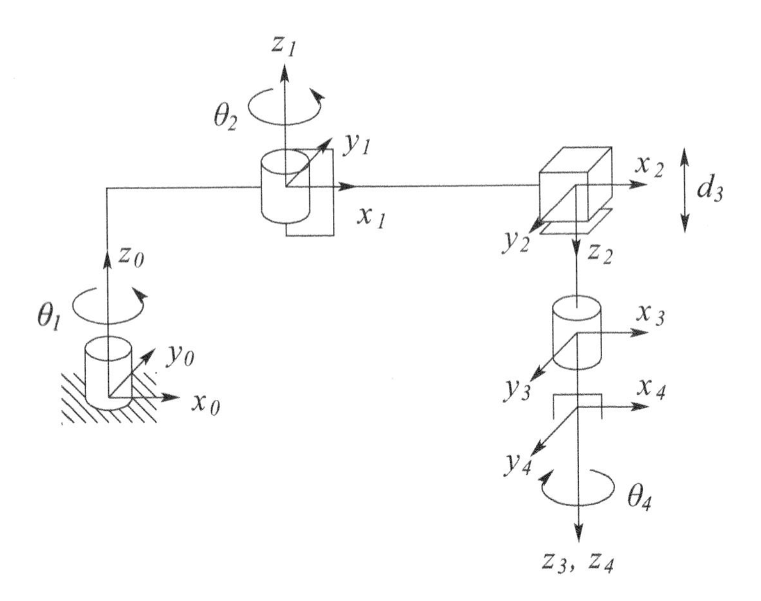

Figure 3.11: DH coordinate frame assignment for the SCARA manipulator.

Table 3.5: DH parameters for the SCARA manipulator.

Link	a_i	α_i	d_i	θ_i
1	a_1	0	0	$\theta_1^\star$
2	a_2	180	0	$\theta_2^\star$
3	0	0	$d_3^\star$	0
4	0	0	d_4	$\theta_4^\star$

* joint variable

AdeptOne robot of Figure 1.14, consists of an RRP arm and a one degree-of-freedom wrist, whose motion is a roll about the vertical axis. The first step is to locate and label the joint axes as shown. Since all joint axes are parallel we have some freedom in the placement of the origins. The origins are placed as shown for convenience. We establish the x_0 axis in the plane of the page as shown. This choice is completely arbitrary, but it does determine the **home position** *of the manipulator, which is defined relative to the zero configuration of the manipulator, that is, the position of the manipulator when the joint variables are equal to zero. The DH parameters are given in Table 3.5.*

and the A matrices are as follows.

$$A_1 = \begin{bmatrix} c_1 & -s_1 & 0 & a_1c_1 \\ s_1 & c_1 & 0 & a_1s_1 \\ 0 & 0 & 1 & 0 \\ 0 & 0 & 0 & 1 \end{bmatrix} \qquad A_2 = \begin{bmatrix} c_2 & s_2 & 0 & a_2c_2 \\ s_2 & -c_2 & 0 & a_2s_2 \\ 0 & 0 & -1 & 0 \\ 0 & 0 & 0 & 1 \end{bmatrix} \tag{3.22}$$

$$A_3 = \begin{bmatrix} 1 & 0 & 0 & 0 \\ 0 & 1 & 0 & 0 \\ 0 & 0 & 1 & d_3 \\ 0 & 0 & 0 & 1 \end{bmatrix} \qquad A_4 = \begin{bmatrix} c_4 & -s_4 & 0 & 0 \\ s_4 & c_4 & 0 & 0 \\ 0 & 0 & 1 & d_4 \\ 0 & 0 & 0 & 1 \end{bmatrix} \tag{3.23}$$

The forward kinematic equations are therefore given by

$$\begin{aligned} T_4^0 &= A_1 \cdots A_4 \\ &= \begin{bmatrix} c_{12}c_4 + s_{12}s_4 & -c_{12}s_4 + s_{12}c_4 & 0 & a_1c_1 + a_2c_{12} \\ s_{12}c_4 - c_{12}s_4 & -s_{12}s_4 - c_{12}c_4 & 0 & a_1s_1 + a_2s_{12} \\ 0 & 0 & -1 & -d_3 - d_4 \\ 0 & 0 & 0 & 1 \end{bmatrix} \end{aligned} \tag{3.24}$$

◇

3.3 INVERSE KINEMATICS

In the previous section we showed how to determine the end effector's position and orientation in terms of the joint variables. This section is concerned with the inverse problem of finding the joint variables in terms of the end effector's position and orientation. This is the problem of **inverse kinematics**, and it is, in general, more difficult than the forward kinematics problem.

We begin by formulating the general inverse kinematics problem. Following this, we describe the principle of kinematic decoupling and how it can be used to simplify the inverse kinematics of most modern manipulators. Using kinematic decoupling, we can consider the position and orientation problems independently. We describe a geometric approach for solving the positioning problem, while we exploit the Euler angle parameterization to solve the orientation problem.

3.3.1 The General Inverse Kinematics Problem

The general problem of inverse kinematics can be stated as follows. Given a 4×4 homogeneous transformation

$$H = \begin{bmatrix} R & o \\ 0 & 1 \end{bmatrix} \in SE(3) \tag{3.25}$$

find a solution, or possibly multiple solutions, of the equation

$$T_n^0(q_1, \ldots, q_n) = H \tag{3.26}$$

where

$$T_n^0(q_1, \ldots, q_n) = A_1(q_1) \cdots A_n(q_n) \tag{3.27}$$

Here, H represents the desired position and orientation of the end effector, and our task is to find the values for the joint variables $q_1, \ldots, q_n$ so that $T_n^0(q_1, \ldots, q_n) = H$.

Equation (3.26) results in twelve nonlinear equations in n unknown variables, which can be written as

$$T_{ij}(q_1, \ldots, q_n) = h_{ij}, \qquad i = 1, 2, 3, \quad j = 1, \ldots, 4 \tag{3.28}$$

where T_{ij}, h_{ij} refer to the twelve nontrivial entries of T_n^0 and H, respectively. Since the bottom row of both T_n^0 and H are (0,0,0,1), four of the sixteen equations represented by Equation (3.26) are trivial.

Example 3.7

Recall the Stanford manipulator of Example 3.5. Suppose that the desired position and orientation of the final frame are given by

$$H = \begin{bmatrix} 0 & 1 & 0 & -0.154 \\ 0 & 0 & 1 & 0.763 \\ 1 & 0 & 0 & 0 \\ 0 & 0 & 0 & 1 \end{bmatrix} \tag{3.29}$$

To find the corresponding joint variables θ_1, θ_2, d_3, θ_4, θ_5, and θ_6 we must solve the following simultaneous set of nonlinear trigonometric equations:

$$\begin{aligned}
c_1[c_2(c_4c_5c_6 - s_4s_6) - s_2s_5c_6] - s_1(s_4c_5c_6 + c_4s_6) &= 0 \\
s_1[c_2(c_4c_5c_6 - s_4s_6) - s_2s_5c_6] + c_1(s_4c_5c_6 + c_4s_6) &= 0 \\
-s_2(c_4c_5c_6 - s_4s_6) - c_2s_5c_6 &= 1 \\
c_1[-c_2(c_4c_5s_6 + s_4c_6) + s_2s_5s_6] - s_1(-s_4c_5s_6 + c_4c_6) &= 1 \\
s_1[-c_2(c_4c_5s_6 + s_4c_6) + s_2s_5s_6] + c_1(-s_4c_5s_6 + c_4c_6) &= 0 \\
s_2(c_4c_5s_6 + s_4c_6) + c_2s_5s_6 &= 0 \\
c_1(c_2c_4s_5 + s_2c_5) - s_1s_4s_5 &= 0 \\
s_1(c_2c_4s_5 + s_2c_5) + c_1s_4s_5 &= 1 \\
-s_2c_4s_5 + c_2c_5 &= 0 \\
c_1s_2d_3 - s_1d_2 + d_6(c_1c_2c_4s_5 + c_1c_5s_2 - s_1s_4s_5) &= -0.154 \\
s_1s_2d_3 + c_1d_2 + d_6(c_1s_4s_5 + c_2c_4s_1s_5 + c_5s_1s_2) &= 0.763 \\
c_2d_3 + d_6(c_2c_5 - c_4s_2s_5) &= 0
\end{aligned}$$

If the values of the nonzero DH parameters are $d_2 = 0.154$ *and* $d_6 = 0.263$, *one solution to this set of equations is given by:*

$$\theta_1 = \pi/2, \quad \theta_2 = \pi/2, \quad d_3 = 0.5, \quad \theta_4 = \pi/2, \quad \theta_5 = 0, \quad \theta_6 = \pi/2.$$

Even though we have not yet seen how one might derive this solution, it is not difficult to verify that it satisfies the forward kinematics equations for the Stanford arm.

◇

The equations in the preceding example are, of course, much too difficult to solve directly in closed form. This is the case for most robot arms. Therefore, we need to develop efficient and systematic techniques that exploit the particular kinematic structure of the manipulator. Whereas the forward kinematics problem always has a unique solution that can be obtained simply by evaluating the forward equations, the inverse kinematics problem may or may not have a solution. Even if a solution exists, it may or may not be unique. Furthermore, because these forward kinematic equations are in general complicated nonlinear functions of the joint variables, the solutions may be difficult to obtain even when they exist.

In solving the inverse kinematics problem we are most interested in finding a closed-form solution of the equations rather than a numerical solution. Finding a closed-form solution means finding an explicit relationship

$$q_k = f_k(h_{11}, \ldots, h_{34}), \qquad k = 1, \ldots, n \tag{3.30}$$

Closed-form solutions are preferable for two reasons. First, in certain applications, such as tracking a welding seam whose location is provided by a vision system, the inverse kinematic equations must be solved at a rapid rate, say every 20 milliseconds, and having closed-form expressions rather than an iterative search is a practical necessity. Second, the kinematic equations in general have multiple solutions. Having closed-form solutions allows one to develop rules for choosing a particular solution among several.

The practical question of the existence of solutions to the inverse kinematics problem depends on engineering as well as mathematical considerations. For example, the motion of the revolute joints may be restricted to less than a full 360 degrees of rotation so that not all mathematical solutions of the kinematic equations will correspond to physically realizable configurations of the manipulator. We will assume that the given position and orientation is such that at least one solution of Equation (3.26) exists. Once a solution to the mathematical equations is identified, it must be further checked to see whether or not it satisfies all constraints on the ranges of possible joint motions.

3.3.2 Kinematic Decoupling

Although the general problem of inverse kinematics is quite difficult, it turns out that for manipulators having six joints with the last three joint axes intersecting at a point (such as the Stanford Manipulator above), it is possible to decouple the inverse kinematics problem into two simpler problems, known respectively as **inverse position kinematics**, and **inverse orientation kinematics**. To put it another way, for a six-DOF manipulator with a spherical wrist, the inverse kinematics problem may be separated into two simpler problems, namely first finding the position of the intersection of the wrist axes, hereafter called the **wrist center**, and then finding the orientation of the wrist.

For concreteness let us suppose that there are exactly six degrees of freedom and that the last three joint axes intersect at a point o_c. We express Equation (3.26) as two sets of equations representing the rotational and positional equations

$$R_6^0(q_1, \ldots, q_6) = R \tag{3.31}$$

$$o_6^0(q_1, \ldots, q_6) = o \tag{3.32}$$

where o and R are the desired position and orientation of the tool frame, expressed with respect to the world coordinate system. Thus, we are given o and R, and the inverse kinematics problem is to solve for $q_1, \ldots, q_6$.

The assumption of a spherical wrist means that the axes z_3, z_4, and z_5 intersect at o_c and hence the origins o_4 and o_5 assigned by the DH convention will always be at the wrist center o_c. Often o_3 will also be at o_c, but this is not necessary for our subsequent development. The important point of this assumption for the inverse kinematics is that motion of the final three joints about these axes will not change the position of o_c, and thus the position of the wrist center is a function of only the first three joint variables.

The origin of the tool frame (whose desired coordinates are given by o) is simply obtained by a translation of distance d_6 along z_5 from o_c (see Table 3.3). In our case, z_5 and z_6 are the same axis, and the third column of R expresses the direction of z_6 with respect to the base frame. Therefore, we have

$$o = o_c^0 + d_6 R \begin{bmatrix} 0 \\ 0 \\ 1 \end{bmatrix} \tag{3.33}$$

Thus, in order to have the end effector of the robot at the point with coordinates given by o and with the orientation of the end effector given by

$R = (r_{ij})$, it is necessary and sufficient that the wrist center o_c have coordinates given by

$$o_c^0 \;=\; o - d_6 R \begin{bmatrix} 0 \\ 0 \\ 1 \end{bmatrix} \tag{3.34}$$

and that the orientation of the frame $o_6x_6y_6z_6$ with respect to the base be given by R. If the components of the end effector's position o are denoted o_x, o_y, o_z and the components of the wrist center o_c^0 are denoted x_c, y_c, z_c then Equation (3.34) gives the relationship

$$\begin{bmatrix} x_c \\ y_c \\ z_c \end{bmatrix} \;=\; \begin{bmatrix} o_x - d_6 r_{13} \\ o_y - d_6 r_{23} \\ o_z - d_6 r_{33} \end{bmatrix} \tag{3.35}$$

Using Equation (3.35) we may find the values of the first three joint variables. This determines the orientation transformation R_3^0 which depends only on these first three joint variables. We can now determine the orientation of the end effector relative to the frame $o_3x_3y_3z_3$ from the expression

$$R \;=\; R_3^0 R_6^3 \tag{3.36}$$

as

$$R_6^3 \;=\; (R_3^0)^{-1} R = (R_3^0)^T R \tag{3.37}$$

As we shall see in Section 3.3.6, the final three joint angles can then be found as a set of Euler angles corresponding to R_6^3. Note that the right-hand side of Equation (3.37) is completely known since R is given and R_3^0 can be calculated once the first three joint variables are known. The idea of kinematic decoupling is illustrated in Figure 3.12.

3.3.3 Inverse Position: A Geometric Approach

For the common kinematic arrangements that we consider, we can use a geometric approach to find the variables q_1, q_2, q_3 corresponding to o_c^0 given by Equation (3.34). We restrict our treatment to the geometric approach for two reasons. First, as we have said, most manipulator designs are kinematically simple, usually consisting of one of the five basic configurations of Chapter 1 with a spherical wrist. Indeed, it is partly due to the difficulty of the general inverse kinematics problem that manipulator designs have evolved to their present state. Second, there are few techniques that can

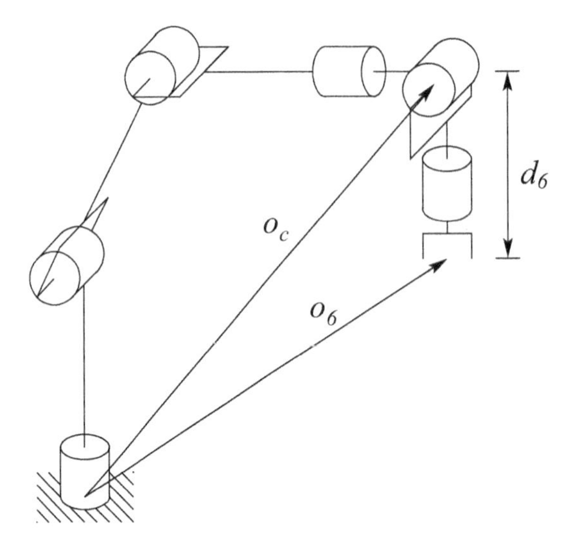

Figure 3.12: Kinematic decoupling.

handle the general inverse kinematics problem for arbitrary configurations. Since the reader is most likely to encounter robot configurations of the type considered here, the added difficulty involved in treating the general case seems unjustified.

In general, the complexity of the inverse kinematics problem increases with the number of nonzero DH parameters. For most manipulators, many of the a_i, d_i are zero, the α_i are 0 or $\pm\pi/2$, etc. In these cases especially, a geometric approach is the simplest and most natural. The general idea of the geometric approach is to solve for joint variable q_i by projecting the manipulator onto the $x_{i-1} - y_{i-1}$ plane and solving a simple trigonometry problem. For example, to solve for θ_1, we project the arm onto the $x_0 - y_0$ plane and use trigonometry to find θ_1. We will illustrate this method with two important examples: the articulated and spherical arms.

3.3.4 Articulated Configuration

Consider the elbow manipulator shown in Figure 3.13, with the components of o_c^0 denoted by x_c, y_c, z_c. We project o_c onto the $x_0 - y_0$ plane as shown in Figure 3.14. We see from this projection that

$$\theta_1 = \text{Atan2}(x_c, y_c) \tag{3.38}$$

in which $\text{Atan2}(x, y)$ denotes the two argument arctangent function defined in Appendix A.

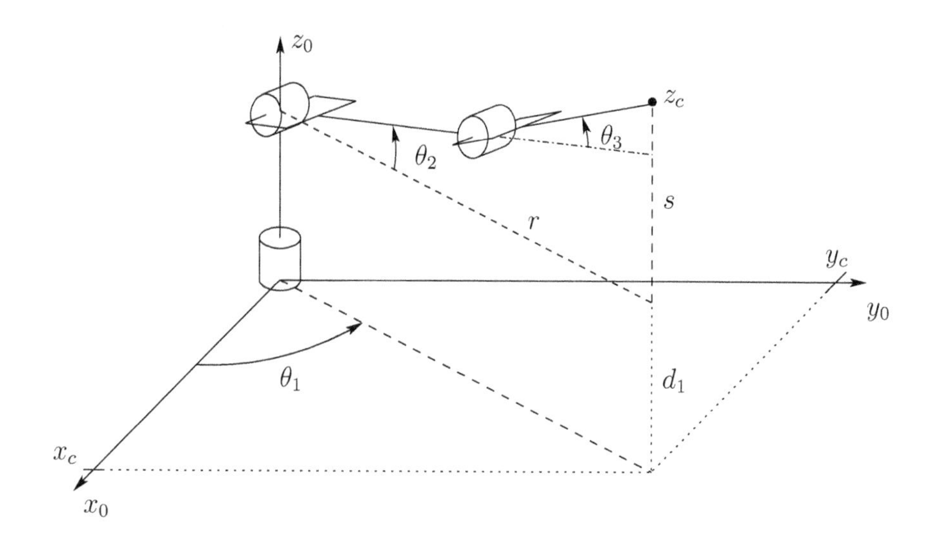

Figure 3.13: Elbow manipulator.

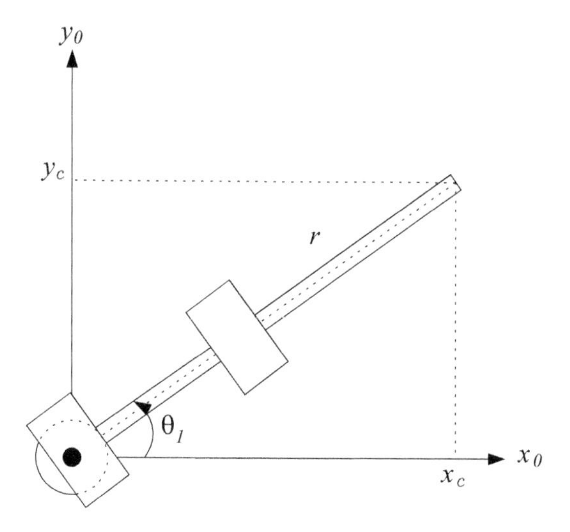

Figure 3.14: Projection of the wrist center onto $x_0 - y_0$ plane.

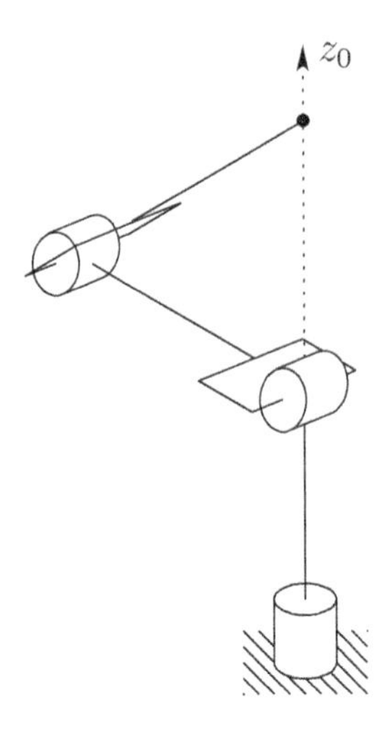

Figure 3.15: Singular configuration in which the wrist center lies on the z_0 axis.

Note that a second valid solution for θ_1 is

$$\theta_1 = \pi + \text{Atan2}(x_c, y_c) \tag{3.39}$$

Of course this will, in turn, lead to different solutions for θ_2 and θ_3, as we will see below.

These solutions for θ_1, are valid unless $x_c = y_c = 0$. In this case Equation (3.38) is undefined and the manipulator is in a singular configuration, shown in Figure 3.15. In this position the wrist center o_c intersects z_0; hence any value of θ_1 leaves o_c fixed. There are thus infinitely many solutions for θ_1 when o_c intersects z_0.

If there is an offset $d \neq 0$ as shown in Figure 3.16 then the wrist center cannot intersect z_0. In this case, depending on how the DH parameters have been assigned, we will have $d_2 = d$ or $d_3 = d$, and there will, in general, be only two solutions for θ_1. These correspond to the so-called **left arm** and **right arm** configurations as shown in Figures 3.17 and 3.18.

Figure 3.17 shows the left arm configuration. From this figure, we see geometrically that

$$\theta_1 = \phi - \alpha \tag{3.40}$$

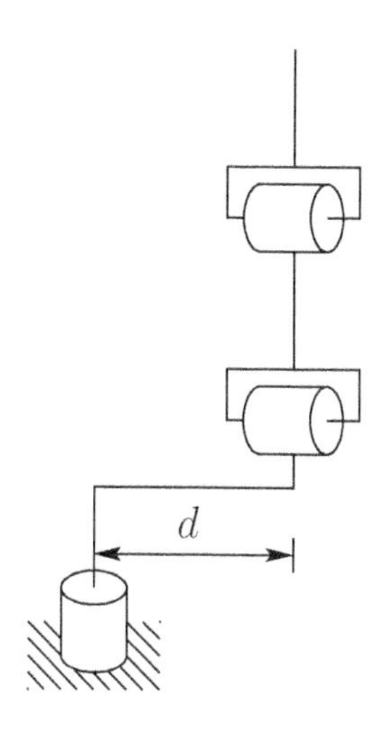

Figure 3.16: Elbow manipulator with shoulder offset.

in which

$$\begin{aligned} \phi &= \text{Atan2}(x_c, y_c) && (3.41) \\ \alpha &= \text{Atan2}\left(\sqrt{r^2 - d^2}, d\right) && (3.42) \\ &= \text{Atan2}\left(\sqrt{x_c^2 + y_c^2 - d^2}, d\right) \end{aligned}$$

The second solution, given by the right arm configuration shown in Figure 3.18 is given by

$$\theta_1 = \text{Atan2}(x_c, y_c) + \text{Atan2}\left(-\sqrt{r^2 - d^2}, -d\right) \tag{3.43}$$

To see this, note that

$$\begin{aligned} \theta_1 &= \alpha + \beta \\ \alpha &= \text{Atan2}(x_c, y_c) \\ \beta &= \gamma + \pi \\ \gamma &= \text{Atan2}\left(\sqrt{r^2 - d^2}, d\right) \end{aligned}$$

which together imply that

$$\beta = \text{Atan2}\left(-\sqrt{r^2 - d^2}, -d\right)$$

since $\cos(\theta + \pi) = -\cos(\theta)$ and $\sin(\theta + \pi) = -\sin(\theta)$.

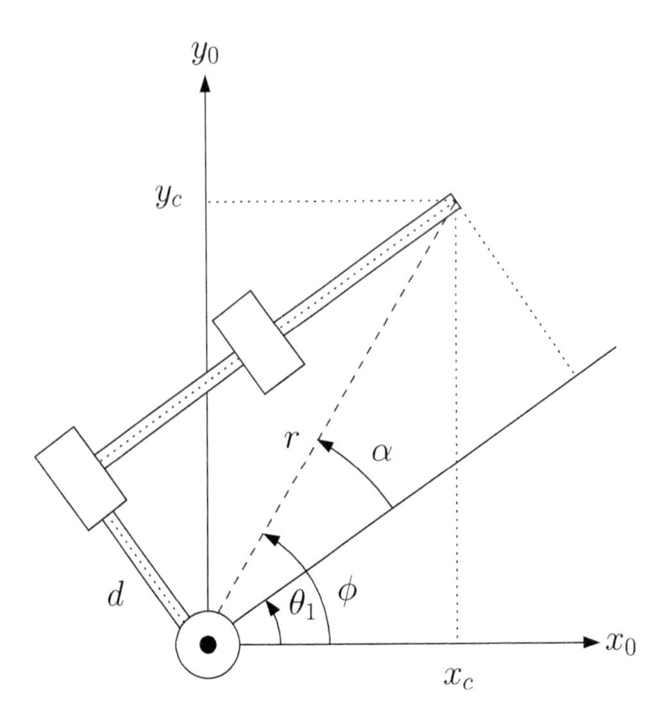

Figure 3.17: Left arm configuration.

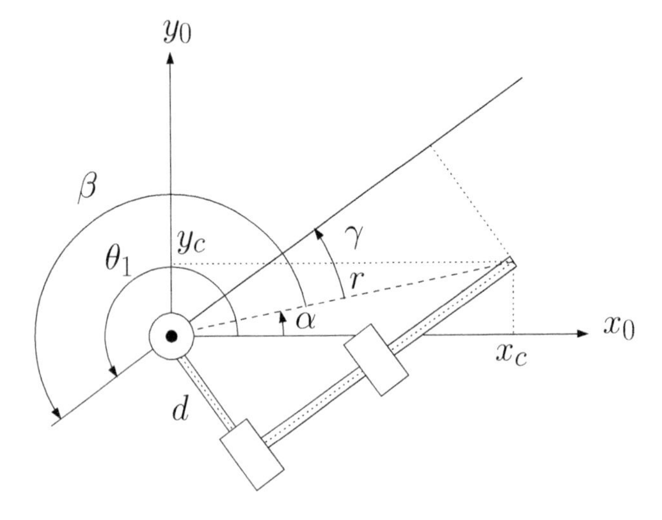

Figure 3.18: Right arm configuration.

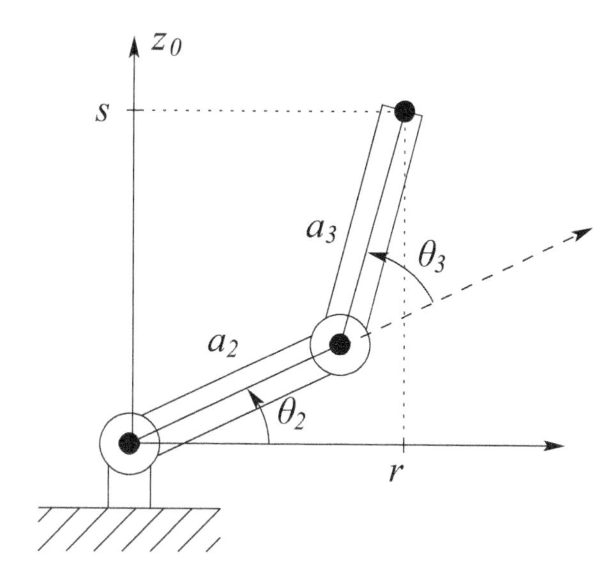

Figure 3.19: Projecting onto the plane formed by links 2 and 3.

To find the angles θ_2, θ_3 for the elbow manipulator given θ_1, we consider the plane formed by the second and third links as shown in Figure 3.19. Since the motion of second and third links is planar, the solution is analogous to that of the two-link manipulator of Chapter 1. As in our previous derivation (cf. Equations (1.7) and (1.8)) we can apply the law of cosines to obtain

$$\begin{aligned} \cos\theta_3 &= \frac{r^2 + s^2 - a_2^2 - a_3^2}{2a_2a_3} \\ &= \frac{x_c^2 + y_c^2 - d^2 + (z_c - d_1)^2 - a_2^2 - a_3^2}{2a_2a_3} := D \end{aligned} \tag{3.44}$$

since $r^2 = x_c^2 + y_c^2 - d^2$ and $s = z_c - d_1$. Hence, θ_3 is given by

$$\theta_3 = \text{Atan2}\left(D, \pm\sqrt{1 - D^2}\right) \tag{3.45}$$

The two solutions for θ_3 correspond to the elbow-down position and elbow-up position, respectively. Similarly θ_2 is given as

$$\begin{aligned} \theta_2 &= \text{Atan2}(r, s) - \text{Atan2}(a_2 + a_3c_3, a_3s_3) \\ &= \text{Atan2}\left(\sqrt{x_c^2 + y_c^2 - d^2}, z_c - d_1\right) - \text{Atan2}(a_2 + a_3c_3, a_3s_3) \end{aligned} \tag{3.46}$$

An example of an elbow manipulator with offsets is the PUMA shown in Figure 3.20. There are four solutions to the inverse position kinematics

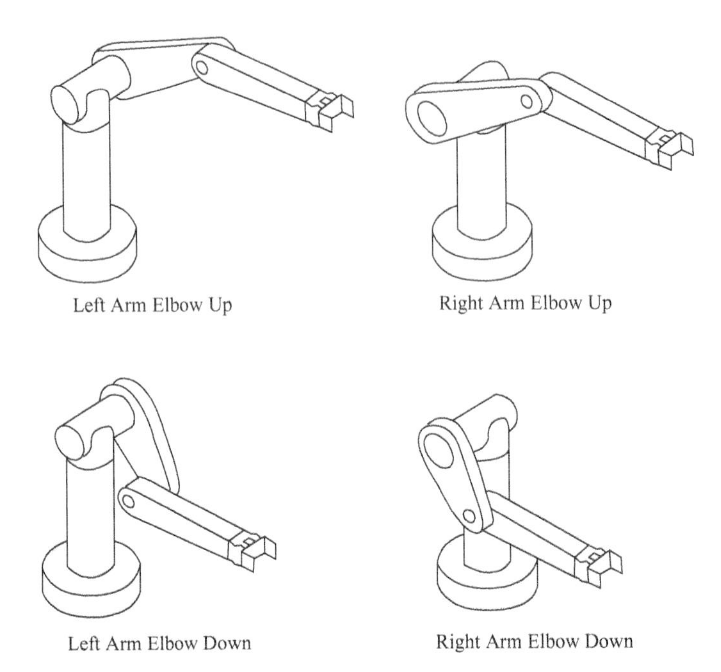

Figure 3.20: Four solutions of the inverse position kinematics for the PUMA manipulator.

as shown. These correspond to the situations left arm-elbow up, left arm–elbow down, right arm–elbow up and right arm–elbow down. We will see that there are two solutions for the wrist orientation thus giving a total of eight solutions of the inverse kinematics for the PUMA manipulator.

3.3.5 Spherical Configuration

Next, we solve the inverse position kinematics for a three degree of freedom spherical manipulator shown in Figure 3.21. As in the case of the elbow manipulator the first joint variable is the base rotation and a solution is given as

$$\theta_1 = \text{Atan2}(x_c, y_c) \tag{3.47}$$

provided x_c and y_c are not both zero. If both x_c and y_c are zero, the configuration is singular as before and θ_1 may take on any value. As in the case of the elbow manipulator, a second solution for θ_1 is given by

$$\theta_1 = \pi + \text{Atan2}(x_c, y_c). \tag{3.48}$$

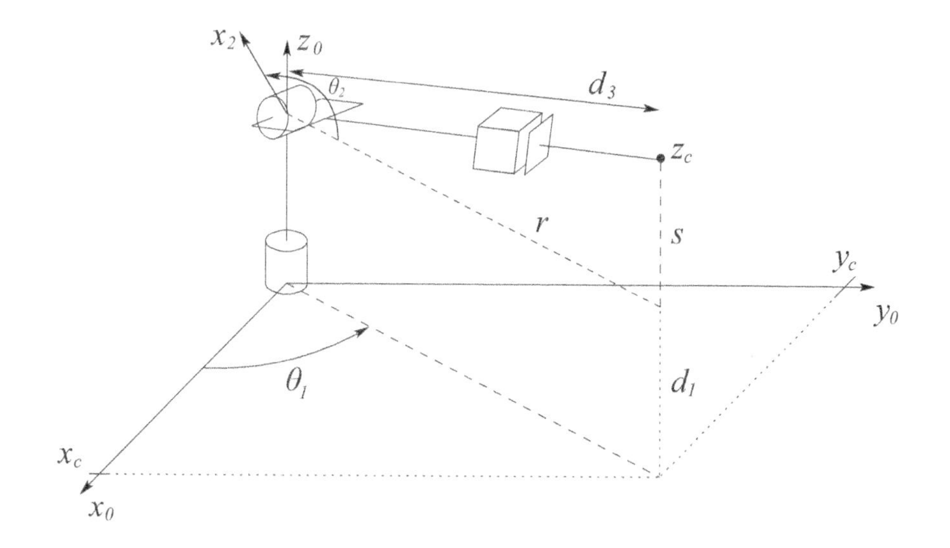

Figure 3.21: Spherical manipulator.

The angle θ_2 is given from Figure 3.21 as

$$\theta_2 = \operatorname{Atan2}(r, s) + \frac{\pi}{2} \tag{3.49}$$

where $r^2 = x_c^2 + y_c^2$, $s = z_c - d_1$.

The linear distance d_3 is found as

$$d_3 = \sqrt{r^2 + s^2} = \sqrt{x_c^2 + y_c^2 + (z_c - d_1)^2} \tag{3.50}$$

The negative square root solution for d_3 is disregarded and thus in this case we obtain two solutions to the inverse position kinematics as long as the wrist center does not intersect z_0. If there is an offset then there will be left and right arm configurations as in the case of the elbow manipulator (Problem 3-21).

3.3.6 Inverse Orientation

In the previous section we used a geometric approach to solve the inverse position problem. This gives the values of the first three joint variables corresponding to a given position of the wrist center. The inverse orientation problem is now one of finding the values of the final three joint variables corresponding to a given orientation with respect to the frame $o_3x_3y_3z_3$. For a spherical wrist, this can be interpreted as the problem of finding a set of Euler angles corresponding to a given rotation matrix R. Recall that

Table 3.6: DH parameters for the articulated manipulator of Figure 3.13.

Link	a_i	α_i	d_i	θ_i
1	0	90	d_1	θ_1^*
2	a_2	0	0	θ_2^*
3	a_3	0	0	θ_3^*

* variable

Equation (3.15) shows that the rotation matrix obtained for the spherical wrist has the same form as the rotation matrix for the Euler transformation given in Equation (2.26). Therefore, we can use the method developed in Section 2.5.1 to solve for the three joint angles of the spherical wrist. In particular, we solve for the three Euler angles, ϕ, θ, ψ, using Equations (2.28) – (2.33), and then use the mapping

$$\begin{aligned} \theta_4 &= \phi \\ \theta_5 &= \theta \\ \theta_6 &= \psi \end{aligned}$$

Example 3.8 Articulated Manipulator with Spherical Wrist

The DH parameters for the frame assignment shown in Figure 3.13 are summarized in Table 3.6. Multiplying the corresponding A_i matrices gives the matrix R_3^0 for the articulated or elbow manipulator as

$$R_3^0 = \begin{bmatrix} c_1 c_{23} & -c_1 s_{23} & s_1 \\ s_1 c_{23} & -s_1 s_{23} & -c_1 \\ s_{23} & c_{23} & 0 \end{bmatrix} \tag{3.51}$$

The matrix R_6^3 is the upper left 3×3 submatrix of the matrix product $A_4 A_5 A_6$ given by

$$R_6^3 = \begin{bmatrix} c_4 c_5 c_6 - s_4 s_6 & -c_4 c_5 s_6 - s_4 c_6 & c_4 s_5 \\ s_4 c_5 c_6 + c_4 s_6 & -s_4 c_5 s_6 + c_4 c_6 & s_4 s_5 \\ -s_5 c_6 & s_5 s_6 & c_5 \end{bmatrix} \tag{3.52}$$

The equation to be solved for the final three variables is therefore

$$R_6^3 = (R_3^0)^T R \tag{3.53}$$

and the Euler angle solution can be applied to this equation. For example, the three equations given by the third column in the above matrix equation are given by

$$c_4 s_5 = c_1 c_{23} r_{13} + s_1 c_{23} r_{23} + s_{23} r_{33} \quad (3.54)$$
$$s_4 s_5 = -c_1 s_{23} r_{13} - s_1 s_{23} r_{23} + c_{23} r_{33} \quad (3.55)$$
$$c_5 = s_1 r_{13} - c_1 r_{23} \quad (3.56)$$

Hence, if not both of Equations (3.54) and (3.55) are zero, we obtain θ_5 *from Equations (2.28) and (2.29) as*

$$\theta_5 = \text{Atan2}\left(s_1 r_{13} - c_1 r_{23}, \pm\sqrt{1 - (s_1 r_{13} - c_1 r_{23})^2}\right) \quad (3.57)$$

If the positive square root is chosen in Equation (3.57), then θ_4 *and* θ_6 *are given by Equations (2.30) and (2.31), respectively, as*

$$\begin{aligned} \theta_4 &= \text{Atan2}(c_1 c_{23} r_{13} + s_1 c_{23} r_{23} + s_{23} r_{33}, \\ &\qquad -c_1 s_{23} r_{13} - s_1 s_{23} r_{23} + c_{23} r_{33}) \quad (3.58) \\ \theta_6 &= \text{Atan2}(-s_1 r_{11} + c_1 r_{21}, s_1 r_{12} - c_1 r_{22}) \quad (3.59) \end{aligned}$$

The other solutions are obtained analogously. If $s_5 = 0$*, then joint axes* z_3 *and* z_5 *are collinear. This is a singular configuration and only the sum* $\theta_4 + \theta_6$ *can be determined. One solution is to choose* θ_4 *arbitrarily and then determine* θ_6 *using Equation (2.35) or (2.37).*

◇

Example 3.9 Elbow Manipulator - Complete Solution

To summarize the geometric approach for solving the inverse kinematics equations, we give here one solution to the inverse kinematics of the six degree-of-freedom elbow manipulator shown in Figure 3.13 which has no joint offsets and a spherical wrist.

Given

$$o = \begin{bmatrix} o_x \\ o_y \\ o_z \end{bmatrix}, \quad R = \begin{bmatrix} r_{11} & r_{12} & r_{13} \\ r_{21} & r_{22} & r_{23} \\ r_{31} & r_{32} & r_{33} \end{bmatrix} \quad (3.60)$$

then with

$$x_c = o_x - d_6 r_{13} \quad (3.61)$$
$$y_c = o_y - d_6 r_{23} \quad (3.62)$$
$$z_c = o_z - d_6 r_{33} \quad (3.63)$$

a set of DH joint variables is given by

$$\theta_1 = \text{Atan2}(x_c, y_c) \tag{3.64}$$

$$\theta_2 = \text{Atan2}\left(\sqrt{x_c^2 + y_c^2 - d^2}, z_c - d_1\right) - \text{Atan2}(a_2 + a_3 c_3, a_3 s_3) \tag{3.65}$$

$$\theta_3 = \text{Atan2}\left(D, \pm\sqrt{1 - D^2}\right), \quad \text{with } D = \frac{x_c^2 + y_c^2 - d^2 + (z_c - d_1)^2 - a_2^2 - a_3^2}{2a_2 a_3} \tag{3.66}$$

$$\theta_4 = \text{Atan2}(c_1 c_{23} r_{13} + s_1 c_{23} r_{23} + s_{23} r_{33}, -c_1 s_{23} r_{13} - s_1 s_{23} r_{23} + c_{23} r_{33}) \tag{3.67}$$

$$\theta_5 = \text{Atan2}\left(s_1 r_{13} - c_1 r_{23}, \pm\sqrt{1 - (s_1 r_{13} - c_1 r_{23})^2}\right) \tag{3.68}$$

$$\theta_6 = \text{Atan2}(-s_1 r_{11} + c_1 r_{21}, s_1 r_{12} - c_1 r_{22}) \tag{3.69}$$

The other possible solutions are left as an exercise (Problem 3-20).

◇

Example 3.10 SCARA Manipulator

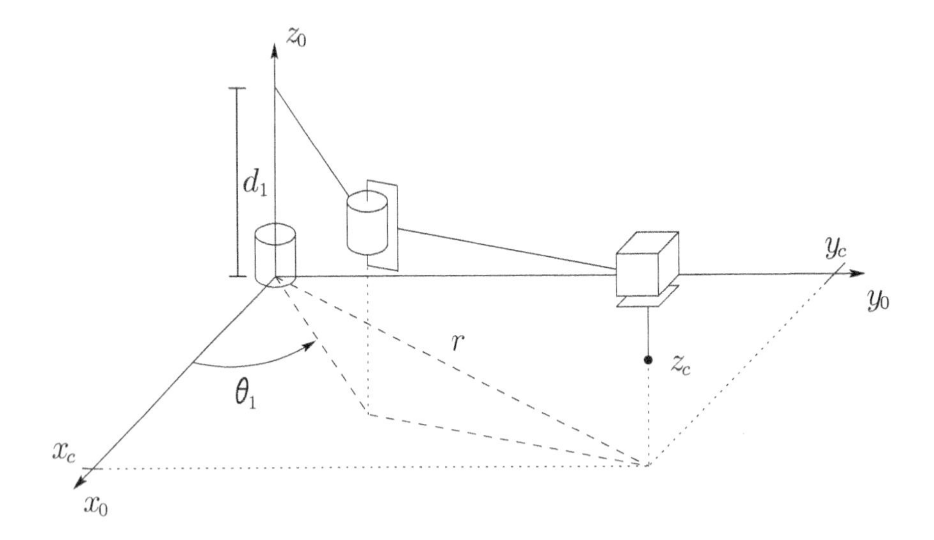

Figure 3.22: SCARA manipulator.

As another example, consider the SCARA manipulator illustrated in Figure 3.22, with forward kinematics defined by T_4^0 from Equation (3.24). The inverse kinematics solution is then given as the set of solutions of the equa-

tion

$$
\begin{aligned}
T_4^1 &= \begin{bmatrix} R & o \\ 0 & 1 \end{bmatrix} \\
&= \begin{bmatrix} c_{12}c_4 + s_{12}s_4 & s_{12}c_4 - c_{12}s_4 & 0 & a_1c_1 + a_2c_{12} \\ s_{12}c_4 - c_{12}s_4 & -c_{12}c_4 - s_{12}s_4 & 0 & a_1s_1 + a_2s_{12} \\ 0 & 0 & -1 & -d_3 - d_4 \\ 0 & 0 & 0 & 1 \end{bmatrix} \qquad (3.70)
\end{aligned}
$$

We first note that, since the SCARA has only four degrees of freedom, not every possible H from $SE(3)$ allows a solution of Equation (3.70). In fact we can easily see that there is no solution of Equation (3.70) unless R is of the form

$$
R = \begin{bmatrix} c_\alpha & s_\alpha & 0 \\ s_\alpha & -c_\alpha & 0 \\ 0 & 0 & -1 \end{bmatrix} \qquad (3.71)
$$

and if this is the case, the sum $\theta_1 + \theta_2 - \theta_4$ is determined by

$$
\theta_1 + \theta_2 - \theta_4 = \alpha = \mathrm{Atan2}(r_{11}, r_{12}) \qquad (3.72)
$$

Projecting the manipulator configuration onto the $x_0 - y_0$ plane yields the geometry shown in Figure 3.22. Using the law of cosines

$$
c_2 = \frac{o_x^2 + o_y^2 - a_1^2 - a_2^2}{2a_1a_2} \qquad (3.73)
$$

$$
\qquad (3.74)
$$

and

$$
\theta_2 = \mathrm{Atan2}\left(c_2, \pm\sqrt{1 - c_2}\right) \qquad (3.75)
$$

The value for θ_1 is then obtained as

$$
\theta_1 = \mathrm{Atan2}(o_x, o_y) - \mathrm{Atan2}(a_1 + a_2c_2, a_2s_2) \qquad (3.76)
$$

We may now determine θ_4 from Equation (3.72) as

$$
\begin{aligned}
\theta_4 &= \theta_1 + \theta_2 - \alpha \qquad (3.77) \\
&= \theta_1 + \theta_2 - \mathrm{Atan2}(r_{11}, r_{12})
\end{aligned}
$$

Finally d_3 is given as

$$
d_3 = o_z + d_4 \qquad (3.78)
$$

◇

3.4 SUMMARY

In this chapter we studied the relationships between joint variables q_i and the position and orientation of the end effector. We began by introducing the Denavit-Hartenberg convention for assigning coordinate frames to the links of a serial manipulator. We may summarize the procedure based on the DH convention in the following algorithm for deriving the forward kinematics for any manipulator.

Step l: Locate and label the joint axes $z_0, \ldots, z_{n-1}$.

Step 2: Establish the base frame. Set the origin anywhere on the z_0-axis. The x_0 and y_0 axes are chosen conveniently to form a right-handed frame.

For $i = 1, \ldots, n-1$ perform Steps 3 to 5.

Step 3: Locate the origin o_i where the common normal to z_i and z_{i-1} intersects z_i. If z_i intersects z_{i-1} locate o_i at this intersection. If z_i and z_{i-1} are parallel, locate o_i in any convenient position along z_i.

Step 4: Establish x_i along the common normal between z_{i-1} and z_i through o_i, or in the direction normal to the $z_{i-1} - z_i$ plane if z_{i-1} and z_i intersect.

Step 5: Establish y_i to complete a right-handed frame.

Step 6: Establish the end-effector frame $o_n x_n y_n z_n$. Assuming the nth joint is revolute, set $z_n = a$ parallel to z_{n-1}. Establish the origin o_n conveniently along z_n, preferably at the center of the gripper or at the tip of any tool that the manipulator may be carrying. Set $y_n = s$ in the direction of the gripper closure and set $x_n = n$ as $s \times a$. If the tool is not a simple gripper set x_n and y_n conveniently to form a right-handed frame.

Step 7: Create a table of DH parameters a_i, d_i, α_i, θ_i.

$a_i =$ distance along x_i from the intersection of the x_i and z_{i-1} axes to o_i.

$d_i =$ distance along z_{i-1} from o_{i-1} to the intersection of the x_i and z_{i-1} axes. If joint i is prismatic, d_i is variable.

$\alpha_i =$ the angle from z_{i-1} to z_i measured about x_i.

$\theta_i =$ the angle from x_{i-1} to x_i measured about z_{i-1}. If joint i is revolute, θ_i is variable.

Step 8: Form the homogeneous transformation matrices A_i by substituting the above parameters into Equation (3.10).

Step 9: Form $T_n^0 = A_1 \cdots A_n$. This then gives the position and orientation of the tool frame expressed in base coordinates.

This DH convention defines the forward kinematics equations for a manipulator, i.e., the mapping from joint variables to end effector position and orientation. To control a manipulator, it is necessary to solve the inverse problem, i.e., given a position and orientation for the end effector, solve for the corresponding set of joint variables.

In this chapter we considered the special case of manipulators for which kinematic decoupling can be used (e.g., a manipulator with a spherical wrist). For this class of manipulators the determination of the inverse kinematics can be summarized by the following algorithm.

Step 1: Find q_1, q_2, q_3 such that the wrist center o_c has coordinates given by

$$o_c^0 = o - d_6 R \begin{bmatrix} 0 \\ 0 \\ 1 \end{bmatrix} \tag{3.79}$$

Step 2: Using the joint variables determined in Step 1, evaluate R_3^0.

Step 3: Find a set of Euler angles corresponding to the rotation matrix

$$R_6^3 = (R_3^0)^{-1} R = (R_3^0)^T R \tag{3.80}$$

In this chapter we demonstrated a geometric approach for Step 1. In particular, to solve for joint variable q_i, we project the manipulator (including the wrist center) onto the $x_{i-1} - y_{i-1}$ plane and use trigonometry to find q_i.

PROBLEMS

3-1 Verify the statement after Equation (3.14) that the rotation matrix R has the form given by Equation (3.13) provided assumptions (DH1) and (DH2) are satisfied.

3-2 Consider the three-link planar manipulator shown in Figure 3.23. Derive the forward kinematic equations using the DH convention.

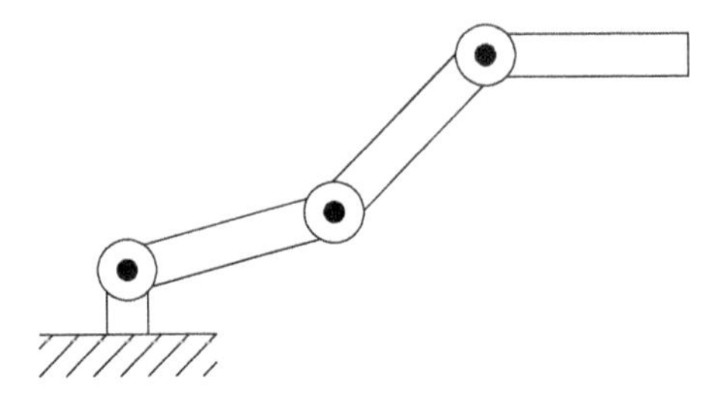

Figure 3.23: Three-link planar arm of Problem 3-2.

3-3 Consider the two-link Cartesian manipulator of Figure 3.24. Derive the forward kinematic equations using the DH convention.

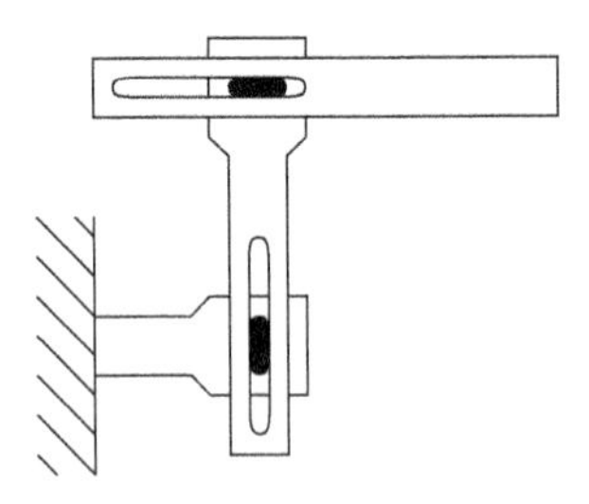

Figure 3.24: Two-link Cartesian robot of Problem 3-3.

3-4 Consider the two-link manipulator of Figure 3.25, which has joint 1 revolute and joint 2 prismatic. Derive the forward kinematic equations using the DH convention.

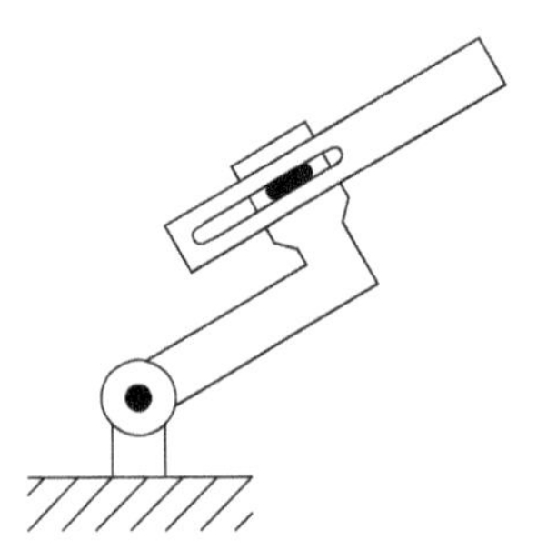

Figure 3.25: Two-link planar arm of Problem 3-4.

3-5 Consider the three-link planar manipulator of Figure 3.26. Derive the forward kinematic equations using the DH convention.

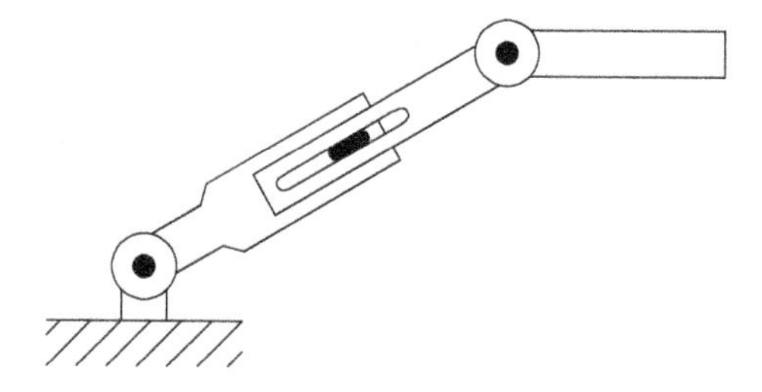

Figure 3.26: Three-link planar arm with prismatic joint of Problem 3-5.

3-6 Consider the three-link articulated robot of Figure 3.27. Derive the forward kinematic equations using the DH convention.

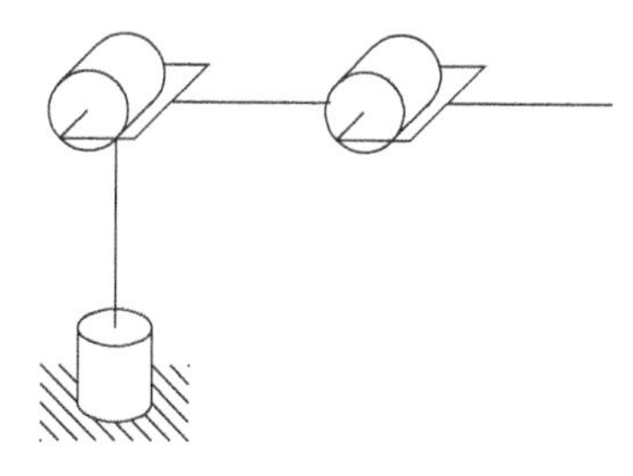

Figure 3.27: Three-link articulated robot.

3-7 Consider the three-link Cartesian manipulator of Figure 3.28. Derive the forward kinematic equations using the DH convention.

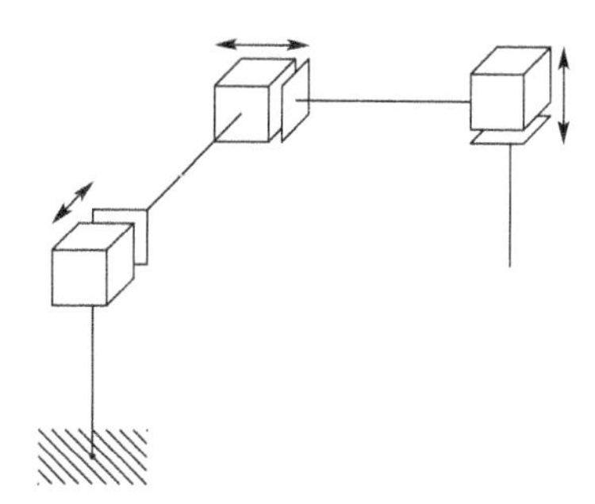

Figure 3.28: Three-link Cartesian robot.

3-8 Attach a spherical wrist to the three-link articulated manipulator of Problem 3-6 as shown in Figure 3.29. Derive the forward kinematic equations for this manipulator.

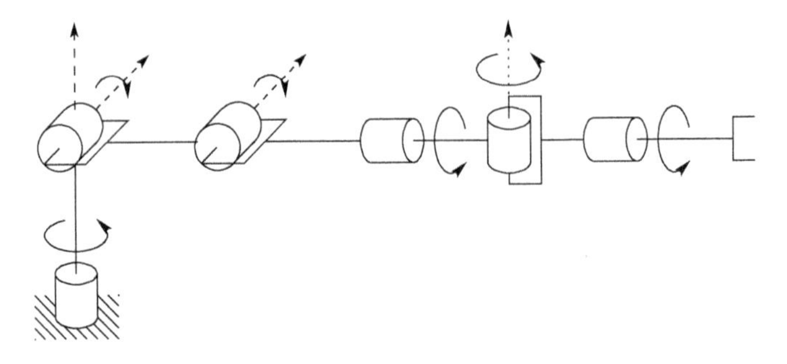

Figure 3.29: Elbow manipulator with spherical wrist.

3-9 Attach a spherical wrist to the three-link Cartesian manipulator of Problem 3-7 as shown in Figure 3.30. Derive the forward kinematic equations for this manipulator.

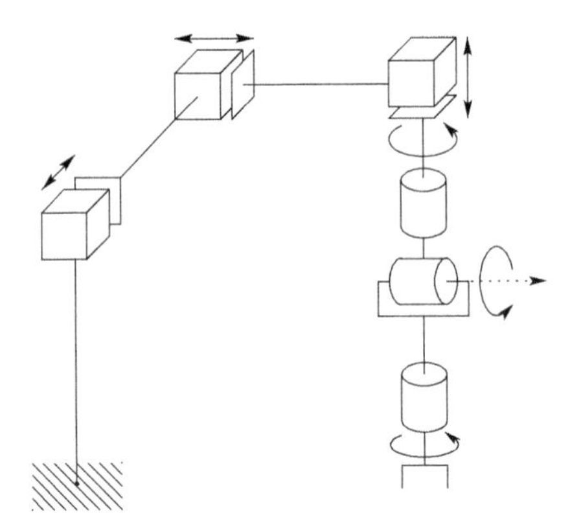

Figure 3.30: Cartesian manipulator with spherical wrist.

3-10 Consider the PUMA 260 manipulator shown in Figure 3.31. Derive the complete set of forward kinematic equations by establishing appropriate DH coordinate frames, constructing a table of DH parameters, forming the A matrices, etc.

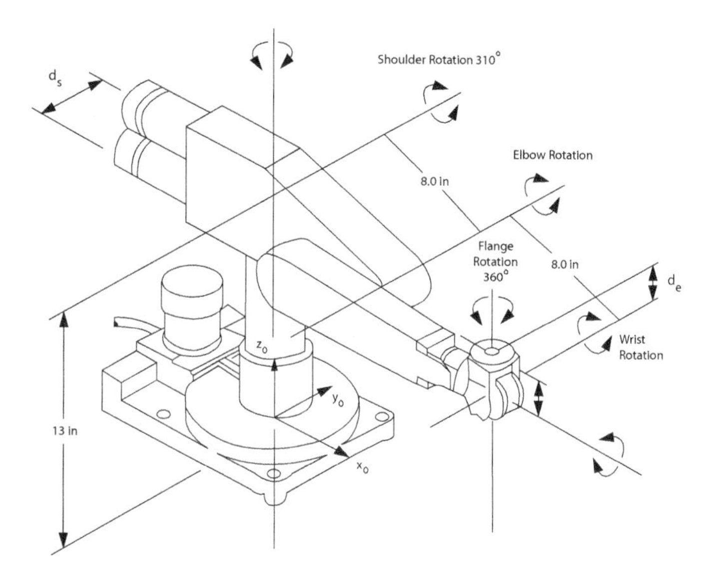

Figure 3.31: PUMA 260 manipulator.

3-11 Given a desired position of the end effector, how many solutions are there to the inverse kinematics of the three-link planar arm shown in Figure 3.32? If the orientation of the end effector is also specified, how many solutions are there? Use the geometric approach to find them.

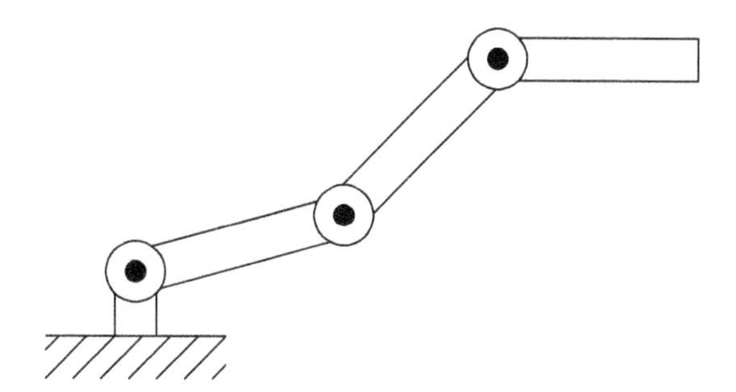

Figure 3.32: Three-link planar robot with revolute joints.

3-12 Repeat Problem 3-11 for the three-link planar arm with prismatic joint of Figure 3.33.

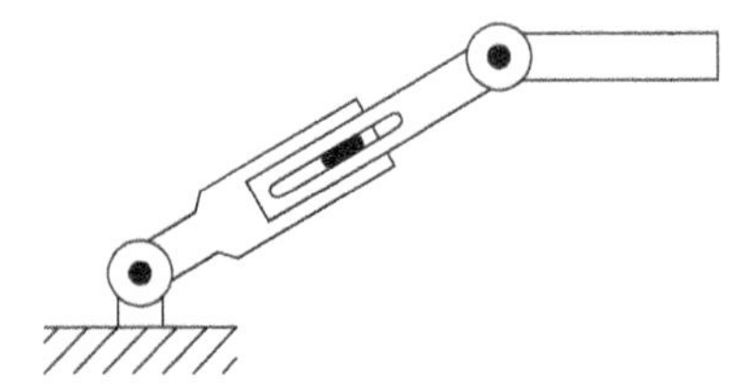

Figure 3.33: Three-link planar robot with prismatic joint.

3-13 Solve the inverse position kinematics for the cylindrical manipulator of Figure 3.34.

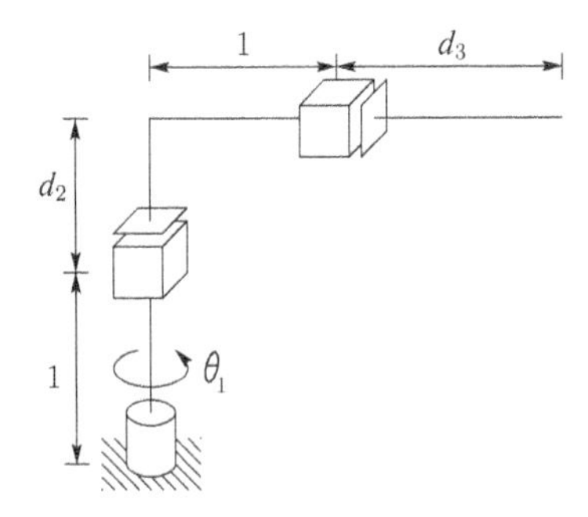

Figure 3.34: Cylindrical configuration.

3-14 Solve the inverse position kinematics for the Cartesian manipulator of Figure 3.35.

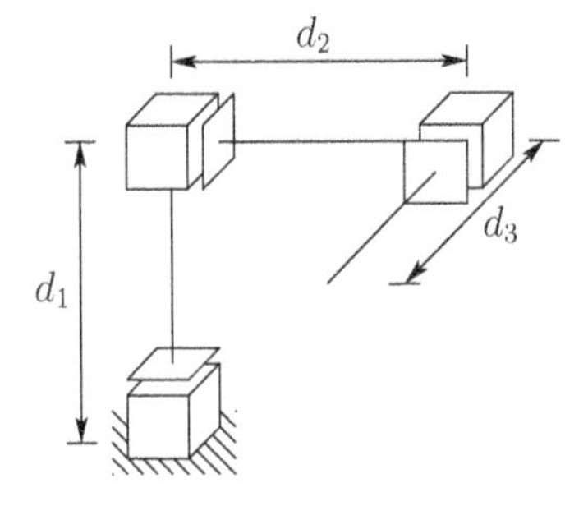

Figure 3.35: Cartesian configuration.

3-15 Add a spherical wrist to the three-link cylindrical arm of Problem 3-13 and write the complete inverse kinematics solution.

3-16 Add a spherical wrist to the Cartesian manipulator of Problem 3-14 and write the complete inverse kinematics solution.

3-17 Write a computer program to compute the inverse kinematic equations for the elbow manipulator using Equations (3.64)–(3.69). Include procedures for identifying singular configurations and choosing a particular solution when the configuration is not singular. Test your routine for various special cases, including singular configurations.

3-18 The Stanford manipulator of Example 3.5 has a spherical wrist. Given a desired position o and orientation R of the end effector,

1. Compute the desired coordinates of the wrist center o_c^0.
2. Solve the inverse position kinematics, that is, find values of the first three joint variables that will place the wrist center at o_c. Is the solution unique? How many solutions did you find?
3. Compute the rotation matrix R_3^0. Solve the inverse orientation problem for this manipulator by finding a set of Euler angles corresponding to R_6^3 given by Equation (3.52).

3-19 Repeat Problem 3-18 for the PUMA 260 manipulator of Problem 3-10, which also has a spherical wrist. How many total solutions did you find?

3-20 Find all other solutions to the inverse kinematics of the elbow manipulator of Example 3.9.

3-21 Modify the solutions θ_1 and θ_2 for the spherical manipulator given by Equations (3.47) and (3.49) for the case of a shoulder offset.

NOTES AND REFERENCES

The Denavit-Hartenberg convention for assigning coordinate frames was introduced in the fifties, and is described in [57] and [27]. Since then, many articles have been written on the topics of forward and inverse kinematics. Seminal articles that deal with forward kinematics include [19], [29], [74], [75], [103], [57], and [138]. Inverse kinematics problems are considered in [6], [45], [53], [75], [76], [103], [105], [113], and [134]. In the late seventies and early eighties, several robotics books were published that covered topics related to robot kinematics, such as, [13], [102], [128], [140]. Since then, most robotics texts have included as standard material descriptions of the DH convention and the inverse kinematics problem, including [41], [93], [110]. More detailed treatment of the general inverse kinematics problem can be found in [45], [53], [105], [134].

Chapter 4

VELOCITY KINEMATICS – THE JACOBIAN

In the previous chapter we derived the forward and inverse position equations relating joint positions to end-effector positions and orientations. In this chapter we derive the velocity relationships, relating the linear and angular velocities of the end effector to the joint velocities.

Mathematically, the forward kinematic equations define a function between the space of Cartesian positions and orientations and the space of joint positions. The velocity relationships are then determined by the **Jacobian** of this function. The Jacobian is a matrix that generalizes the notion of the ordinary derivative of a scalar function. The Jacobian is one of the most important quantities in the analysis and control of robot motion. It arises in virtually every aspect of robotic manipulation: in the planning and execution of smooth trajectories, in the determination of singular configurations, in the execution of coordinated anthropomorphic motion, in the derivation of the dynamic equations of motion, and in the transformation of forces and torques from the end effector to the manipulator joints.

We begin this chapter with an investigation of velocities and how to represent them. We first consider angular velocity about a fixed axis and then generalize this to rotation about an arbitrary, possibly moving axis with the aid of skew symmetric matrices. Equipped with this general representation of angular velocities, we are able to derive equations for both the angular velocity and the linear velocity for the origin of a moving frame.

We then proceed to the derivation of the manipulator Jacobian. For an n-link manipulator we first derive the Jacobian representing the instantaneous transformation between the n-vector of joint velocities and the 6-vector consisting of the linear and angular velocities of the end effector. This Jacobian is then a $6 \times n$ matrix. The same approach is used to determine the transformation between the joint velocities and the linear and angular velocity of any point on the manipulator. This will be important when we discuss the derivation of the dynamic equations of motion in Chapter 7. We then show how the end-effector velocity is related to the velocity of an attached tool. Following this, we describe the **analytic Jacobian**, which uses an alternative parameterization of end-effector velocity. We then discuss the notion of **singular configurations**. These are configurations in which the manipulator loses one or more degrees of freedom of motion. We show how the singular configurations are determined geometrically and give several examples. Following this, we briefly discuss the inverse problems of determining joint velocities and accelerations for specified end-effector velocities and accelerations. We then discuss how the manipulator Jacobian can be used to relate forces at the end effector to joint torques. We end the chapter by considering redundant manipulators. This includes discussions of the inverse velocity problem, singular value decomposition and manipulability.

4.1 ANGULAR VELOCITY: THE FIXED AXIS CASE

When a rigid body moves in a pure rotation about a fixed axis, every point of the body moves in a circle. The centers of these circles lie on the axis of rotation. As the body rotates, a perpendicular from any point of the body to the axis sweeps out an angle θ, and this angle is the same for every point of the body. If k is a unit vector in the direction of the axis of rotation, then the angular velocity is given by

$$\omega = \dot{\theta} k \tag{4.1}$$

in which $\dot{\theta}$ is the time derivative of θ.

Given the angular velocity of the body, one learns in introductory dynamics courses that the linear velocity of any point on the body is given by the equation

$$v = \omega \times r \tag{4.2}$$

in which r is a vector from the origin (which in this case is assumed to lie on the axis of rotation) to the point. In fact, the computation of this velocity v is normally the goal in introductory dynamics courses, and therefore, the main role of an angular velocity is to induce linear velocities of points in a rigid body. In our applications, we are interested in describing the motion of a moving frame, including the motion of the origin of the frame through space and also the rotational motion of the frame's axes. Therefore, for our purposes, the angular velocity will hold equal status with linear velocity.

As in previous chapters, in order to specify the orientation of a rigid object, we attach a coordinate frame rigidly to the object, and then specify the orientation of the attached frame. Since every point on the object experiences the same angular velocity (each point sweeps out the same angle θ in a given time interval), and since each point of the body is in a fixed geometric relationship to the body-attached frame, we see that the angular velocity is a property of the attached coordinate frame itself. Angular velocity is not a property of individual points. Individual points may experience a *linear velocity* that is induced by an angular velocity, but it makes no sense to speak of a point itself rotating. Thus, in Equation (4.2) v corresponds to the linear velocity of a point, while ω corresponds to the angular velocity associated with a rotating coordinate frame.

In this fixed axis case, the problem of specifying angular displacements is really a planar problem, since each point traces out a circle, and since every circle lies in a plane. Therefore, it is tempting to use $\dot{\theta}$ to represent the angular velocity. However, as we have already seen in Chapter 2, this choice does not generalize to the three-dimensional case, either when the axis of rotation is not fixed, or when the angular velocity is the result of multiple rotations about distinct axes. For this reason, we will develop a more general representation for angular velocities. This is analogous to our development of rotation matrices in Chapter 2 to represent orientation in three dimensions. The key tool that we will need to develop this representation is the skew symmetric matrix, which is the topic of the next section.

4.2 SKEW SYMMETRIC MATRICES

In Section 4.3 we will derive properties of rotation matrices that can be used to compute relative velocity transformations between coordinate frames. Such transformations involve derivatives of rotation matrices. By introducing the notion of a skew symmetric matrix it is possible to simplify many of the computations involved.

An $n \times n$ matrix S is said to be **skew symmetric** if and only if

$$S^T + S = 0 \tag{4.3}$$

We denote the set of all 3×3 skew symmetric matrices by $so(3)$.

If $S \in so(3)$ has components s_{ij}, $i, j = 1, 2, 3$ then Equation (4.3) is equivalent to the nine equations

$$s_{ij} + s_{ji} = 0 \qquad i, j = 1, 2, 3 \tag{4.4}$$

From Equation (4.4) we see that $s_{ii} = 0$; that is, the diagonal terms of S are zero and the off diagonal terms s_{ij}, $i \neq j$ satisfy $s_{ij} = -s_{ji}$. Thus,

S contains only three independent entries and every 3×3 skew symmetric matrix has the form

$$S = \begin{bmatrix} 0 & -s_3 & s_2 \\ s_3 & 0 & -s_1 \\ -s_2 & s_1 & 0 \end{bmatrix} \tag{4.5}$$

If $a = [a_x, a_y, a_z]^T$ is a 3-vector, we define the skew symmetric matrix $S(a)$ as

$$S(a) = \begin{bmatrix} 0 & -a_z & a_y \\ a_z & 0 & -a_x \\ -a_y & a_x & 0 \end{bmatrix}$$

Example 4.1

Denote by i, j, and k the three unit basis coordinate vectors,

$$i = \begin{bmatrix} 1 \\ 0 \\ 0 \end{bmatrix}; \quad j = \begin{bmatrix} 0 \\ 1 \\ 0 \end{bmatrix}; \quad k = \begin{bmatrix} 0 \\ 0 \\ 1 \end{bmatrix}$$

The skew symmetric matrices $S(i)$, $S(j)$, and $S(k)$ are given by

$$S(i) = \begin{bmatrix} 0 & 0 & 0 \\ 0 & 0 & -1 \\ 0 & 1 & 0 \end{bmatrix} \qquad S(j) = \begin{bmatrix} 0 & 0 & 1 \\ 0 & 0 & 0 \\ -1 & 0 & 0 \end{bmatrix}$$

$$S(k) = \begin{bmatrix} 0 & -1 & 0 \\ 1 & 0 & 0 \\ 0 & 0 & 0 \end{bmatrix}$$

◇

4.2.1 Properties of Skew Symmetric Matrices

Skew symmetric matrices possess several properties that will prove useful for subsequent derivations.[1] Among these properties are

1. The operator S is linear, that is,

$$S(\alpha a + \beta b) = \alpha S(a) + \beta S(b) \tag{4.6}$$

for any vectors a and b belonging to $\mathbb{R}^3$ and scalars α and β.

[1]These properties are consequences of the fact that $so(3)$ is a *Lie algebra*, a vector space with a suitably defined product operation [12].

2. For any vectors a and p belonging to $\mathbb{R}^3$,

$$S(a)p = a \times p \tag{4.7}$$

where $a \times p$ denotes the vector cross product. Equation (4.7) can be verified by direct calculation.

3. For $R \in SO(3)$ and $a \in \mathbb{R}^3$

$$RS(a)R^T = S(Ra) \tag{4.8}$$

To show this, we use that fact that if $R \in SO(3)$ and a, b are vectors in $\mathbb{R}^3$

$$R(a \times b) = Ra \times Rb \tag{4.9}$$

This can be shown by direct calculation. Equation (4.9) is **not** true in general unless R is orthogonal. It says that if we first rotate the vectors a and b using the rotation transformation R and then form the cross product of the rotated vectors Ra and Rb, the result is the same as that obtained by first forming the cross product $a \times b$ and then rotating to obtain $R(a \times b)$. Equation (4.8) now follows easily from Equations (4.7) and (4.9) as follows. Let $b \in \mathbb{R}^3$ be an arbitrary vector. Then

$$\begin{aligned} RS(a)R^T b &= R(a \times R^T b) \\ &= (Ra) \times (RR^T b) \\ &= (Ra) \times b \\ &= S(Ra)b \end{aligned}$$

and the result follows. The left-hand side of Equation (4.8) represents a similarity transformation of the matrix $S(a)$. The equation says therefore that the matrix representation of $S(a)$ in a coordinate frame rotated by R is the same as the skew symmetric matrix $S(Ra)$ corresponding to the vector a rotated by R.

4. For an $n \times n$ skew symmetric matrix S and any vector $X \in \mathbb{R}^n$

$$X^T S X = 0 \tag{4.10}$$

4.2.2 The Derivative of a Rotation Matrix

Suppose now that a rotation matrix R is a function of the single variable θ. Hence, $R = R(\theta) \in SO(3)$ for every θ. Since R is orthogonal for all θ it follows that

$$R(\theta)R(\theta)^T = I \tag{4.11}$$

Differentiating both sides of Equation (4.11) with respect to θ using the product rule gives

$$\left[\frac{d}{d\theta}R\right]R(\theta)^T + R(\theta)\left[\frac{d}{d\theta}R^T\right] = 0 \tag{4.12}$$

Let us define the matrix S as

$$S = \left[\frac{d}{d\theta}R\right]R(\theta)^T \tag{4.13}$$

Then the transpose of S is

$$S^T = \left(\left[\frac{d}{d\theta}R\right]R(\theta)^T\right)^T = R(\theta)\left[\frac{d}{d\theta}R^T\right] \tag{4.14}$$

Equation (4.12) says therefore that

$$S + S^T = 0 \tag{4.15}$$

In other words, the matrix S defined by Equation (4.13) is skew symmetric. Multiplying both sides of Equation (4.13) on the right by R and using the fact that $R^TR = I$ yields

$$\frac{d}{d\theta}R = SR(\theta) \tag{4.16}$$

Equation (4.16) is very important. It says that computing the derivative of the rotation matrix R is equivalent to a matrix multiplication by a skew symmetric matrix S. The most commonly encountered situation is the case where R is a basic rotation matrix or a product of basic rotation matrices.

Example 4.2

If $R = R_{x,\theta}$, the basic rotation matrix given by Equation (2.6), then direct computation shows that

$$\begin{aligned} S = \left[\frac{d}{d\theta}R\right]R^T &= \begin{bmatrix} 0 & 0 & 0 \\ 0 & -\sin\theta & -\cos\theta \\ 0 & \cos\theta & -\sin\theta \end{bmatrix}\begin{bmatrix} 1 & 0 & 0 \\ 0 & \cos\theta & \sin\theta \\ 0 & -\sin\theta & \cos\theta \end{bmatrix} \\ &= \begin{bmatrix} 0 & 0 & 0 \\ 0 & 0 & -1 \\ 0 & 1 & 0 \end{bmatrix} = S(i) \end{aligned}$$

Thus, we have shown that

$$\frac{d}{d\theta} R_{x,\theta} = S(i) R_{x,\theta}$$

Similar computations show that

$$\frac{d}{d\theta} R_{y,\theta} = S(j) R_{y,\theta} \quad \text{and} \quad \frac{d}{d\theta} R_{z,\theta} = S(k) R_{z,\theta} \tag{4.17}$$

◇

Example 4.3

Let $R_{k,\theta}$ be a rotation about the axis defined by k as in Equation (2.43). Note that in this example k is not the unit coordinate vector $[0, 0, 1]^T$. It is easy to check that $S(k)^3 = -S(k)$. Using this fact together with Problem 4-8 it follows that

$$\frac{d}{d\theta} R_{k,\theta} = S(k) R_{k,\theta} \tag{4.18}$$

◇

4.3 ANGULAR VELOCITY: THE GENERAL CASE

We now consider the general case of angular velocity about an arbitrary, possibly moving, axis. Suppose that a rotation matrix R is time varying, so that $R = R(t) \in SO(3)$ for every $t \in \mathbb{R}$. Assuming that $R(t)$ is continuously differentiable as a function of t, an argument identical to the one in the previous section shows that the time derivative $\dot{R}(t)$ of $R(t)$ can be written as

$$\dot{R}(t) = S(\omega(t)) R(t) \tag{4.19}$$

where the matrix $S(\omega(t))$ is skew symmetric. The vector $\omega(t)$ is the **angular velocity** of the rotating frame with respect to the fixed frame at time t. To see that ω is the angular velocity vector, consider a point p rigidly attached to a moving frame. The coordinates of p relative to the fixed frame are given by $p^0 = R_1^0 p^1$. Differentiating this expression we obtain

$$\begin{aligned} \frac{d}{dt} p^0 &= \dot{R}_1^0 p^1 \\ &= S(\omega) R_1^0 p^1 \\ &= \omega \times R_1^0 p^1 \\ &= \omega \times p^0 \end{aligned}$$

which shows that ω is indeed the traditional angular velocity vector.

Equation (4.19) shows the relationship between angular velocity and the derivative of a rotation matrix. In particular, if the instantaneous orientation of a frame $o_1x_1y_1z_1$ with respect to a frame $o_0x_0y_0z_0$ is given by R_1^0, then the angular velocity of frame $o_1x_1y_1z_1$ is directly related to the derivative of R_1^0 by Equation (4.19). When there is a possibility of ambiguity, we will use the notation $\omega_{i,j}$ to denote the angular velocity that corresponds to the derivative of the rotation matrix R_j^i. Since ω is a free vector, we can express it with respect to any coordinate system of our choosing. As usual we use a superscript to denote the reference frame. For example, $\omega_{1,2}^0$ would give the angular velocity that corresponds to the derivative of R_2^1, expressed in coordinates relative to frame $o_0x_0y_0z_0$. In cases where the angular velocities specify rotation relative to the base frame, we will often simplify the subscript, for example, using ω_2 to represent the angular velocity that corresponds to the derivative of R_2^0.

Example 4.4

Suppose that $R(t) = R_{x,\theta(t)}$. Then $\dot{R}(t)$ is computed using the chain rule as

$$\dot{R} = \frac{dR}{dt} = \frac{dR}{d\theta}\frac{d\theta}{dt} = \dot{\theta}S(i)R(t) = S(\omega(t))R(t) \tag{4.20}$$

in which $\omega = i\dot{\theta}$ is the angular velocity. Note, $i = [1,0,0]^T$.

◇

4.4 ADDITION OF ANGULAR VELOCITIES

We are often interested in finding the resultant angular velocity due to the relative rotation of several coordinate frames. We now derive the expressions for the composition of angular velocities of two moving frames $o_1x_1y_1z_1$ and $o_2x_2y_2z_2$ relative to the fixed frame $o_0x_0y_0z_0$. For now, we assume that the three frames share a common origin. Let the relative orientations of the frames $o_1x_1y_1z_1$ and $o_2x_2y_2z_2$ be given by the rotation matrices $R_1^0(t)$ and $R_2^1(t)$ (both time varying).

In the derivations that follow, we will use the notation $\omega_{i,j}$ to denote the angular velocity vector that corresponds to the time derivative of the rotation matrix R_j^i. Since we can express this vector relative to the coordinate frame of our choosing, we again use a superscript to define the reference frame. Thus, $\omega_{i,j}^k$ would denote the angular velocity vector corresponding to the derivative of R_j^i, expressed relative to frame k.

As in Chapter 2,

$$R_2^0(t) = R_1^0(t)R_2^1(t) \tag{4.21}$$

Taking derivatives of both sides of Equation (4.21) with respect to time yields

$$\dot{R}_2^0 = \dot{R}_1^0 R_2^1 + R_1^0 \dot{R}_2^1 \tag{4.22}$$

Using Equation (4.19), the term $\dot{R}_2^0$ on the left-hand side of Equation (4.22) can be written

$$\dot{R}_2^0 = S(\omega_{0,2}^0)R_2^0 \tag{4.23}$$

In this expression, $\omega_{0,2}^0$ denotes the total angular velocity experienced by frame $o_2x_2y_2z_2$. This angular velocity results from the combined rotations expressed by R_1^0 and R_2^1.

The first term on the right-hand side of Equation (4.22) is simply

$$\dot{R}_1^0 R_2^1 = S(\omega_{0,1}^0)R_1^0 R_2^1 = S(\omega_{0,1}^0)R_2^0 \tag{4.24}$$

Note that in this equation, $\omega_{0,1}^0$ denotes the angular velocity of frame $o_1x_1y_1z_1$ that results from the changing R_1^0, and this angular velocity vector is expressed relative to the coordinate system $o_0x_0y_0z_0$.

Let us examine the second term on the right-hand side of Equation (4.22). Using Equation (4.8) we have

$$\begin{aligned} R_1^0 \dot{R}_2^1 &= R_1^0 S(\omega_{1,2}^1)R_2^1 \\ &= R_1^0 S(\omega_{1,2}^1){R_1^0}^T R_1^0 R_2^1 = S(R_1^0 \omega_{1,2}^1)R_1^0 R_2^1 \\ &= S(R_1^0 \omega_{1,2}^1)R_2^0 \end{aligned} \tag{4.25}$$

Note that in this equation, $\omega_{1,2}^1$ denotes the angular velocity of frame $o_2x_2y_2z_2$ that corresponds to the changing R_2^1, expressed relative to the coordinate system $o_1x_1y_1z_1$. Thus, the product $R_1^0\omega_{1,2}^1$ expresses this angular velocity relative to the coordinate system $o_0x_0y_0z_0$. In other words, $R_1^0\omega_{1,2}^1$ gives the coordinates of the free vector $\omega_{1,2}$ with respect to frame 0.

Now, combining the above expressions we have shown that

$$S(\omega_2^0)R_2^0 = \{S(\omega_{0,1}^0) + S(R_1^0\omega_{1,2}^1)\}R_2^0 \tag{4.26}$$

Since $S(a) + S(b) = S(a + b)$, we see that

$$\omega_2^0 = \omega_{0,1}^0 + R_1^0\omega_{1,2}^1 \tag{4.27}$$

In other words, the angular velocities can be added once they are expressed relative to the same coordinate frame, in this case $o_0x_0y_0z_0$.

The above reasoning can be extended to any number of coordinate systems. In particular, suppose that we are given

$$R_n^0 = R_1^0 R_2^1 \cdots R_n^{n-1} \tag{4.28}$$

Extending the above reasoning we obtain

$$\dot{R}_n^0 = S(\omega_{0,n}^0) R_n^0 \tag{4.29}$$

in which

$$\omega_{0,n}^0 = \omega_{0,1}^0 + R_1^0 \omega_{1,2}^1 + R_2^0 \omega_{2,3}^2 + R_3^0 \omega_{3,4}^3 + \cdots + R_{n-1}^0 \omega_{n-1,n}^{n-1} \tag{4.30}$$

$$= \omega_{0,1}^0 + \omega_{1,2}^0 + \omega_{2,3}^0 + \omega_{3,4}^0 + \cdots + \omega_{n-1,n}^0 \tag{4.31}$$

4.5 LINEAR VELOCITY OF A POINT ATTACHED TO A MOVING FRAME

We now consider the linear velocity of a point that is rigidly attached to a moving frame. Suppose the point p is rigidly attached to the frame $o_1x_1y_1z_1$, and that $o_1x_1y_1z_1$ is rotating relative to the frame $o_0x_0y_0z_0$. Then the coordinates of p with respect to the frame $o_0x_0y_0z_0$ are given by

$$p^0 = R_1^0(t) p^1 \tag{4.32}$$

The velocity $\dot{p}^0$ is then given by the product rule for differentiation as

$$\begin{aligned} \dot{p}^0 &= \dot{R}_1^0(t) p^1 + R_1^0(t) \dot{p}^1 \\ &= S(\omega^0) R_1^0(t) p^1 \\ &= S(\omega^0) p^0 = \omega^0 \times p^0 \end{aligned} \tag{4.33}$$

which is the familiar expression for the velocity in terms of the vector cross product. Note that Equation (4.33) follows from the fact that p is rigidly attached to frame $o_1x_1y_1z_1$, and therefore its coordinates relative to frame $o_1x_1y_1z_1$ do not change, giving $\dot{p}^1 = 0$.

Now, suppose that the motion of the frame $o_1x_1y_1z_1$ relative to $o_0x_0y_0z_0$ is more general. Suppose that the homogeneous transformation relating the two frames is time-dependent, so that

$$H_1^0(t) = \begin{bmatrix} R_1^0(t) & o_1^0(t) \\ 0 & 1 \end{bmatrix} \tag{4.34}$$

For simplicity we omit the argument t and the subscripts and superscripts on R_1^0 and o_1^0, and write

$$p^0 = Rp^1 + o \tag{4.35}$$

Differentiating the above expression using the product rule gives

$$\begin{aligned} \dot{p}^0 &= \dot{R}p^1 + \dot{o} \\ &= S(\omega)Rp^1 + \dot{o} \\ &= \omega \times r + v \end{aligned} \tag{4.36}$$

where $r = Rp^1$ is the vector from o_1 to p expressed in the orientation of the frame $o_0x_0y_0z_0$, and v is the rate at which the origin o_1 is moving.

If the point p is moving relative to the frame $o_1x_1y_1z_1$, then we must add to the term v the term $R(t)\dot{p}^1$, which is the rate of change of the coordinates p^1 expressed in the frame $o_0x_0y_0z_0$.

4.6 DERIVATION OF THE JACOBIAN

Consider an n-link manipulator with joint variables $q_1, \ldots, q_n$. Let

$$T_n^0(q) = \begin{bmatrix} R_n^0(q) & o_n^0(q) \\ 0 & 1 \end{bmatrix} \tag{4.37}$$

denote the transformation from the end-effector frame to the base frame, where $q = [q_1, \ldots, q_n]^T$ is the vector of joint variables. As the robot moves about, both the joint variables q_i and the end-effector position o_n^0 and orientation R_n^0 will be functions of time. The objective of this section is to relate the linear and angular velocity of the end effector to the vector of joint velocities $\dot{q}(t)$. Let

$$S(\omega_n^0) = \dot{R}_n^0(R_n^0)^T \tag{4.38}$$

define the angular velocity vector ω_n^0 of the end effector, and let

$$v_n^0 = \dot{o}_n^0 \tag{4.39}$$

denote the linear velocity of the end effector. We seek expressions of the form

$$v_n^0 = J_v\dot{q} \tag{4.40}$$
$$\omega_n^0 = J_\omega\dot{q} \tag{4.41}$$

where J_v and J_ω are $3 \times n$ matrices. We may write Equations (4.40) and (4.41) together as

$$\xi = J\dot{q} \tag{4.42}$$

in which ξ and J are given by

$$\xi = \begin{bmatrix} v_n^0 \\ \omega_n^0 \end{bmatrix} \quad \text{and} \quad J = \begin{bmatrix} J_v \\ J_\omega \end{bmatrix} \tag{4.43}$$

The vector ξ is sometimes called a body velocity. Note that this velocity vector is *not* the derivative of a position variable, since the angular velocity vector is not the derivative of any particular time varying quantity. The matrix J is called the **manipulator Jacobian** or **Jacobian** for short. Note that J is a $6 \times n$ matrix where n is the number of links. We next derive a simple expression for the Jacobian of any manipulator.

4.6.1 Angular Velocity

Recall from Equation (4.30) that angular velocities can be added as free vectors, provided that they are expressed relative to a common coordinate frame. Thus, we can determine the angular velocity of the end effector relative to the base by expressing the angular velocity contributed by each joint in the orientation of the base frame and then summing these.

If the ith joint is revolute, then the ith joint variable q_i equals θ_i and the axis of rotation is z_{i-1}. Slightly abusing notation, let ω_i^{i-1} represent the angular velocity of link i that is imparted by the rotation of joint i, expressed relative to frame $o_{i-1}x_{i-1}y_{i-1}z_{i-1}$. This angular velocity is expressed in the frame $i-1$ by

$$\omega_i^{i-1} = \dot{q}_i z_{i-1}^{i-1} = \dot{q}_i k \tag{4.44}$$

in which, as above, k is the unit coordinate vector $[0, 0, 1]^T$.

If the ith joint is prismatic, then the motion of frame i relative to frame $i-1$ is a translation and

$$\omega_i^{i-1} = 0 \tag{4.45}$$

Thus, if joint i is prismatic, the angular velocity of the end effector does not depend on q_i, which now equals d_i.

Therefore, the overall angular velocity of the end effector, ω_n^0, in the base frame is determined by Equation (4.30) as

$$\omega_n^0 = \rho_1\dot{q}_1 k + \rho_2\dot{q}_2 R_1^0 k + \cdots + \rho_n\dot{q}_n R_{n-1}^0 k = \sum_{i-1}^{n} \rho_i\dot{q}_i z_{i-1}^0 \tag{4.46}$$

in which ρ_i is equal to 1 if joint i is revolute and 0 if joint i is prismatic, since

$$z_{i-1}^0 = R_{i-1}^0 k \tag{4.47}$$

Of course $z_0^0 = k = [0,0,1]^T$.

The lower half of the Jacobian J_ω, in Equation (4.43) is thus given as

$$J_\omega = [\rho_1 z_0 \cdots \rho_n z_{n-1}] \tag{4.48}$$

Note that in this equation, we have omitted the superscripts for the unit vectors along the z-axes, since these are all referenced to the world frame. In the remainder of the chapter we will follow this convention when there is no ambiguity concerning the reference frame.

4.6.2 Linear Velocity

The linear velocity of the end effector is just $\dot{o}_n^0$. By the chain rule for differentiation

$$\dot{o}_n^0 = \sum_{i=1}^{n} \frac{\partial o_n^0}{\partial q_i} \dot{q}_i \tag{4.49}$$

Thus, we see that the ith column of J_v, which we denote as J_{v_i} is given by

$$J_{v_i} = \frac{\partial o_n^0}{\partial q_i} \tag{4.50}$$

Furthermore, this expression is just the linear velocity of the end effector that would result if $\dot{q}_i$ were equal to one and the other $\dot{q}_j$ were zero. In other words, the ith column of the Jacobian can be generated by holding all joints fixed but the ith and actuating the ith at unit velocity. This observation leads to a simple and intuitive derivation of the linear velocity Jacobian as we now show. We now consider the two cases of prismatic and revolute joints separately.

Case 1: Prismatic Joints

Figure 4.1 illustrates the case when all joints are fixed except a single prismatic joint. Since joint i is prismatic, it imparts a pure translation to the end effector. The direction of translation is parallel to the axis z_{i-1} and the

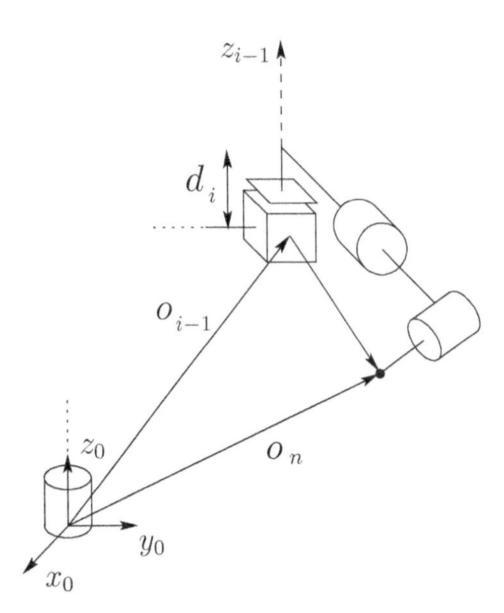

Figure 4.1: Motion of the end effector due to primsmatic joint i.

magnitude of the translation is $\dot{d}_i$, where d_i is the DH joint variable. Thus, in the orientation of the base frame we have

$$\dot{o}_n^0 = \dot{d}_i R_{i-1}^0 \begin{bmatrix} 0 \\ 0 \\ 1 \end{bmatrix} = \dot{d}_i z_{i-1}^0 \tag{4.51}$$

in which d_i is the joint variable for prismatic joint i. Thus, for the case of prismatic joints, after dropping the superscripts we have

$$J_{v_i} = z_{i-1} \tag{4.52}$$

Case 2: Revolute Joints

Figure 4.2 illustrates the case when all joints are fixed except a single revolute joint. Since joint i is revolute, we have $q_i = \theta_i$. Referring to Figure 4.2 and assuming that all joints are fixed except joint i, we see that the linear velocity of the end effector is simply of the form $\omega \times r$, where

$$\omega = \dot{\theta}_i z_{i-1} \tag{4.53}$$

and

$$r = o_n - o_{i-1} \tag{4.54}$$

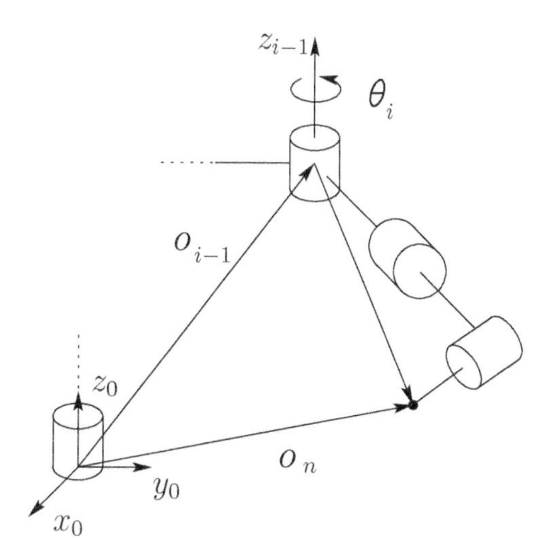

Figure 4.2: Motion of the end effector due to revolute joint i.

Thus, putting these together and expressing the coordinates relative to the base frame, for a revolute joint we obtain

$$J_{v_i} = z_{i-1} \times (o_n - o_{i-1}) \tag{4.55}$$

in which we have omitted the zero superscripts following our convention.

4.6.3 Combining the Linear and Angular Velocity Jacobians

As we have seen in the preceding section, the upper half of the Jacobian J_v is given as

$$J_v = [J_{v_1} \cdots J_{v_n}] \tag{4.56}$$

in which the ith column J_{v_i} is

$$J_{v_i} = \begin{cases} z_{i-1} \times (o_n - o_{i-1}) & \text{for revolute joint } i \\ z_{i-1} & \text{for prismatic joint } i \end{cases} \tag{4.57}$$

The lower half of the Jacobian is given as

$$J_\omega = [J_{\omega_1} \cdots J_{\omega_n}] \tag{4.58}$$

in which the ith column J_{ω_i} is

$$J_{\omega_i} = \begin{cases} z_{i-1} & \text{for revolute joint } i \\ 0 & \text{for prismatic joint } i \end{cases} \tag{4.59}$$

The above formulas make the determination of the Jacobian of any manipulator simple since all of the quantities needed are available once the forward kinematics are worked out. Indeed, the only quantities needed to compute the Jacobian are the unit vectors z_i and the coordinates of the origins $o_1, \ldots, o_n$. A moment's reflection shows that the coordinates for z_i with respect to the base frame are given by the first three elements in the third column of T_i^0 while o_i is given by the first three elements of the fourth column of T_i^0. Thus, only the third and fourth columns of the T matrices are needed in order to evaluate the Jacobian according to the above formulas.

The above procedure works not only for computing the velocity of the end effector but also for computing the velocity of any point on the manipulator. This will be important in Chapter 7 when we will need to compute the velocity of the center of mass of the various links in order to derive the dynamic equations of motion.

Example 4.5 Two-Link Planar Manipulator

Consider the two-link planar manipulator of Example 3.1. Since both joints are revolute the Jacobian matrix, which in this case is 6×2, *is of the form*

$$J(q) = \begin{bmatrix} z_0 \times (o_2 - o_0) & z_1 \times (o_2 - o_1) \\ z_0 & z_1 \end{bmatrix} \tag{4.60}$$

The various quantities above are easily seen to be

$$o_0 = \begin{bmatrix} 0 \\ 0 \\ 0 \end{bmatrix} \quad o_1 = \begin{bmatrix} a_1 c_1 \\ a_1 s_1 \\ 0 \end{bmatrix} \quad o_2 = \begin{bmatrix} a_1 c_1 + a_2 c_{12} \\ a_1 s_1 + a_2 s_{12} \\ 0 \end{bmatrix} \tag{4.61}$$

$$z_0 = z_1 = \begin{bmatrix} 0 \\ 0 \\ 1 \end{bmatrix} \tag{4.62}$$

Performing the required calculations then yields

$$J = \begin{bmatrix} -a_1 s_1 - a_2 s_{12} & -a_2 s_{12} \\ a_1 c_1 + a_2 c_{12} & a_2 c_{12} \\ 0 & 0 \\ 0 & 0 \\ 0 & 0 \\ 1 & 1 \end{bmatrix} \tag{4.63}$$

It is easy to see how the above Jacobian compares with Equation (1.1) derived in Chapter 1. The first two rows of Equation (4.62) are exactly the 2×2 *Jacobian of Chapter 1 and give the linear velocity of the origin* o_2 *relative to the base. The third row in Equation (4.63) is the linear velocity in the direction of* z_0*, which is of course always zero in this case. The last three rows represent the angular velocity of the final frame, which is simply a rotation about the vertical axis at the rate* $\dot{\theta}_1 + \dot{\theta}_2$.

◇

Example 4.6 Jacobian for an Arbitrary Point on a Link

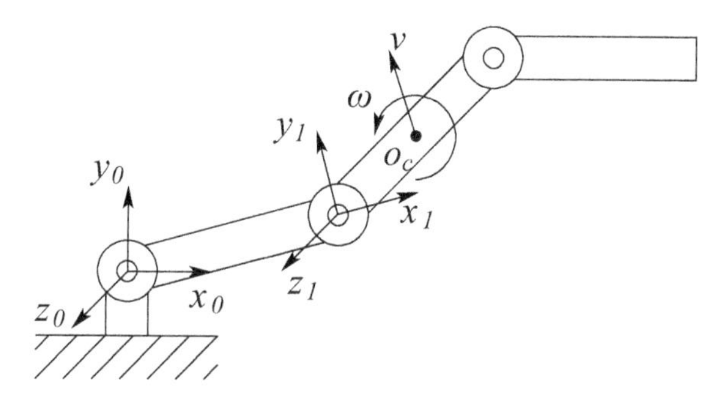

Figure 4.3: Finding the velocity of link 2 of a 3-link planar robot.

Consider the three-link planar manipulator of Figure 4.3. Suppose we wish to compute the linear velocity v *and the angular velocity* ω *of the center of link 2 as shown. In this case we have that*

$$J(q) = \begin{bmatrix} z_0 \times (o_c - o_0) & z_1 \times (o_c - o_1) & 0 \\ z_0 & z_1 & 0 \end{bmatrix} \tag{4.64}$$

which is merely the usual Jacobian with o_c *in place of* o_n*. Note that the third column of the Jacobian is zero, since the velocity of the second link is unaffected by motion of the third link.*[2] *Note that in this case the vector* o_c *must be computed as it is not given directly by the* T *matrices (Problem 4-16).*

◇

Example 4.7 Stanford Manipulator

Consider the Stanford manipulator of Example 3.5 with its associated Denavit-Hartenberg coordinate frames. Note that joint 3 is prismatic and

[2] Note that we are treating only kinematic effects here. Reaction forces on link 2 due to the motion of link 3 will influence the motion of link 2. These dynamic effects are treated by the methods of Chapter 7.

that $o_3 = o_4 = o_5$ as a consequence of the spherical wrist and the frame assignment. Denoting this common origin by o we see that the columns of the Jacobian have the form

$$J_i = \begin{bmatrix} z_{i-1} \times (o_6 - o_{i-1}) \\ z_{i-1} \end{bmatrix} \qquad i = 1, 2$$

$$J_3 = \begin{bmatrix} z_2 \\ 0 \end{bmatrix}$$

$$J_i = \begin{bmatrix} z_{i-1} \times (o_6 - o) \\ z_{i-1} \end{bmatrix} \qquad i = 4, 5, 6$$

Now, using the A matrices given by Equations (3.18)–(3.20) and the T matrices formed as products of the A matrices, these quantities are easily computed as follows. First, o_j is given by the first three entries of the last column of $T_j^0 = A_1 \cdots A_j$, with $o_0 = [0, 0, 0]^T = o_1$. The vector z_j is given as $z_j = R_j^0 k$ where R_j^0 is the rotational part of T_j^0. Thus, it is only necessary to compute the matrices T_j^0 to calculate the Jacobian. Carrying out these calculations one obtains the following expressions for the Stanford manipulator:

$$o_6 = \begin{bmatrix} c_1 s_2 d_3 - s_1 d_2 + d_6(c_1 c_2 c_4 s_5 + c_1 c_5 s_2 - s_1 s_4 s_5) \\ s_1 s_2 d_3 - c_1 d_2 + d_6(c_1 s_4 s_5 + c_2 c_4 s_1 s_5 + c_5 s_1 s_2) \\ c_2 d_3 + d_6(c_2 c_5 - c_4 s_2 s_5) \end{bmatrix} \quad (4.65)$$

$$o_3 = \begin{bmatrix} c_1 s_2 d_3 - s_1 d_2 \\ s_1 s_2 d_3 + c_1 d_2 \\ c_2 d_3 \end{bmatrix} \quad (4.66)$$

The z_i are given as

$$z_0 = \begin{bmatrix} 0 \\ 0 \\ 1 \end{bmatrix} \qquad z_1 = \begin{bmatrix} -s_1 \\ c_1 \\ 0 \end{bmatrix} \quad (4.67)$$

$$z_2 = \begin{bmatrix} c_1 s_2 \\ s_1 s_2 \\ c_2 \end{bmatrix} \qquad z_3 = \begin{bmatrix} c_1 s_2 \\ s_1 s_2 \\ c_2 \end{bmatrix} \quad (4.68)$$

$$z_4 = \begin{bmatrix} -c_1c_2s_4 - s_1c_4 \\ -s_1c_2s_4 + c_1c_4 \\ s_2s_4 \end{bmatrix} \quad (4.69)$$

$$z_5 = \begin{bmatrix} c_1c_2c_4s_5 - s_1s_4s_5 + c_1s_2c_5 \\ s_1c_2c_4s_5 + c_1s_4s_5 + s_1s_2c_5 \\ -s_2c_4s_5 + c_2c_5 \end{bmatrix} \quad (4.70)$$

The Jacobian of the Stanford Manipulator is now given by combining these expressions according to the given formulas (Problem 4-22).

◇

Example 4.8 SCARA Manipulator

We will now derive the Jacobian of the SCARA manipulator of Example 3.6. This Jacobian is a 6×4 *matrix since the SCARA has only four degrees of freedom. As before we need only compute the matrices* $T_j^0 = A_1 \dots A_j$*, where the A-matrices are given by Equations (3.22) and (3.23).*

Since joints 1,2, and 4 are revolute and joint 3 is prismatic, and since $o_4 - o_3$ *is parallel to* z_3 *(and thus,* $z_3 \times (o_4 - o_3) = 0$*), the Jacobian is of the form*

$$J = \begin{bmatrix} z_0 \times (o_4 - o_0) & z_1 \times (o_4 - o_1) & z_2 & 0 \\ z_0 & z_1 & 0 & z_3 \end{bmatrix} \quad (4.71)$$

The origins of the DH frames are given by

$$o_1 = \begin{bmatrix} a_1c_1 \\ a_1s_1 \\ 0 \end{bmatrix} \qquad o_2 = \begin{bmatrix} a_1c_1 + a_2c_{12} \\ a_1s_1 + a_2s_{12} \\ 0 \end{bmatrix} \quad (4.72)$$

$$o_4 = \begin{bmatrix} a_1c_1 + a_2c_{12} \\ a_1s_2 + a_2s_{12} \\ d_3 - d_4 \end{bmatrix} \quad (4.73)$$

Similarly $z_0 = z_1 = k$*, and* $z_2 = z_3 = -k$*. Therefore the Jacobian of the SCARA Manipulator is*

$$J = \begin{bmatrix} -a_1s_1 - a_2s_{12} & -a_2s_{12} & 0 & 0 \\ a_1c_1 + a_2c_{12} & a_2c_{12} & 0 & 0 \\ 0 & 0 & -1 & 0 \\ 0 & 0 & 0 & 0 \\ 0 & 0 & 0 & 0 \\ 1 & 1 & 0 & -1 \end{bmatrix} \quad (4.74)$$

◇

4.7 THE TOOL VELOCITY

Many tasks require that a tool be attached to the end effector. In such cases, it is necessary to relate the velocity of the tool frame to the velocity of the end-effector frame. Suppose that the tool is rigidly attached to the end effector, and the fixed spatial relationship between the end effector and the tool frame is given by the constant homogeneous transformation matrix

$$T_{\text{tool}}^{6} = \begin{bmatrix} R & d \\ 0 & 1 \end{bmatrix} \tag{4.75}$$

We will assume that the end effector velocity is given and expressed in coordinates relative to the end-effector frame, that is, we are given ξ_6^6. In this section we will derive the velocity of the tool expressed in coordinates relative to the tool frame, that is, we will derive $\xi_{\text{tool}}^{\text{tool}}$.

Since the two frames are rigidly attached, the angular velocity of the tool frame is the same as the angular velocity of the end-effector frame. To see this, simply compute the angular velocities of each frame by taking the derivatives of the appropriate rotation matrices. Since R is constant and $R_{\text{tool}}^0 = R_6^0 R$, we have

$$\begin{aligned} & \dot{R}_{\text{tool}}^0 = \dot{R}_6^0 R \\ \Longrightarrow \quad & S(\omega_{\text{tool}}^0) R_{\text{tool}}^0 = S(\omega_6^0) R_6^0 R \\ \Longrightarrow \quad & S(\omega_{\text{tool}}^0) = S(\omega_6^0) \end{aligned}$$

Thus, $\omega_{\text{tool}} = \omega_6$, and to obtain the tool angular velocity relative to the tool frame we apply a rotational transformation

$$\omega_{\text{tool}}^{\text{tool}} = \omega_6^{\text{tool}} = R^T \omega_6^6 \tag{4.76}$$

If the end-effector frame is moving with body velocity $\xi = [v_6^T, \omega_6^T]^T$, then the linear velocity of the origin of the tool frame, which is rigidly attached to the end-effector frame, is given by

$$v_{\text{tool}} = v_6 + \omega_6 \times r \tag{4.77}$$

where r is the vector from the origin of the end-effector frame to the origin of the tool frame. From Equation (4.75), we see that d gives the coordinates of the origin of the tool frame with respect to the end-effector frame, and therefore we can express r in coordinates relative to the tool frame as $r^{\text{tool}} =$

$R^T d$. Thus, we write $\omega_6 \times r$ in coordinates with respect to the tool frame as

$$\begin{aligned} \omega_6^{\text{tool}} \times r^{\text{tool}} &= R^T \omega_6^6 \times (R^T d) \\ &= -R^T d \times R^T \omega_6^6 \\ &= -S(R^T d) R^T \omega_6^6 \\ &= -R^T S(d) R R^T \omega_6^6 \\ &= -R^T S(d) \omega_6^6 \end{aligned} \tag{4.78}$$

To express the free vector v_6^6 in coordinates relative to the tool frame, we apply the rotational transformation

$$v_6^{\text{tool}} = R^T v_6^6 \tag{4.79}$$

Combining Equations (4.77), (4.78), and (4.79) to obtain the linear velocity of the tool frame and using Equation (4.76) for the angular velocity of the tool frame, we have

$$\begin{aligned} v_{\text{tool}}^{\text{tool}} &= R^T v_6^6 - R^T S(d) \omega_6^6 \\ \omega_{\text{tool}}^{\text{tool}} &= R^T \omega_6^6 \end{aligned}$$

which can be written as the matrix equation

$$\xi_{\text{tool}}^{\text{tool}} = \begin{bmatrix} R^T & -R^T S(d) \\ 0_{3\times3} & R^T \end{bmatrix} \xi_6^6 \tag{4.80}$$

In many cases, it is useful to solve the inverse problem: compute the required end effector velocity to produce a desired tool velocity. Since

$$\begin{bmatrix} R & S(d)R \\ 0_{3\times3} & R \end{bmatrix} = \begin{bmatrix} R^T & -R^T S(d) \\ 0_{3\times3} & R^T \end{bmatrix}^{-1} \tag{4.81}$$

(Problem 4-23) we can solve Equation (4.80) for ξ_6^6, obtaining

$$\xi_6^6 = \begin{bmatrix} R & S(d)R \\ 0_{3\times3} & R \end{bmatrix} \xi_{\text{tool}}^{\text{tool}}$$

This gives the general expression for transforming velocities between two rigidly attached moving frames

$$\xi_A^A = \begin{bmatrix} R_B^A & S(d_B^A) R_B^A \\ 0_{3\times3} & R_B^A \end{bmatrix} \xi_B^B \tag{4.82}$$

4.8 THE ANALYTICAL JACOBIAN

The Jacobian matrix derived above is sometimes called the *Geometric Jacobian* to distinguish it from the *Analytical Jacobian*, denoted $J_a(q)$, which is based on a minimal representation for the orientation of the end-effector frame. Let

$$X = \begin{bmatrix} d(q) \\ \alpha(q) \end{bmatrix} \tag{4.83}$$

denote the end-effector pose, where $d(q)$ is the usual vector from the origin of the base frame to the origin of the end-effector frame and α denotes a minimal representation for the orientation of the end-effector frame relative to the base frame. For example, let $\alpha = [\phi, \theta, \psi]^T$ be a vector of Euler angles as defined in Chapter 2. Then we seek an expression of the form

$$\dot{X} = \begin{bmatrix} \dot{d} \\ \dot{\alpha} \end{bmatrix} = J_a(q)\dot{q} \tag{4.84}$$

to define the analytical Jacobian.

It can be shown (Problem 4-13) that, if $R = R_{z,\psi}R_{y,\theta}R_{z,\phi}$ is the Euler angle transformation then

$$\dot{R} = S(\omega)R \tag{4.85}$$

in which ω defining the angular velocity is given by

$$\omega = \begin{bmatrix} c_\psi s_\theta \dot{\phi} - s_\psi \dot{\theta} \\ s_\psi s_\theta \dot{\phi} + c_\psi \dot{\theta} \\ \dot{\psi} + c_\theta \dot{\phi} \end{bmatrix} \tag{4.86}$$

$$= \begin{bmatrix} c_\psi s_\theta & -s_\psi & 0 \\ s_\psi s_\theta & c_\psi & 0 \\ c_\theta & 0 & 1 \end{bmatrix} \begin{bmatrix} \dot{\phi} \\ \dot{\theta} \\ \dot{\psi} \end{bmatrix} = B(\alpha)\dot{\alpha} \tag{4.87}$$

The components of ω are called **nutation**, **spin**, and **precession**, respectively. Combining the above relationship with the previous definition of the Jacobian

$$\begin{bmatrix} v \\ \omega \end{bmatrix} = \begin{bmatrix} \dot{d} \\ \omega \end{bmatrix} = J(q)\dot{q} \tag{4.88}$$

yields

$$J(q)\dot{q} = \begin{bmatrix} v \\ \omega \end{bmatrix} = \begin{bmatrix} \dot{d} \\ B(\alpha)\dot{\alpha} \end{bmatrix} = \begin{bmatrix} I & 0 \\ 0 & B(\alpha) \end{bmatrix} \begin{bmatrix} \dot{d} \\ \dot{\alpha} \end{bmatrix} = \begin{bmatrix} I & 0 \\ 0 & B(\alpha) \end{bmatrix} J_a(q)\dot{q}$$

Thus, the analytical Jacobian, $J_a(q)$, may be computed from the geometric Jacobian as

$$J_a(q) = \begin{bmatrix} I & 0 \\ 0 & B^{-1}(\alpha) \end{bmatrix} J(q) \tag{4.89}$$

provided $\det B(\alpha) \neq 0$.

In the next section we discuss the notion of Jacobian singularities, which are configurations where the Jacobian loses rank. Singularities of the matrix $B(\alpha)$ are called **representational singularities**. It can easily be shown (Problem 4-24) that $B(\alpha)$ is invertible provided $s_\theta \neq 0$. This means that the singularities of the analytical Jacobian include the singularities of the geometric Jacobian, J, as defined in the next section, together with the representational singularities.

4.9 SINGULARITIES

The $6 \times n$ Jacobian $J(q)$ defines a mapping

$$\xi = J(q)\dot{q} \tag{4.90}$$

between the vector $\dot{q}$ of joint velocities and the vector $\xi = [v^T, \omega^T]^T$ of end-effector velocities. This implies that the all possible end-effector velocities are linear combinations of the columns of the Jacobian matrix,

$$\xi = J_1\dot{q}_1 + J_2\dot{q}_2 \cdots + J_n\dot{q}_n$$

For example, for the two-link planar arm, the Jacobian matrix given in Equation (4.63) has two columns. It is easy to see that the linear velocity of the end effector must lie in the xy-plane, since neither column has a nonzero entry for the third row. Since $\xi \in \mathbb{R}^6$, it is necessary that J have six linearly independent columns for the end effector to be able to achieve any arbitrary velocity (see Appendix B).

The rank of a matrix is the number of linearly independent columns (or rows) in the matrix. Thus, when $\operatorname{rank} J = 6$, the end effector can execute any arbitrary velocity. For a matrix $J \in \mathbb{R}^{6\times n}$, it is always the case that $\operatorname{rank} J \leq \min(6, n)$. For example, for the two-link planar arm, we always have $\operatorname{rank} J \leq 2$, while for an anthropomorphic arm with spherical wrist we always have $\operatorname{rank} J \leq 6$.

The rank of a matrix is not necessarily constant. Indeed, the rank of the manipulator Jacobian matrix will depend on the configuration q. Configurations for which $\operatorname{rank} J(q)$ is less than its maximum value are called **singularities** or **singular configurations**.

Identifying manipulator singularities is important for several reasons.

- Singularities represent configurations from which certain directions of motion may be unattainable.
- At singularities, bounded end-effector velocities may correspond to unbounded joint velocities.
- At singularities, bounded joint torques may correspond to unbounded end-effector forces and torques. (We will see this in Chapter 9.)
- Singularities often correspond to points on the boundary of the manipulator workspace, that is, to points of maximum reach of the manipulator.
- Singularities correspond to points in the manipulator workspace that may be unreachable under small perturbations of the link parameters, such as length, offset, etc.

There are a number of methods that can be used to determine the singularities of the Jacobian. In this chapter, we will exploit the fact that a square matrix is singular when its determinant is equal to zero. In general, it is difficult to solve the nonlinear equation $\det J(q) = 0$. Therefore, we now introduce the method of decoupling singularities, which is applicable whenever, for example, the manipulator is equipped with a spherical wrist.

4.9.1 Decoupling of Singularities

We saw in Chapter 3 that a set of forward kinematic equations can be derived for any manipulator by attaching a coordinate frame rigidly to each link in any manner that we choose, computing a set of homogeneous transformations relating the coordinate frames, and multiplying them together as needed. The DH convention is merely a systematic way to do this. Although the resulting equations are dependent on the coordinate frames chosen, the manipulator configurations themselves are geometric quantities, independent of the frames used to describe them. Recognizing this fact allows us to decouple the determination of singular configurations, for those manipulators with spherical wrists, into two simpler problems. The first is to determine so-called **arm singularities**, that is, singularities resulting from motion of the arm, which consists of the first three or more links, while the second is to determine the **wrist singularities** resulting from motion of the spherical wrist.

Consider the case that $n = 6$, that is, the manipulator consists of a 3-DOF arm with a 3-DOF spherical wrist. In this case the Jacobian is a 6×6 matrix and a configuration q is singular if and only if

$$\det J(q) \quad = \quad 0 \tag{4.91}$$

If we now partition the Jacobian J into 3×3 blocks as

$$J = [J_P \mid J_O] \quad = \quad \left[\begin{array}{c|c} J_{11} & J_{12} \\ \hline J_{21} & J_{22} \end{array} \right] \tag{4.92}$$

then, since the final three joints are always revolute

$$J_O \quad = \quad \left[\begin{array}{ccc} z_3 \times (o_6 - o_3) & z_4 \times (o_6 - o_4) & z_5 \times (o_6 - o_5) \\ z_3 & z_4 & z_5 \end{array} \right] \tag{4.93}$$

Since the wrist axes intersect at a common point o, if we choose the coordinate frames so that $o_3 = o_4 = o_5 = o_6 = o$, then J_O becomes

$$J_O \quad = \quad \left[\begin{array}{ccc} 0 & 0 & 0 \\ z_3 & z_4 & z_5 \end{array} \right] \tag{4.94}$$

In this case the Jacobian matrix has the block triangular form

$$J \quad = \quad \left[\begin{array}{cc} J_{11} & 0 \\ J_{21} & J_{22} \end{array} \right] \tag{4.95}$$

with determinant

$$\det J \quad = \quad \det J_{11} \det J_{22} \tag{4.96}$$

where J_{11} and J_{22} are each 3×3 matrices. J_{11} has i^{th} column $z_{i-1} \times (o - o_{i-1})$ if joint i is revolute, and z_{i-1} if joint i is prismatic, while

$$J_{22} \quad = \quad [z_3 \; z_4 \; z_5] \tag{4.97}$$

Therefore the set of singular configurations of the manipulator is the union of the set of arm configurations satisfying $\det J_{11} = 0$ and the set of wrist configurations satisfying $\det J_{22} = 0$. *Note that this form of the Jacobian does not necessarily give the correct relation between the velocity of the end effector and the joint velocities. It is intended only to simplify the determination of singularities.*

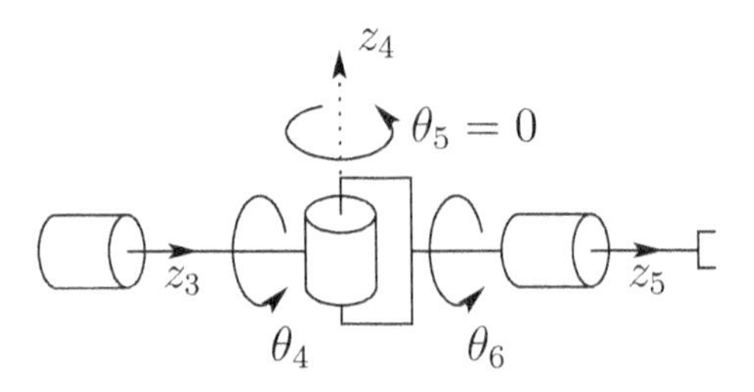

Figure 4.4: Spherical wrist singularity.

4.9.2 Wrist Singularities

We can now see from Equation (4.97) that a spherical wrist is in a singular configuration whenever the vectors z_3, z_4, and z_5 are linearly dependent. Referring to Figure 4.4 we see that this happens when the joint axes z_3 and z_5 are collinear, that is, when $\theta_4 = 0$ or π. These are the only singularities of the spherical wrist, and the are unavoidable without imposing mechanical limits on the wrist design to restrict its motion in such a way that z_3 and z_5 are prevented from lining up. In fact, when *any* two revolute joint axes are collinear a singularity results, since an equal and opposite rotation about the axes results in no net motion of the end effector.

4.9.3 Arm Singularities

To investigate arm singularities we need only to compute $\det J_{11}$, which is done using Equation (4.57) but with the wrist center o in place of o_n. In the remainder of this section, we will determine the singularities for three common arms, the elbow manipulator, the spherical manipulator and the SCARA manipulator.

Example 4.9 Elbow Manipulator Singularities

Consider the three-link articulated manipulator with coordinate frames attached as shown in Figure 4.5. It is left as an exercise (Problem 4-17) to show that

$$J_{11} = \begin{bmatrix} -a_2 s_1 c_2 - a_3 s_1 c_{23} & -a_2 s_2 c_1 - a_3 s_{23} c_1 & -a_3 c_1 s_{23} \\ a_2 c_1 c_2 + a_3 c_1 c_{23} & -a_2 s_1 s_2 - a_3 s_1 s_{23} & -a_3 s_1 s_{23} \\ 0 & a_2 c_2 + a_3 c_{23} & a_3 c_{23} \end{bmatrix} \tag{4.98}$$

and that the determinant of J_{11} is

$$\det J_{11} = a_2 a_3 s_3 (a_2 c_2 + a_3 c_{23}) \tag{4.99}$$

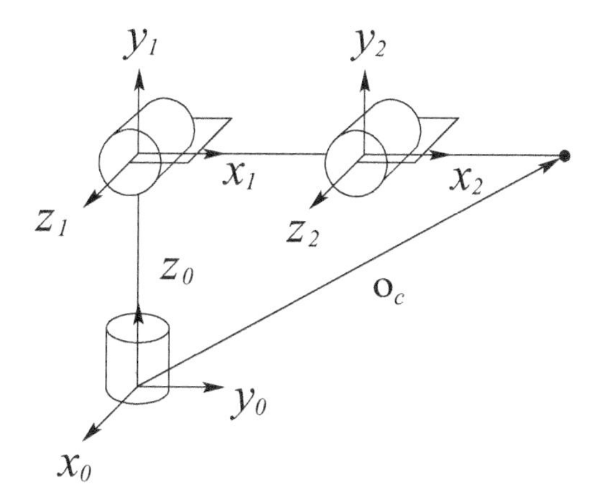

Figure 4.5: Elbow manipulator.

We see from Equation (4.99) that the elbow manipulator is in a singular configuration when

$$s_3 \quad = \quad 0, \quad \textit{that is, } \theta_3 = 0 \textit{ or } \pi \tag{4.100}$$

and whenever

$$a_2c_2 + a_3c_{23} \quad = \quad 0 \tag{4.101}$$

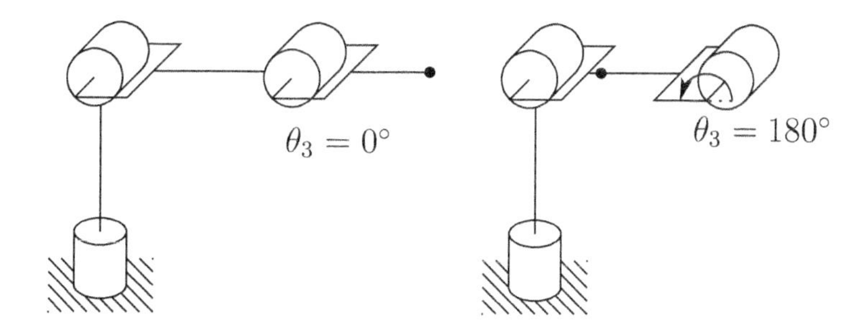

Figure 4.6: Elbow singularities of the elbow manipulator.

The situation of Equation (4.100) is shown in Figure 4.6 and arises when the elbow is fully extended or fully retracted as shown.

The second situation of Equation (4.101) is shown in Figure 4.7. This configuration occurs when the wrist center intersects the axis of the base rotation, z_0. As we saw in Chapter 3, there are infinitely many singular

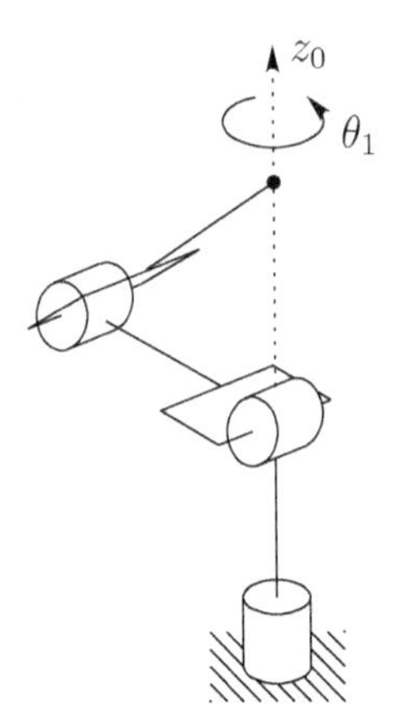

Figure 4.7: Singularity of the elbow manipulator with no offsets.

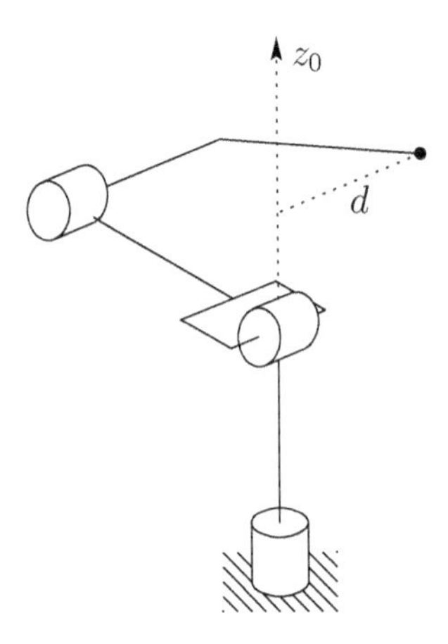

Figure 4.8: Elbow manipulator with an offset at the elbow.

configurations and infinitely many solutions to the inverse position kinematics when the wrist center is along this axis.

For an elbow manipulator with an offset, as shown in Figure 4.8, the wrist center cannot intersect z_0*, which corroborates our earlier statement that points reachable at singular configurations may not be reachable under arbitrarily small perturbations of the manipulator parameters, in this case an offset in either the elbow or the shoulder.*

◇

Example 4.10 Spherical Manipulator

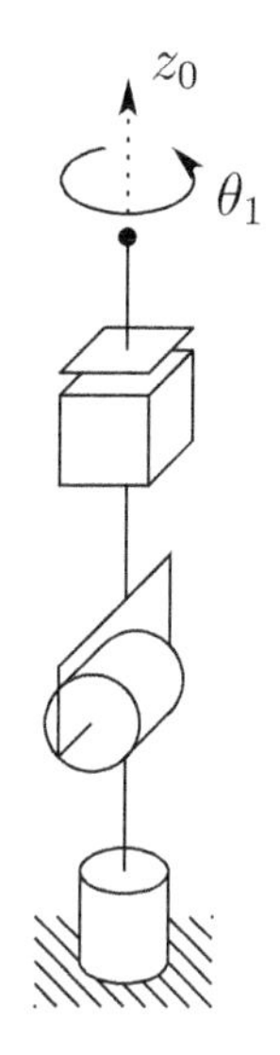

Figure 4.9: Singularity of spherical manipulator with no offsets.

Consider the spherical arm of Figure 4.9. This manipulator is in a singular configuration when the wrist center intersects z_0 *as shown since, as before, any rotation about the base leaves this point fixed.*

◇

Example 4.11 SCARA Manipulator

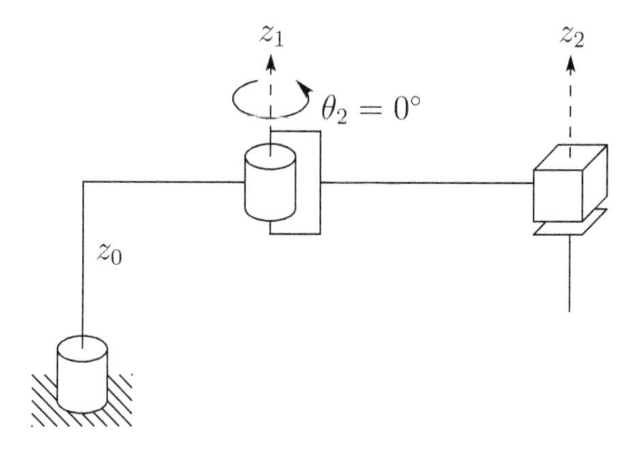

Figure 4.10: SCARA manipulator singularity.

We have already derived the complete Jacobian for the the SCARA manipulator. This Jacobian is simple enough to be used directly rather than deriving the modified Jacobian as we have done above. Referring to Figure 4.10 we can see geometrically that the only singularity of the SCARA arm is when the elbow is fully extended or fully retracted. Indeed, since the portion of the Jacobian of the SCARA governing arm singularities is given as

$$J_{11} = \begin{bmatrix} \alpha_1 & \alpha_3 & 0 \\ \alpha_2 & \alpha_4 & 0 \\ 0 & 0 & -1 \end{bmatrix} \tag{4.102}$$

where

$$\begin{aligned} \alpha_1 &= -a_1 s_1 - a_2 s_{12} \\ \alpha_2 &= a_1 c_1 + a_2 c_{12} \\ \alpha_3 &= -a_1 s_{12} \\ \alpha_4 &= a_1 c_{12} \end{aligned} \tag{4.103}$$

we see that the rank of J_{11} will be less than three precisely whenever $\alpha_1\alpha_4 - \alpha_2\alpha_3 = 0$. It is easy to compute this quantity and show that it is equivalent to (Problem 4-19)

$$s_2 = 0, \quad \textit{which implies} \quad \theta_2 = 0, \pi \tag{4.104}$$

Note the similarities between this case and the singularities for the elbow manipulator shown in Figure 4.6. In each case, the relevant portion of the arm is merely a two-link planar manipulator. As can be seen from Equation (4.63), the Jacobian for the two-link planar manipulator loses rank when $\theta_2 = 0$ or π.

◇

4.10 STATIC FORCE/TORQUE RELATIONSHIPS

Interaction of the manipulator with the environment produces forces and moments at the end effector or tool. These, in turn, produce torques at the joints of the robot.[3] In this section we discuss the role of the manipulator Jacobian in the quantitative relationship between the end-effector forces and joint torques. This relationship is important for the development of path planning methods in Chapter 5, the derivation of the dynamic equations in

[3] Here, we consider the case of revolute joints. If the joints are prismatic, forces and moments at the end effector produce forces at the joints.

Chapter 7 and in the design of force control algorithms in Chapter 9. We can state the main result concisely as follows.

Let $F = [F_x, F_y, F_z, n_x, n_y, n_z]^T$ represent the vector of forces and moments at the end effector. Let τ denote the corresponding vector of joint torques. Then F and τ are related by

$$\tau = J^T(q)F \tag{4.105}$$

where $J^T(q)$ is the transpose of the manipulator Jacobian.

An easy way to derive this relationship is through the so-called **principle of virtual work**. We will defer a detailed discussion of the principle of virtual work until Chapter 7, where we will introduce the notions of generalized coordinates, holonomic constraints, and virtual constraints in more detail. However, we can give a somewhat informal justification in this section as follows. Let δX and δq represent infinitesimal displacements in the task space and joint space, respectively. These displacements are called **virtual displacements** if they are consistent with any constraints imposed on the system. For example, if the end effector is in contact with a rigid wall, then the virtual displacements in position are tangent to the wall. These virtual displacements are related through the manipulator Jacobian $J(q)$ according to

$$\delta X = J(q)\delta q \tag{4.106}$$

The virtual work δw of the system is

$$\delta w = F^T \delta X - \tau^T \delta q \tag{4.107}$$

Substituting Equation (4.106) into Equation (4.107) yields

$$\delta w = (F^T J - \tau^T)\delta q \tag{4.108}$$

The principle of virtual work says, in effect, that the quantity given by Equation (4.108) is equal to zero if the manipulator is in equilibrium. This leads to the relationship given by Equation (4.105). In other words, the end-effector forces are related to the joint torques by the *transpose* of the manipulator Jacobian.

Example 4.12

Consider the two-link planar manipulator of Figure 4.11, with a force F applied at the end of link two as shown. The Jacobian of this manipulator is

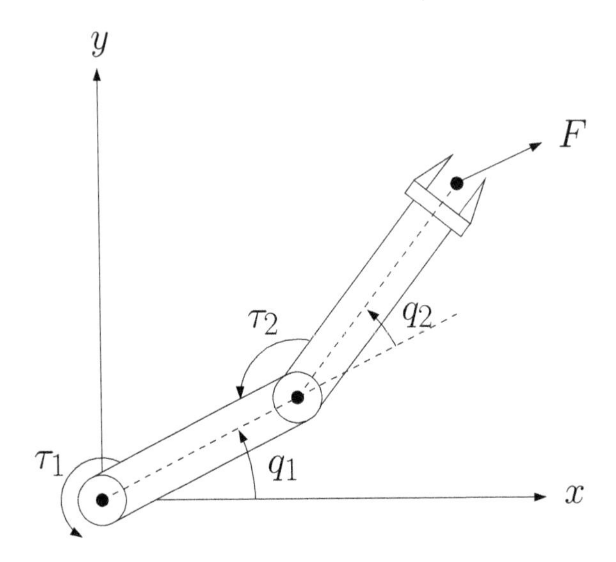

Figure 4.11: Two-link planar robot.

given by Equation (4.63). The resulting joint torques $\tau = (\tau_1, \tau_2)$ are then given as

$$\begin{bmatrix} \tau_1 \\ \tau_2 \end{bmatrix} = \begin{bmatrix} -a_1 s_1 - a_2 s_{12} & a_1 c_1 + a_2 c_{12} & 0 & 0 & 0 & 1 \\ -a_2 s_{12} & a_2 c_{12} & 0 & 0 & 0 & 1 \end{bmatrix} \begin{bmatrix} F_x \\ F_y \\ F_z \\ n_x \\ n_y \\ n_z \end{bmatrix} \tag{4.109}$$

◇

4.11 INVERSE VELOCITY AND ACCELERATION

The Jacobian relationship

$$\xi = J\dot{q} \tag{4.110}$$

specifies the end-effector velocity that will result when the joints move with velocity $\dot{q}$. The inverse velocity problem is the problem of finding the joint velocities $\dot{q}$ that produce the desired end-effector velocity. It is perhaps a bit surprising that the inverse velocity relationship is conceptually simpler than the inverse position relationship. When the Jacobian is square and nonsingular, this problem can be solved by simply inverting the Jacobian matrix to give

$$\dot{q} = J^{-1}\xi \tag{4.111}$$

For manipulators that do not have exactly six joints, the Jacobian cannot be inverted. In this case there will be a solution to Equation (4.110) if and only if ξ lies in the range space of the Jacobian. This can be determined by the following simple rank test. A vector ξ belongs to the range of J if and only if

$$\operatorname{rank} J(q) = \operatorname{rank}\left[J(q) \mid \xi\right] \tag{4.112}$$

In other words, Equation (4.110) may be solved for $\dot{q} \in \mathbb{R}^n$ provided that the rank of the *augmented matrix* $[J(q) \mid \xi]$ is the same as the rank of the Jacobian $J(q)$. This is a standard result from linear algebra, and several algorithms exist, such as Gaussian elimination, for solving such systems of linear equations.

For the case when $n > 6$ we can solve for $\dot{q}$ using the right pseudoinverse of J. To construct this psuedoinverse, we use the fact that when $J \in \mathbb{R}^{m\times n}$, if $m < n$ and rank $J = m$, then $(JJ^T)^{-1}$ exists. In this case $(JJ^T) \in \mathbb{R}^{m\times m}$, and has rank m. Using this result, we can regroup terms to obtain

$$\begin{aligned} I &= (JJ^T)(JJ^T)^{-1} \\ &= J\left[J^T(JJ^T)^{-1}\right] \\ &= JJ^+ \end{aligned}$$

Here, $J^+ = J^T(JJ^T)^{-1}$ is called a right pseudoinverse of J, since $JJ^+ = I$. Note that $J^+J \in \mathbb{R}^{n\times n}$ and that in general $J^+J \neq I$ (recall that matrix multiplication is not commutative).

It is now easy to demonstrate (Problem 4-25) that a solution to Equation (4.110) is given by

$$\dot{q} = J^+\xi + (I - J^+J)b \tag{4.113}$$

in which $b \in \mathbb{R}^n$ is an arbitrary vector.

In general, for $m < n$, $(I - J^+J) \neq 0$, and all vectors of the form $(I - J^+J)b$ lie in the null space of J. This means that, if $\dot{q}'$ is a joint velocity vector such that $\dot{q}' = (I - J^+J)b$, then when the joints move with velocity $\dot{q}'$, the end effector will remain fixed since $J\dot{q}' = 0$. Thus, if $\dot{q}$ is a solution to Equation (4.110), then so is $\dot{q} + \dot{q}'$ with $\dot{q}' = (I - J^+J)b$, for any value of b. If the goal is to minimize the resulting joint velocities, we choose $b = 0$ (Problem 4-25).

We can construct the right pseudoinverse of J using its singular value decomposition (see Appendix B). In particular, we can write J as the matrix product

$$J = U\Sigma V^T$$

in which $U \in \mathbb{R}^{m\times m}$ and $V \in \mathbb{R}^{n\times n}$ are orthogonal (rotation) matrices and $\Sigma \in \mathbb{R}^{m\times n}$ is given by

$$\Sigma = \left[\begin{array}{ccccc|c} \sigma_1 & & & & & \\ & \sigma_2 & & & & \\ & & . & & & 0 \\ & & & . & & \\ & & & & \sigma_m & \end{array}\right]$$

in which the σ_i are the **singular values** of the Jacobian. This matrix product is known as the **singular value decomposition** of the Jacobian.

It is now a simple matter to construct the right pseudoinverse of J using its singular value decomposition.

$$J^+ = V\Sigma^+U^T \tag{4.114}$$

in which

$$\Sigma^+ = \left[\begin{array}{ccccc|c} \sigma_1^{-1} & & & & & \\ & \sigma_2^{-1} & & & & \\ & & . & & & 0 \\ & & & . & & \\ & & & & \sigma_m^{-1} & \end{array}\right]^T$$

We can apply a similar approach when the analytical Jacobian is used in place of the manipulator Jacobian. Recall from Equation (4.84) that the joint velocities and the end-effector velocities are related by the analytical Jacobian as

$$\dot{X} \quad = \quad J_a(q)\dot{q} \tag{4.115}$$

Thus, the inverse velocity problem becomes one of solving the linear system given by Equation (4.115), which can be accomplished as above for the manipulator Jacobian.

Differentiating Equation (4.115) yields an expression for the acceleration

$$\ddot{X} \quad = \quad J_a(q)\ddot{q} + \left(\frac{d}{dt}J_a(q)\right)\dot{q} \tag{4.116}$$

Thus, given a vector $\ddot{X}$ of end-effector accelerations, the instantaneous joint acceleration vector $\ddot{q}$ is given as a solution of

$$J_a(q)\ddot{q} = \ddot{X} - \left(\frac{d}{dt}J_a(q)\right)\dot{q}$$

For 6-DOF manipulators the inverse velocity and acceleration equations can therefore be written as

$$\dot{q} = J_a(q)^{-1}\dot{X} \tag{4.117}$$

and

$$\ddot{q} = J_a(q)^{-1}\left[\ddot{X} - \left(\frac{d}{dt}J_a(q)\right)\dot{q}\right] \tag{4.118}$$

provided $\det J_a(q) \neq 0$.

4.12 MANIPULABILITY

For a specific value of q, the Jacobian relationship defines the linear system given by $\xi = J\dot{q}$. We can think of J as scaling the input $\dot{q}$ to produce the output ξ. It is often useful to characterize quantitatively the effects of this scaling. Often, in systems with a single input and a single output, this kind of characterization is given in terms of the so called impulse response of a system, which essentially characterizes how the system responds to a unit input. In this multidimensional case, the analogous concept is to characterize the output in terms of an input that has unit norm. Consider the set of all robot joint velocities $\dot{q}$ such that

$$||\dot{q}||^2 = \dot{q}_1^2 + \dot{q}_2^2 + \ldots \dot{q}_n^2 \leq 1 \tag{4.119}$$

If we use the minimum norm solution $\dot{q} = J^+\xi$, we obtain

$$\begin{aligned} ||\dot{q}||^2 &= \dot{q}^T\dot{q} \\ &= (J^+\xi)^T J^+\xi \\ &= \xi^T(JJ^T)^{-1}\xi \end{aligned} \tag{4.120}$$

The derivation of this is left as an exercise (Problem 4-27). Equation (4.120) gives us a quantitative characterization of the scaling effected by the Jacobian. In particular, if the manipulator Jacobian is full rank, that is, if rank $J = m$, then Equation (4.120) defines an m-dimensional ellipsoid that is known as the **manipulability ellipsoid**. If the input (joint velocity) vector has unit norm, then the output (end-effector velocity) will lie within the ellipsoid given by Equation (4.120). We can more easily see that Equation (4.120) defines an ellipsoid by replacing the Jacobian by its SVD $J = U\Sigma V^T$ (see Appendix B) to obtain

$$\xi^T(JJ^T)^{-1}\xi^T = (U^T\xi)^T\Sigma_m^{-2}(U^T\xi) \tag{4.121}$$

in which

$$\Sigma_m^{-2} = \begin{bmatrix} \sigma_1^{-2} & & & & \\ & \sigma_2^{-2} & & & \\ & & \cdot & & \\ & & & \cdot & \\ & & & & \sigma_m^{-2} \end{bmatrix}$$

and $\sigma_1 \geq \sigma_2 \cdots \geq \sigma_m \geq 0$. The derivation of Equation (4.121) is left as an exercise (Problem 4-28). If we make the substitution $w = U^T \xi$, then Equation (4.121) can be written as

$$w^T \Sigma_m^{-2} w = \sum \frac{w_i^2}{\sigma_i^2} \leq 1 \tag{4.122}$$

and it is clear that this is the equation for an axis-aligned ellipse in a new coordinate system that is obtained by rotation according to the orthogonal matrix U^T. In the original coordinate system, the axes of the ellipsoid are given by the vectors $\sigma_i u_i$. The volume of the ellipsoid is given by

$$\text{volume} = K \sigma_1 \sigma_2 \cdots \sigma_m$$

in which K is a constant that depends only on the dimension m of the ellipsoid. The manipulability measure is given by

$$\mu = \sigma_1 \sigma_2 \cdots \sigma_m \tag{4.123}$$

Note that the constant K is not included in the definition of manipulability, since it is fixed once the task has been defined, that is, once the dimension of the task space has been fixed.

Now, consider the special case when the robot is not redundant, that is, $J \in \mathbb{R}^{m \times m}$. Recall that the determinant of a product is equal to the product of the determinants, and that a matrix and its transpose have the same determinant. Thus, we have

$$\det JJ^T \quad = \quad \lambda_1^2 \lambda_2^2 \cdots \lambda_m^2 \tag{4.124}$$

in which $\lambda_1 \geq \lambda_2 \cdots \leq \lambda_m$ are the eigenvalues of J (see Appendix B). This leads to

$$\mu = \sqrt{\det JJ^T} = |\lambda_1 \lambda_2 \cdots \lambda_m| = |\det J| \tag{4.125}$$

The manipulability μ has the following properties.

- In general, $\mu = 0$ holds if and only if $\text{rank}(J) < m$, that is, when J is not full rank.

- Suppose that there is some error in the measured velocity $\Delta\xi$. We can bound the corresponding error in the computed joint velocity $\Delta\dot{q}$ by

$$(\sigma_1)^{-1} \leq \frac{||\Delta\dot{q}||}{||\Delta\xi||} \leq (\sigma_m)^{-1} \tag{4.126}$$

Example 4.13 Two-link Planar Arm

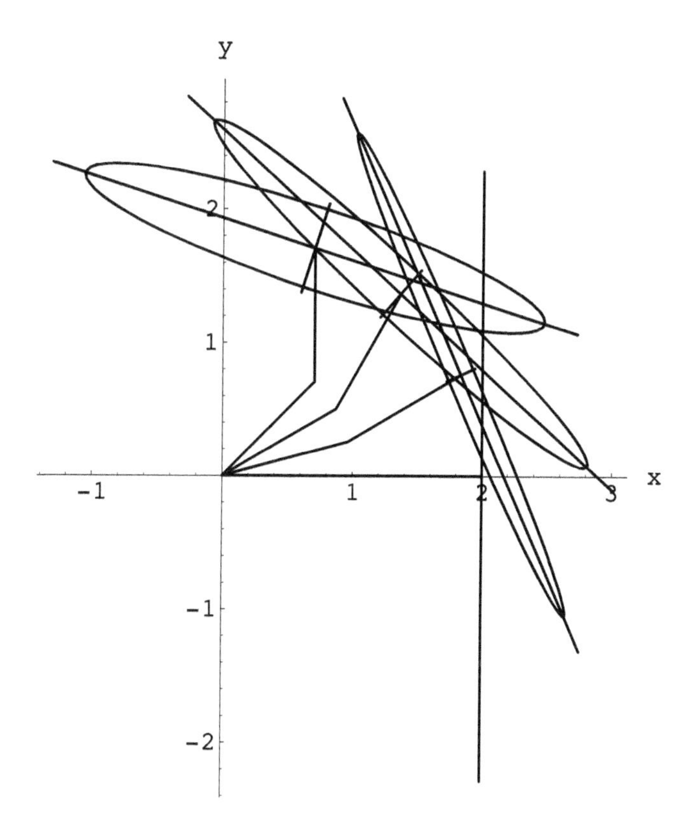

Figure 4.12: Manipulability ellipsoids are shown for several configurations of the two-link arm.

Consider the two-link planar arm and the task of positioning in the plane. The Jacobian is given by

$$J = \begin{bmatrix} -a_1 s_1 - a_2 s_{12} & -a_2 s_{12} \\ a_1 c_1 + a_2 c_{12} & a_2 c_{12} \end{bmatrix} \tag{4.127}$$

and the manipulability is given by

$$\mu = |det\ J| = a_1 a_2 |s_2|$$

Manipulability ellipsoids for several configurations of the two-link arm are shown in Figure 4.12.

We can use manipulability to determine the optimal configurations in which to perform certain tasks. In some cases it is desirable to perform a task in the configuration for which the end effector has the maximum manipulability. For the two-link arm the maximum manipulability is obtained for $\theta_2 = \pm\pi/2$.

Manipulability can also be used to aid in the design of manipulators. For example, suppose that we wish to design a two-link planar arm whose total link length a_1+a_2 *is fixed. What values should be chosen for* a_1 *and* a_2*? If we design the robot to maximize the maximum manipulability, then we need to maximize* $\mu = a_1 a_2 |s_2|$. *We have already seen that the maximum is obtained when* $\theta_2 = \pm\pi/2$, *so we need only find* a_1 *and* a_2 *to maximize the product* $a_1 a_2$. *This is achieved when* $a_1 = a_2$. *Thus, to maximize manipulability, the link lengths should be chosen to be equal.*

◇

4.13 SUMMARY

A moving coordinate frame has both a linear and an angular velocity. Linear velocity is associated to a moving point, while angular velocity is associated to a rotating frame. Thus, the linear velocity of a moving frame is merely the velocity of its origin. The angular velocity for a moving frame is related to the time derivative of the rotation matrix that describes the instantaneous orientation of the frame. In particular, if $R(t) \in SO(3)$, then

$$\dot{R}(t) = S(\omega(t))R(t) \tag{4.128}$$

and the vector $\omega(t)$ is the instantaneous angular velocity of the frame. The operator S gives a skew symmetrix matrix

$$S(\omega) = \begin{bmatrix} 0 & -\omega_z & \omega_y \\ \omega_z & 0 & -\omega_x \\ -\omega_y & \omega_x & 0 \end{bmatrix} \tag{4.129}$$

The manipulator Jacobian relates the vector of joint velocities to the body velocity $\xi = [v^T, \omega^T]^T$ of the end effector

$$\xi = J\dot{q} \tag{4.130}$$

This relationship can be written as two equations, one for linear velocity and one for angular velocity,

$$v = J_v \dot{q} \tag{4.131}$$

$$\omega = J_\omega \dot{q} \tag{4.132}$$

The ith column of the Jacobian matrix corresponds to the ith joint of the robot manipulator, and takes one of two forms depending on whether the ith joint is prismatic or revolute

$$J_i = \begin{cases} \begin{bmatrix} z_{i-1} \times (o_n - o_{i-1}) \\ z_{i-1} \end{bmatrix} & \text{if joint i is revolute} \\ \\ \begin{bmatrix} z_{i-1} \\ 0 \end{bmatrix} & \text{if joint i is prismatic} \end{cases} \tag{4.133}$$

It is often the case that a tool is attached to the end effector. When two frames are rigidly attached, their velocities are related by

$$\xi_A^A = \begin{bmatrix} R_B^A & S(d_B^A)R_B^A \\ 0_{3\times 3} & R_B^A \end{bmatrix} \xi_B^B$$

and this relationship allows us to compute the required end effector velocity to achieve a desired tool velocity.

For a given parameterization of orientation, for example, Euler angles, the analytical Jacobian relates joint velocities to the time derivative of the pose parameters

$$X = \begin{bmatrix} d(q) \\ \alpha(q) \end{bmatrix} \qquad \dot{X} = \begin{bmatrix} \dot{d} \\ \dot{\alpha} \end{bmatrix} = J_a(q)\dot{q}$$

in which $d(q)$ is the usual vector from the origin of the base frame to the origin of the end-effector frame and α denotes a parameterization of the rotation matrix that specifies the orientation of the end-effector frame relative to the base frame. For the Euler angle parameterization, the analytical Jacobian is given by

$$J_a(q) = \begin{bmatrix} I & 0 \\ 0 & B(\alpha)^{-1} \end{bmatrix} J(q) \tag{4.134}$$

in which

$$B(\alpha) = \begin{bmatrix} c_\psi s_\theta & -s_\psi & 0 \\ s_\psi s_\theta & c_\psi & 0 \\ c_\theta & 0 & 1 \end{bmatrix}$$

A configuration at which the Jacobian loses rank, that is, a configuration q such that $\operatorname{rank} J \leq \max_q \operatorname{rank} J(q)$, is called a singularity. For a manipulator with a spherical wrist, the set of singular configurations includes singularites of the wrist (which are merely the singularities in the

Euler angle parameterization) and singularites in the arm. The latter can be found by solving

$$\det J_{11} = 0$$

with J_{11} the upper left 3×3 block of the manipulator Jacobian.

The Jacobian matrix can also be used to relate forces F applied at the end-effector frame to the induced joint torques τ

$$\tau = J^T(q)F$$

For nonsingular configurations, the Jacobian relationship can be used to find the joint velocities $\dot{q}$ necessary to achieve a desired end-effector velocity ξ. The minimum norm solution is given by

$$\dot{q} = J^+\xi$$

in which $J^+ = J^T(JJ^T)^{-1}$ is the right pseudoinverse of J.

Manipulability is defined by $\mu = \sigma_1\sigma_2\cdots\sigma_m$ in which σ_i are the singular values for the manipulator Jacobian. The manipulatibility can be used to characterize the range of possible end-effector velocities for a given configuration q.

PROBLEMS

4-1 Verify Equation (4.6) by direct calculation.

4-2 Verify Equation (4.7) by direct calculation.

4-3 Prove the assertion given in Equation (4.9) that $R(a \times b) = Ra \times Rb$, for $R \in S0(3)$.

4-4 Verify Equation (4.10).

4-5 Verify Equation (4.17) by direct calculation.

4-6 Suppose that $a = [1, -1, 2]^T$ and that $R = R_{x,90}$. Show by direct calculation that $RS(a)R^T = S(Ra)$.

4-7 Given $R = R_{x,\theta}R_{y,\phi}$, compute $\frac{\partial R}{\partial \phi}$. Evaluate $\frac{\partial R}{\partial \phi}$ at $\theta = \frac{\pi}{2}$, $\phi = \frac{\phi}{2}$.

4-8 Use Equation (2.43) to show that

$$R_{k,\theta} = I + S(k)\sin(\theta) + S^2(k)\operatorname{vers}(\theta)$$

4-9 Show that $S^3(k) = -S(k)$. Use this and Problem 4-8 to verify Equation (4.18).

4-10 Given any square matrix A, the exponential of A is a matrix defined as

$$e^A = I + A + \frac{1}{2}A^2 + \frac{1}{3!}A^3 + \cdot$$

Given $S \in so(3)$ show that $e^S \in SO(3)$.

Use the acts that $e^A e^B = e^{A+B}$ provided that A and B commute, that is, $AB = BA$, and the fact that $\det(e^A) = e^{Tr(A)}$.

4-11 Show that $R_{k,\theta} = e^{S(k)\theta}$ for k a unit vector.

Hint: Use the series expansion for the matrix exponential together with Problems 4-8 and 4-9. Alternatively use the fact that $R_{k,\theta}$ satisfies the differential equation

$$\frac{dR}{d\theta} = S(k)R.$$

4-12 Use Problem 4-11 to show the converse of Problem 4-10, that is, if $R \in SO(3)$ then there exists $S \in so(3)$ such that $R = e^S$.

4-13 Given the Euler angle transformation

$$R = R_{z,\psi}R_{y,\theta}R_{z,\phi}$$

show that $\frac{d}{dt}R = S(\omega)R$ where

$$\omega = \{c_\psi s_\theta \dot{\phi} - s_\psi \dot{\theta}\}i + \{s_\psi s_\theta \dot{\phi} + c_\psi \dot{\theta}\}j + \{\dot{\psi} + c_\theta \dot{\phi}\}k$$

The components of i, j, k, respectively, are called the **nutation**, **spin**, and **precession**.

4-14 Repeat Problem 4-13 for the Roll-Pitch-Yaw transformation. In other words, find an explicit expression for ω such that $\frac{d}{dt}R = S(\omega)R$, where R is given by Equation (2.38).

4-15 Two frames $o_0x_0y_0z_0$ and $o_1x_1y_1z_1$ are related by the homogeneous transformation

$$H = \begin{bmatrix} 0 & -1 & 0 & 1 \\ 1 & 0 & 0 & -1 \\ 0 & 0 & 1 & 0 \\ 0 & 0 & 0 & 1 \end{bmatrix}$$

A particle has velocity $v_1(t) = [3, 1, 0]^T$ relative to frame $o_1x_1y_1z_1$. What is the velocity of the particle in frame $o_0x_0y_0z_0$?

4-16 For the three-link planar manipulator of Example 4.6, compute the vector o_c and derive the manipulator Jacobian matrix.

4-17 Compute the Jacobian J_{11} for the 3-link elbow manipulator of Example 4.9 and show that it agrees with Equation (4.98). Show that the determinant of this matrix agrees with Equation (4.99).

4-18 Compute the Jacobian J_{11} for the three-link spherical manipulator of Example 4.10.

4-19 Use Equation (4.102) to show that the singularities of the SCARA manipulator are given by Equation (4.104).

4-20 Find the 6×3 Jacobian for the three links of the cylindrical manipulator of Figure 3.7. Find the singular configurations for this arm.

4-21 Repeat Problem 4-20 for the Cartesian manipulator of Figure 3.28.

4-22 Complete the derivation of the Jacobian for the Stanford manipulator from Example 4.7.

4-23 Verify Equation (4.81) by direct computation.

4-24 Show that $B(\alpha)$ given by Equation (4.87) is invertible provided $s_\theta \neq 0$.

4-25 Suppose that $\dot{q}$ is a solution to Equation (4.110) for $m < n$.

1. Show that $\dot{q} + (I - J^+J)b$ is also a solution to Equation (4.110) for any $b \in \mathbb{R}^n$.
2. Show that $b = 0$ gives the solution that minimizes the resulting joint velocities.

4-26 Verify Equation (4.114).

4-27 Verify Equation (4.120).

4-28 Verify Equation (4.121).

NOTES AND REFERENCES

Angular velocity is fundamentally related to the derivative of a rotation matrix, and therefore to the Lie algebra $so(3)$. This relationship, and more generally the geometry of $so(n)$, is explored in differential geometry texts, as well as in more advanced robotics texts such as [93].

The concept of a Jacobian matrix as a linear mapping from the tangent space of one manifold to the tangent space of a second manifold is also dealt with in differential geometry texts, and even in advanced calculus texts when one or both of the manifolds is a Euclidean space. The use of the geometric Jacobian matrix of Equation (4.42) to map joint velocities (which lie in the tangent space to the configuration space) to a velocity $\xi = [v^T, \omega^T]^T$ is not commonly found in mathematics texts (note that ξ is *not* itself the derivative of any quantity). However, most robotics texts include some description of the geometric Jacobian, including [102], [110], and [93].

Since the relationship between the end effector velocity and the joint velocities is defined by a linear map, the inverse velocity problem is a special case of the more general problem of solving linear systems, a problem that is the subject of linear algebra. Algorithms for solving this problem can be found in a variety of texts, including [106] and [47].

The manipulability measure discussed in Section 4.12 is due to Yoshikawa [142].

Chapter 5

PATH AND TRAJECTORY PLANNING

In previous chapters we studied the geometry of robot arms, developing solutions for both the forward and inverse kinematics problems. The solutions to these problems depend only on the intrinsic geometry of the robot, and they do not reflect any constraints imposed by the workspace in which the robot operates. In particular, they do not take into account the possiblity of collision between the robot and objects in the workspace. In this chapter we address the problem of planning collision free paths for the robot. We will assume that the initial and final configurations of the robot are specified and that the problem is to find a collision free path connecting these configurations.

The description of this problem is deceptively simple, yet the path planning problem is among the most difficult problems in computer science. The computational complexity of the best known complete[1] path planning algorithm grows exponentially with the number of internal degrees of freedom of the robot. For this reason, for robot systems with more than a few degrees of freedom, complete algorithms are not used in practice.

In this chapter we treat the problem of planning paths for a high degree-of-freedom robot as a search problem. The algorithms we describe are not guaranteed to find solutions to all problems, but they are quite effective in a wide range of practical applications. Furthermore, these algorithms are fairly easy to implement, and require only moderate computation time for most problems.

[1]An algorithm is said to be complete if it finds a solution whenever one exists, and signals failure in finite time when a solution does not exist.

Path planning provides a geometric description of robot motion, but it does not specify any dynamic aspects of the motion. For example, what should be the joint velocities and accelerations while traversing the path? These questions are addressed by a trajectory planner. The trajectory planner computes a function $q(t)$ that completely specifies the desired motion of the robot as it traverses the path.

We begin in Section 5.1 by discussing, in more detail, the notion of **configuration space** that was first introduced in Chapter 1. We give a brief description of the geometry of the configuration space and describe how obstacles in the workspace can be mapped into the configuration space. We then introduce path planning methods that use so-called **artificial potential fields** in Section 5.2. The corresponding algorithms use gradient descent search to find a collision-free path to the goal and, as with all gradient descent methods, these algorithms can become trapped in local minima in the potential field. Therefore, in Section 5.3 we describe how random motions can be used to escape local minima. In Section 5.4 we describe another randomized method known as the **Probabilistic Roadmap (PRM)** method. Finally, since each of these methods generates a sequence of configurations, we describe how polynomial splines can be used to generate smooth trajectories from a sequence of configurations in Section 5.5.

5.1 THE CONFIGURATION SPACE

In Chapter 3 we learned that the forward kinematic map can be used to determine the position and orientation of the end effector frame given the vector of joint variables. Furthermore, the A matrices can be used to infer the position and orientation of any link of the robot. Since each link of the robot is assumed to be a rigid body, the A matrices can be used to infer the position of any point on the robot, given the values of the joint variables. In the path planning literature a complete specification of the location of every point on the robot is referred to as a **configuration**, and the set of all possible configurations is referred to as the **configuration space**. For our purposes, the vector of joint variables q provides a convenient representation of a configuration. We will denote the configuration space by $\mathcal{Q}$.

For a one link revolute arm the configuration space is merely the set of orientations of the link, and thus $\mathcal{Q} = S^1$, where S^1 represents the unit circle. We could also say $\mathcal{Q} = SO(2)$. In fact, the choice of S^1 or $SO(2)$ is not particularly important, since these two are equivalent representations. In either case we can parameterize $\mathcal{Q}$ by a single parameter, the joint angle θ_1. For the two-link planar arm we have $\mathcal{Q} = S^1 \times S^1 = T^2$, in which T^2

represents the torus, and we can represent a configuration by $q = (\theta_1, \theta_2)$. For a Cartesian arm, we have $\mathcal{Q} = \mathbb{R}^3$ and we can represent a configuration by $q = (d_1, d_2, d_3)$.

Although we have chosen to represent a configuration by a vector of joint variables, the notion of a configuration is more general than this. For example, as we saw in Chapter 2, for any rigid two-dimensional object we can specify the location of every point on the object by rigidly attaching a coordinate frame to the object and then specifying the position and orientation of this frame. Thus, for a rigid object moving in the plane we can represent a configuration by the triple $q = (x, y, \theta)$, and the configuration space can be represented by $\mathcal{Q} = \mathbb{R}^2 \times SO(2)$. Again, this is merely one possible representation of the configuration space, but it is a convenient one given the representations of position and orientation that we learned in previous chapters.

A collision occurs when the robot contacts an obstacle in the workspace. To describe collisions we introduce some additional notation. We denote the workspace, that is, the Cartesian space in which the robot moves, by $\mathcal{W}$ and we denote the obstacles in the workspace by $\mathcal{O}_i$. We will denote the robot by $\mathcal{A}$ and the subset of the workspace that is occupied by the robot at configuration q by $\mathcal{A}(q)$. To plan a collision free path we must ensure that the robot never reaches a configuration q that causes it to make contact with an obstacle. The set of configurations for which the robot collides with an obstacle is referred to as the **configuration space obstacle** and it is defined by

$$\mathcal{QO} = \{q \in \mathcal{Q} \mid \mathcal{A}(q) \cap \mathcal{O} \neq \emptyset\}$$

in which $\mathcal{O} = \cup \mathcal{O}_i$. The set of collision-free configurations, referred to as the free configuration space, is then simply the set difference

$$\mathcal{Q}_{\text{free}} = \mathcal{Q} \setminus \mathcal{QO}$$

Example 5.1 A Rigid Body that Translates in the Plane

Consider a gantry robot with two prismatic joints and forward kinematics given by $x = d_1, y = d_2$. For this case, the robot's configuration space is $\mathcal{Q} = \mathbb{R}^2$, so it is particularly easy to visualize both the configuration space and the configuration space obstacle region. If there is only one obstacle in the workspace and both the end effector and the obstacle are convex polygons, it is a simple matter to compute the configuration space obstacle region $\mathcal{QO}$. We assume here that the arm itself is positioned above the workspace, so that the only possible collisions are between the end effector and the obstacle.

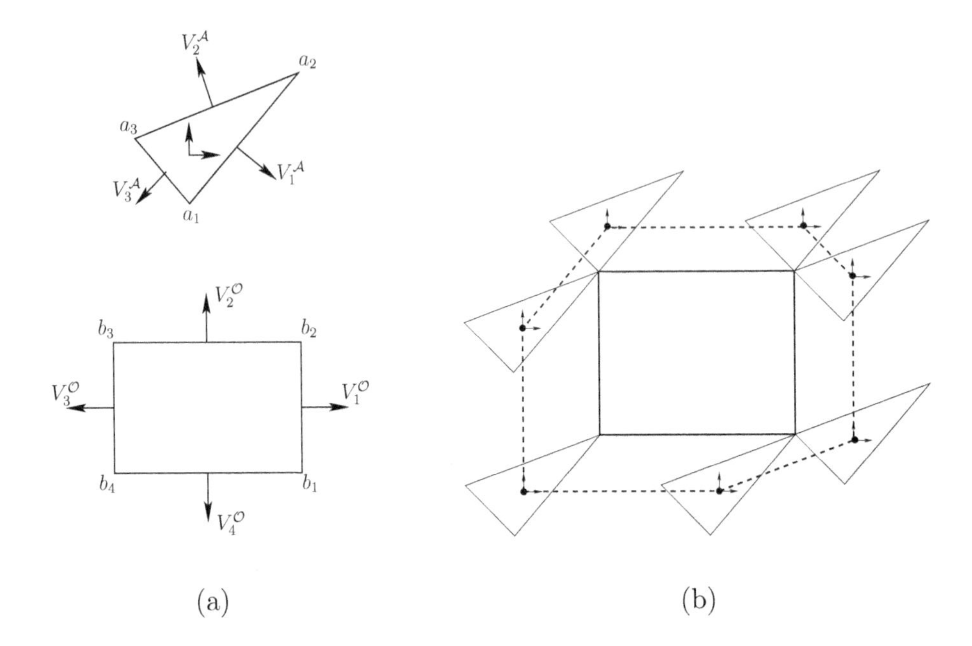

Figure 5.1: (a) The robot end effector is a triangle-shaped rigid object in a workspace that contains a single rectangular obstacle. (b) The boundary of the configuration space obstacle $\mathcal{QO}$ (shown as a dashed line) can be obtained by computing the convex hull of the configurations at which the end effector makes vertex-to-vertex contact with the single convex obstacle.

Let $V_i^{\mathcal{A}}$ denote the vector that is normal to the ith edge of the end effector and $V_i^{\mathcal{O}}$ denote the vector that is normal to the ith edge of the obstacle. Define a_i to be the vector from the origin of the robot's coordinate frame to the ith vertex of the end effector and b_j to be the vector from the origin of the world coordinate frame to the jth vertex of the obstacle. An example is shown in Figure 5.1(a). The vertices of $\mathcal{QO}$ can be determined as follows.

- *For each pair $V_j^{\mathcal{O}}$ and $V_{j-1}^{\mathcal{O}}$, if $V_i^{\mathcal{A}}$ points between $-V_j^{\mathcal{O}}$ and $-V_{j-1}^{\mathcal{O}}$, then add to $\mathcal{QO}$ the vertices $b_j - a_i$ and $b_j - a_{i+1}$.*
- *For each pair $V_i^{\mathcal{A}}$ and $V_{i-1}^{\mathcal{A}}$, if $V_j^{\mathcal{O}}$ points between $-V_i^{\mathcal{A}}$ and $-V_{i-1}^{\mathcal{A}}$, then add to $\mathcal{QO}$ the vertices $b_j - a_i$ and $b_{j+1} - a_i$.*

This is illustrated in Figure 5.1(b). Note that this algorithm essentially places the end effector at all positions where vertex-to-vertex contact between robot and obstacle are possible. The origin of the robot's local coordinate frame at each such configuration defines a vertex of $\mathcal{QO}$. *The polygon defined by these vertices is* $\mathcal{QO}$.

If there are multiple convex obstacles $\mathcal{O}_i$, *then the configuration space obstacle region is merely the union of the obstacle regions* $\mathcal{QO}_i$ *for the individual obstacles. For a nonconvex obstacle, the configuration space obstacle region can be computed by first decomposing the nonconvex obstacle into convex pieces* $\mathcal{O}_i$ *computing the configuration space obstacle region* $\mathcal{QO}_i$ *for each piece, and finally, computing the union of the* $\mathcal{QO}_i$.

This example also illustrates how the configuration space can be constructed for a robot with polygonal shape that translates in the plane, such as a mobile robot moving on a factory floor.

◇

Example 5.2 A Two Link Planar Arm

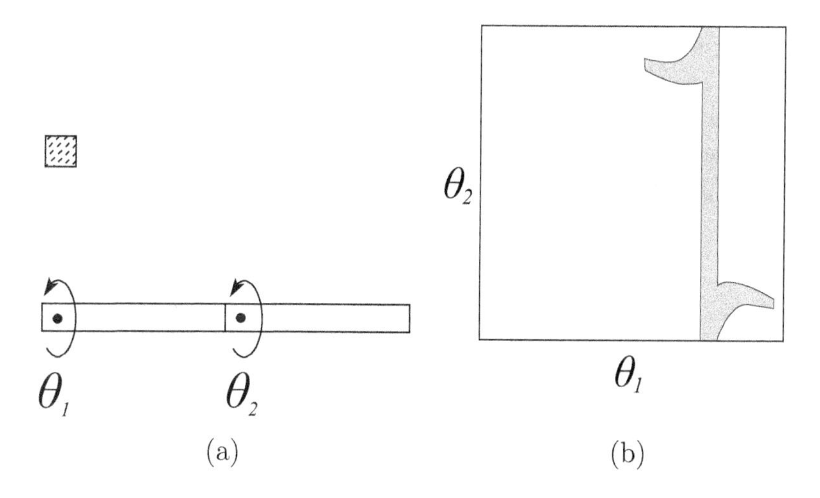

Figure 5.2: (a) The robot is a two-link planar arm and the workspace contains a single, small polygonal obstacle. (b) The corresponding configuration space obstacle region contains all configurations $q = (\theta_1, \theta_2)$ such that the arm at configuration q intersects the obstacle.

The computation of $\mathcal{QO}$ *is more difficult for robots with revolute joints. Consider a two-link planar arm in a workspace containing a single obstacle as shown in Figure 5.2(a). The configuration space obstacle region is illustrated in Figure 5.2(b). For values of* θ_1 *very near* $\pi/2$, *the first link of the arm collides with the obstacle. When the first link is near the obstacle (*θ_1

near $\pi/2$), for some values of θ_2 the second link of the arm collides with the obstacle. The region $\mathcal{QO}$ shown in Figure 5.2(b) was computed using a discrete grid on the configuration space. For each cell in the grid, a collision test was performed, and the cell was shaded when a collision occured. This is only an approximate representation of $\mathcal{QO}$; however, for robots with revolute joints, exact representations are very expensive to compute, and therefore such approximate representations are often used for robots with a few degrees of freedom.

◇

Computing $\mathcal{QO}$ for the two-dimensional case of $\mathcal{Q} = \mathbb{R}^2$ and polygonal obstacles is straightforward, but, as can be seen from the two-link planar arm example, computing $\mathcal{QO}$ becomes difficult for even moderately complex configuration spaces. In the general case (for example, articulated arms or rigid bodies that can both translate and rotate), the problem of computing a representation of the configuration space obstacle region is intractable. One of the reasons for this complexity is that the size of the representation of the configuration space tends to grow exponentially with the number of degrees of freedom. This is easy to understand intuitively by considering the number of n-dimensional unit cubes needed to fill a space of size k. For the one-dimensional case k unit intervals will cover the space. For the two-dimensional case k^2 squares are required. For the three-dimensional case k^3 cubes are required, and so on. Therefore, in this chapter we will develop methods that avoid the construction of an explicit representation of $\mathcal{QO}$ or of $\mathcal{Q}_{\text{free}}$.

The path planning problem is to find a path from an initial configuration q_s to a final configuration q_f, such that the robot does not collide with any obstacle as it traverses the path. More formally, a collision-free path from q_s to q_f is a continuous map, $\gamma : [0, 1] \rightarrow \mathcal{Q}_{\text{free}}$, with $\gamma(0) = q_s$ and $\gamma(1) = q_f$. We will develop path planning methods that compute a sequence of discrete configurations (set points) in the configuration space. In Section 5.5 we will show how smooth trajectories can be generated from such a sequence of set points.

5.2 PATH PLANNING USING POTENTIAL FIELDS

As mentioned above, it is typically not feasible to build an explicit representation of $\mathcal{QO}$ or of $\mathcal{Q}_{\text{free}}$. An alternative is to develop a search algorithm that incrementally explores $\mathcal{Q}_{\text{free}}$ while searching for a path. One of the most popular strategies for exploring $\mathcal{Q}_{\text{free}}$ uses an **artificial potential field** to guide the search.

The basic idea behind the potential field approach is to treat the robot as a point particle in the configuration space under the influence of an artificial potential field U. The field U is constructed so that the robot is attracted to the final configuration q_{f} while being repelled from the boundaries of $\mathcal{QO}$. If possible, U is constructed so that there is a single global minimum of U at q_{f} and there are no local minima. Unfortunately it is typically difficult or even impossible to construct such a field.

In general, the field U is an additive field consisting of one component that attracts the robot to q_{f} and a second component that repels the robot from the boundary of $\mathcal{QO}$

$$U(q) = U_{\mathrm{att}}(q) + U_{\mathrm{rep}}(q)$$

Given this formulation path planning can be treated as an optimization problem, that is, the problem of finding the global minimum in U starting from initial configuration q_{s}. One of the simplest algorithms to solve this problem is gradient descent. In this case, the negative gradient of U can be considered as a generalized force acting on the robot in configuration space

$$\tau(q) = -\nabla U(q) = -\nabla U_{\mathrm{att}}(q) - \nabla U_{\mathrm{rep}}(q)$$

in which τ is a vector of joint torques (for a revolute arm). Allowing this force to act on the robot will cause it to move toward its goal configuration along the path of steepest descent of the potential function.

In general, it is difficult to construct a potential field directly on the configuration space, and even more difficult to compute the gradient of the field on the configuration space. The reasons for this include the difficulty of computing shortest distances to configuration space obstacles (a computation that is required when computing the value of the repulsive field, as we will see below) and the complex geometry of the configuration space. For this reason, we will define our potential fields directly on the workspace of the robot. In particular, for an n-link arm, we will define a potential field for each of the origins of the n DH frames (excluding the fixed, frame 0). These **workspace potential fields** will attract the origins of the DH frames to their goal locations while repelling them from obstacles. We will use these fields to define motions in the configuration space using the manipulator Jacobian matrix.

In the remainder of this section we will describe typical choices for the attractive and repulsive potential fields, how the manipulator Jacobian can be used to map these fields to configuration space motions, and a gradient descent algorithm that can be used to plan paths in this field.

5.2.1 The Attractive Field

To attract the robot to its goal configuration, we will define an attractive potential field $U_{\text{att},i}$ for o_i, the origin of the ith DH frame. When all n origins reach their goal positions, the arm will have reached its goal configuration.

There are several criteria that the potential field $U_{\text{att},i}$ should satisfy. First, $U_{\text{att},i}$ should be monotonically increasing with the distance to o_i from its goal position. The simplest choice for such a field is a field that grows linearly with this distance, a so-called **conic well potential**. If we denote the position of the origin of the ith DH frame by $o_i(q)$, then the conic well potential is given by

$$U_{\text{att},i}(q) = ||o_i(q) - o_i(q_{\text{f}})||$$

The gradient of such a field has unit magnitude everywhere but at the goal position where it is zero. This can lead to stability problems since there is a discontinuity in the attractive force at the goal position. We prefer a field that is continuously differentiable such that the attractive force decreases as o_i approaches its goal position. The simplest such field is a field that grows quadratically with distance

$$U_{\text{att},i}(q) = \frac{1}{2}\zeta_i||o_i(q) - o_i(q_{\text{f}})||^2 \tag{5.1}$$

in which ζ_i is a parameter used to scale the effects of the attractive potential. This field is sometimes referred to as a **parabolic well potential**. The workspace attractive force for o_i is equal to the negative gradient of $U_{\text{att},i}$, which is given by (Problem 5-7)

$$F_{\text{att},i}(q) = -\nabla U_{\text{att},i}(q) = -\zeta(o_i(q) - o_i(q_{\text{f}})) \tag{5.2}$$

For the parabolic well the attractve force for the origin of the ith DH frame is a vector directed toward $o_i(q_{\text{f}})$ with magnitude linearly related to the distance to $o_i(q)$ from $o_i(q_{\text{f}})$.

Note that while this force converges linearly to zero as q approaches q_{f}, which is a desirable property, it grows without bound as q moves away from q_{f}. If q_{s} is very far from q_{f}, this may produce an initial attractive force that is very large. For this reason we may choose to combine the quadratic and conic potentials so that the conic potential attracts o_i when it is very distant from its goal position, and the quadratic potential attracts o_i when it is near its goal position. Of course it is necessary that the gradient be defined at the boundary between the conic and quadratic fields. Such a field can be defined by

$$U_{\text{att},i}(q) = \begin{cases} \frac{1}{2}\zeta_i ||o_i(q) - o_i(q_{\text{f}})||^2 & ; \quad ||o_i(q) - o_i(q_{\text{f}})|| \leq d \\ \\ d\zeta_i ||o_i(q) - o_i(q_{\text{f}})|| - \dfrac{1}{2}\zeta_i d^2 & ; \quad ||o_i(q) - o_i(q_{\text{f}})|| > d \end{cases} \tag{5.3}$$

in which d is the distance that defines the transition from conic to parabolic well. In this case the workspace force for o_i is given by

$$F_{\text{att},i}(q) = \begin{cases} -\zeta_i(o_i(q) - o_i(q_{\text{f}})) & : \quad ||o_i(q) - o_i(q_{\text{f}})|| \leq d \\ \\ -d\zeta_i \dfrac{(o_i(q) - o_i(q_{\text{f}}))}{||o_i(q) - o_i(q_{\text{f}})||} & : \quad ||o_i(q) - o_i(q_{\text{f}})|| > d \end{cases} \tag{5.4}$$

The gradient is well defined at the boundary of the two fields since at the boundary $d = ||o_i(q) - o_i(q_{\text{f}})||$ and the gradient of the quadratic potential is equal to the gradient of the conic potential $F_{\text{att},i}(q) = -\zeta_i(o_i(q) - o_i(q_{\text{f}}))$.

Example 5.3 Two-link Planar Arm

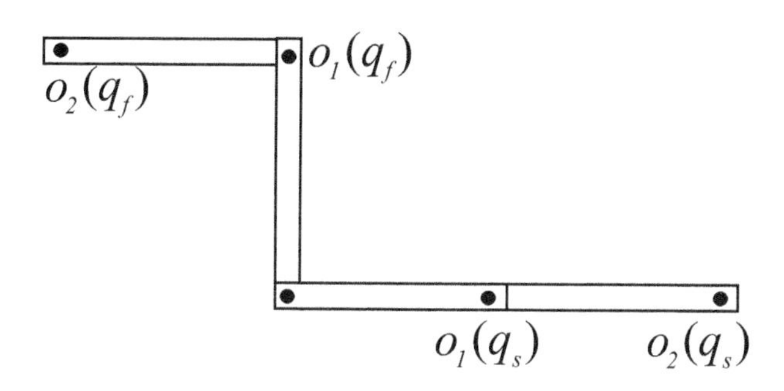

Figure 5.3: The initial configuration for the two-link arm is given by $\theta_1 = \theta_2 = 0$ and the final configuration is given by $\theta_1 = \theta_2 = \pi/2$. The origins for DH frames 1 and 2 are shown at both q_{s} and q_{f}.

Consider the two-link planar arm shown in Figure 5.3 with $a_1 = a_2 = 1$ *and with initial and final configurations given by*

$$q_{\text{s}} = \begin{bmatrix} 0 \\ 0 \end{bmatrix} \qquad q_{\text{f}} = \begin{bmatrix} \frac{\pi}{2} \\ \frac{\pi}{2} \end{bmatrix}$$

Using the forward kinematic equations for this arm (see Example 3.1) we obtain

$$o_1(q_s) = \begin{bmatrix} 1 \\ 0 \end{bmatrix} \quad o_1(q_f) = \begin{bmatrix} 0 \\ 1 \end{bmatrix} \quad o_2(q_s) = \begin{bmatrix} 2 \\ 0 \end{bmatrix} \quad o_2(q_f) = \begin{bmatrix} -1 \\ 1 \end{bmatrix}$$

Using these coordinates for the origins of the two DH frames at their initial and goal configurations, assuming that d is sufficiently large, we obtain the attractive forces

$$\begin{aligned} F_{\text{att},1}(q_s) &= -\zeta_1(o_1(q_s) - o_1(q_f)) = \zeta_1 \begin{bmatrix} -1 \\ 1 \end{bmatrix} \\ F_{\text{att},2}(q_s) &= -\zeta_2(o_2(q_s) - o_2(q_f)) = \zeta_2 \begin{bmatrix} -3 \\ 1 \end{bmatrix} \end{aligned}$$

◇

5.2.2 The Repulsive Field

In order to prevent collisions between the robot and obstacles we will define a **workspace repulsive potential field** for the origin of each DH frame (excluding frame 0). There are several criteria that these repulsive fields should satisfy. They should repel the robot from obstacles, never allowing the robot to collide with an obstacle, and, when the robot is far away from an obstacle, that obstacle should exert little or no influence on the motion of the robot. One way to achieve this is to define a potential function whose value approaches infinity as the configuration approaches an obstacle boundary, and whose value decreases to zero at a specified distance from the obstacle boundary. Note that by defining repulsive potentials only for the origins of the DH frames we cannot ensure that collisions never occur (for example, the middle portion of a long link might collide with an obstacle), but it is fairly easy to modify the method to prevent such collsions as we will see below. For now, we will deal only with the origins of the DH frames.

We define ρ_0 to be the distance of influence of an obstacle. This means that an obstacle will not repel o_i if the distance from o_i to the obstacle is greater than ρ_0. One potential function that meets the criteria described above is given by

$$U_{\text{rep},i}(q) = \begin{cases} \frac{1}{2}\eta_i \left(\frac{1}{\rho(o_i(q))} - \frac{1}{\rho_0} \right)^2 & ; \quad \rho(o_i(q)) \le \rho_0 \\ 0 & ; \quad \rho(o_i(q)) > \rho_0 \end{cases} \tag{5.5}$$

in which $\rho(o_i(q))$ is the shortest distance between o_i and any workspace obstacle. The workspace repulsive force is equal to the negative gradient of $U_{\text{rep},i}$. For $\rho(o_i(q)) \leq \rho_0$, this force is given by (Problem 5-11)

$$F_{\text{rep},i}(q) = \eta_i \left(\frac{1}{\rho(o_i(q))} - \frac{1}{\rho_0} \right) \frac{1}{\rho^2(o_i(q))} \nabla\rho(o_i(q)) \tag{5.6}$$

in which the notation $\nabla\rho(o_i(q))$ indicates the gradient $\nabla\rho(x)$ evaluated at $x = o_i(q)$. If the obstacle region is convex and b is the point on the obstacle boundary that is closest to o_i, then $\rho(o_i(q)) = ||o_i(q) - b||$, and its gradient is

$$\nabla\rho(x)\Bigg|_{x=o_i(q)} = \frac{o_i(q) - b}{||o_i(q) - b||} \tag{5.7}$$

that is, the unit vector directed from b toward $o_i(q)$.

Example 5.4 Two-link Planar Arm

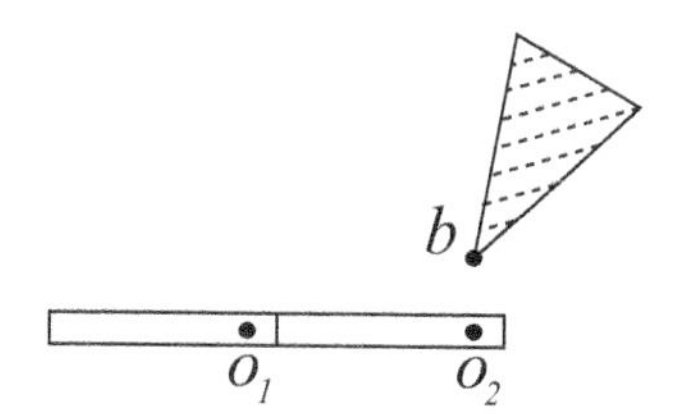

Figure 5.4: The obstacle shown repels o_2, but is outside the distance of influence for o_1. Therefore, it exerts no repulsive force on o_1.

Consider the previous Example 5.3, with a single convex obstacle in the workspace as shown in Figure 5.4. Let $\rho_0 = 1$. This prevents the obstacle from repelling o_1, which is reasonable since link 1 can never contact the obstacle. The nearest obstacle point to o_2 is the vertex b of the polygonal obstacle. Suppose that b has the coordinates $(2, 0.5)$. Then the distance from $o_2(q_s)$ to b is $\rho(o_2(q_s)) = 0.5$ and $\nabla\rho(o_2(q_s)) = [0, -1]^T$. The repulsive force at the initial configuration for o_2 is then given by

$$F_{\text{rep},2}(q_s) = \eta_2 \left(\frac{1}{0.5} - 1 \right) \frac{1}{0.25} \begin{bmatrix} 0 \\ -1 \end{bmatrix} = \eta_2 \begin{bmatrix} 0 \\ -4 \end{bmatrix}$$

This force has no effect on joint 1, but causes joint 2 to rotate slightly in the clockwise direction, moving link 2 away from the obstacle.

◇

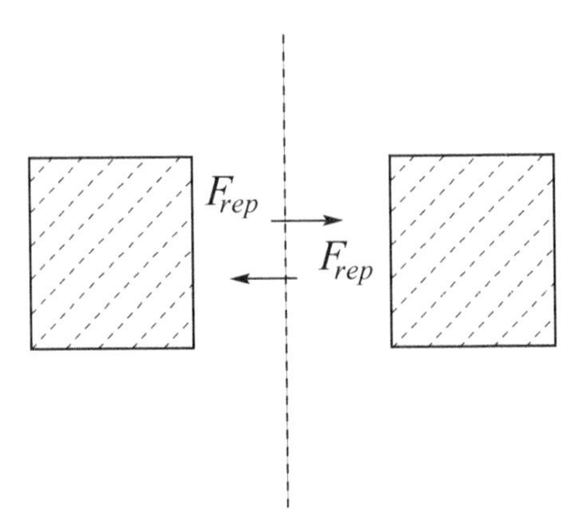

Figure 5.5: In this case the gradient of the repulsive potential given by Equation (5.6) is not continuous. In particular, the gradient changes discontinuously when o_i crosses the line midway between the two obstacles.

If the obstacle is not convex, then the distance function ρ will not necessarily be differentiable everywhere, which implies discontinuity in the force vector. Figure 5.5 illustrates such a case. Here the obstacle region is defined by two rectangular obstacles. For all configurations to the left of the dashed line the force vector points to the right, while for all configurations to the right of the dashed line the force vector points to the left. Thus, when o_i crosses the dashed line, a discontinuity in force occurs. There are various ways to deal with this problem. The simplest of these is merely to ensure that the regions of influence of distinct obstacles do not overlap.

As mentioned above, defining repulsive fields only for the origins of the DH frames does not guarantee that the robot cannot collide with an obstacle. Figure 5.6 shows an example where this is the case. In this figure o_1 and o_2 are very far from the obstacle and therefore the repulsive influence may not be great enough to prevent link 2 from colliding with the obstacle. To cope with this problem we can use a set of **floating repulsive control points** $o_{float,i}$ typically one per link. The floating control points are defined as points on the boundary of a link that are closest to any workspace obstacle. Obviously the choice of the $o_{float,i}$ depends on the configuration q. For the case shown in Figure 5.6, $o_{float,2}$ would be located near the center of link 2, thus repelling the robot from the obstacle. The repulsive force acting on $o_{float,i}$ is defined in the same way as for the other control points using Equation (5.6).

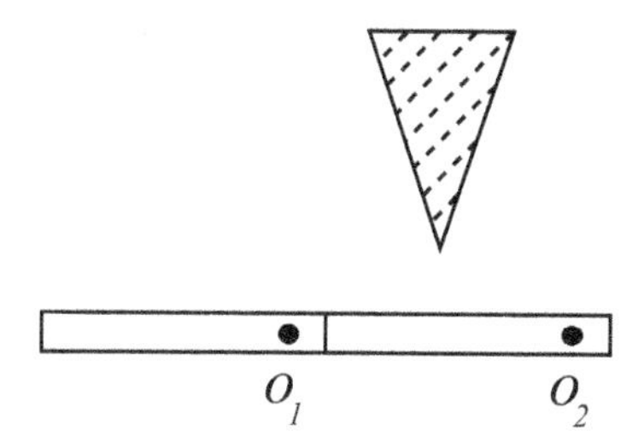

Figure 5.6: The repulsive forces exerted on the origins of the DH frames o_1 and o_2 may not be sufficient to prevent a collision between link 2 and the obstacle.

5.2.3 Mapping Workspace Forces to Joint Torques

We have shown how to construct potential fields in the robot's workspace that induce artificial forces on the origins o_i of the DH frames for the robot arm. In this section we describe how these forces can be mapped to joint torques.

As we derived in Chapter 4 using the principle of virtual work, if τ denotes the vector of joint torques induced by the workspace force F exerted at the end effector, then

$$J_v^T F = \tau \tag{5.8}$$

where J_v inlcudes the top three rows of the manipulator Jacobian. We do not use the lower three rows, since we have considered only attractive and repulsive workspace forces, and not attractive and repulsive workspace torques. Note that for each o_i an appropriate Jacobian must be constructed, but this is straightforward given the techniques described in Chapter 4 and the A matrices for the arm. We denote the Jacobian for o_i by J_{o_i}.

Example 5.5 Two-link Planar Arm

Consider again the two-link arm of Example 5.3 with repulsive workspace forces as given in Example 5.4. The Jacobians that map joint velocities to linear velocities satisfy

$$\dot{o}_i = J_{o_i}(q) \begin{bmatrix} \dot{q}_1 \\ \dot{q}_2 \end{bmatrix}$$

For the two-link arm the Jacobian matrix for o_2 is merely the Jacobian that we derived in Chapter 4, namely

$$J_{o_2}(q_1, q_2) = \begin{bmatrix} -s_1 - s_{12} & -s_{12} \\ c_1 + c_{12} & c_{12} \end{bmatrix} \tag{5.9}$$

The Jacobian matrix for o_1 is similar, but takes into account that motion of joint 2 does not affect the velocity of o_1. Thus

$$J_{o_1}(q_1, q_2) = \begin{bmatrix} -s_1 & 0 \\ c_1 & 0 \end{bmatrix}$$

At $q_s = (0, 0)$ we have

$$J_{o_1}^T(q_s) = \begin{bmatrix} -s_1 & c_1 \\ 0 & 0 \end{bmatrix} = \begin{bmatrix} 0 & 1 \\ 0 & 0 \end{bmatrix}$$

and

$$J_{o_2}^T(q_s) = \begin{bmatrix} -s_1 - s_{12} & c_1 + c_{12} \\ -s_{12} & c_{12} \end{bmatrix} = \begin{bmatrix} 0 & 2 \\ 0 & 1 \end{bmatrix}$$

Using these Jacobians, we can easily map the workspace attractive and repulsive forces to joint torques. If we let $\zeta_1 = \zeta_2 = \eta_2 = 1$ we obtain

$$\tau_{\text{att},1}(q_s) = \begin{bmatrix} 0 & 1 \\ 0 & 0 \end{bmatrix} \begin{bmatrix} -1 \\ 1 \end{bmatrix} = \begin{bmatrix} 1 \\ 0 \end{bmatrix}$$

$$\tau_{\text{att},2}(q_s) = \begin{bmatrix} 0 & 2 \\ 0 & 1 \end{bmatrix} \begin{bmatrix} -3 \\ 1 \end{bmatrix} = \begin{bmatrix} 2 \\ 1 \end{bmatrix}$$

$$\tau_{\text{rep},2}(q_s) = \begin{bmatrix} 0 & 2 \\ 0 & 1 \end{bmatrix} \begin{bmatrix} 0 \\ -4 \end{bmatrix} = \begin{bmatrix} -8 \\ -4 \end{bmatrix}$$

◇

Example 5.6 A Polygonal Robot in the Plane

Artificial potential fields can also be used to plan the motions of a mobile robot moving in the plane. In this case the mobile robot is typically modeled as a polygon that can translate and rotate in the plane.

Consider the polygonal robot shown in Figure 5.7. The vertex a has coordinates (a_x, a_y) in the robot's local coordinate frame. Therefore, if the robot's configuration is given by $q = (x, y, \theta)$, the forward kinematic map for vertex a (that is, the mapping from $q = (x, y, \theta)$ to the global coordinates of the vertex a) is given by

$$a(x, y, \theta) = \begin{bmatrix} x + a_x \cos\theta - a_y \sin\theta \\ y + a_x \sin\theta + a_y \cos\theta \end{bmatrix} \tag{5.10}$$

The corresponding Jacobian matrix is given by

$$J_a(x, y, \theta) = \begin{bmatrix} 1 & 0 & -a_x \sin\theta - a_y \cos\theta \\ 0 & 1 & a_x \cos\theta - a_y \sin\theta \end{bmatrix}$$

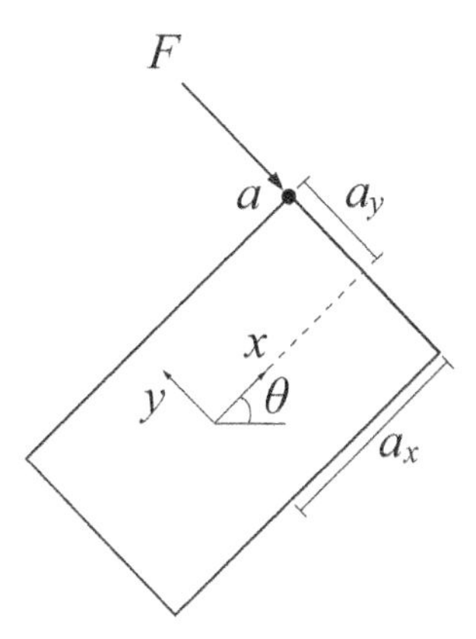

Figure 5.7: In this example, the robot is a polygon whose configuration can be represented as $q = (x, y, \theta)$, in which θ is the angle from the world x-axis to the x-axis of the robot's local frame. A force F is exerted on vertex a with local coordinates (a_x, a_y).

Using the transpose of the Jacobian to map the workspace forces to generalized forces for the configuration space, we obtain

$$J_a^T(x, y, \theta) \begin{bmatrix} F_x \\ F_y \end{bmatrix} = \begin{bmatrix} F_x \\ F_y \\ -F_x(a_x \sin\theta - a_y \cos\theta) + F_y(a_x \cos\theta - a_y \sin\theta) \end{bmatrix}$$

The bottom entry in this vector corresponds to the torque exerted about the origin of the robot frame.

◇

The total artificial joint torque acting on the arm is the sum of the artificial joint torques that result from all attractive and repulsive potentials

$$\tau(q) = \sum_i J_{o_i}^T(q) F_{\text{att},i}(q) + \sum_i J_{o_i}^T(q) F_{\text{rep},i}(q) \tag{5.11}$$

It is essential that we add the joint torques and *not* the workspace forces. In other words, we must use the Jacobians to transform forces to joint torques before we combine the effects of the potential fields. For example, Figure 5.8 shows a case in which two workspace forces F_1 and F_2, act on opposite corners of a rectangle. It is easy to see that $F_1 + F_2 = 0$, but that the combination of these forces produces a pure torque about the center of the rectangle.

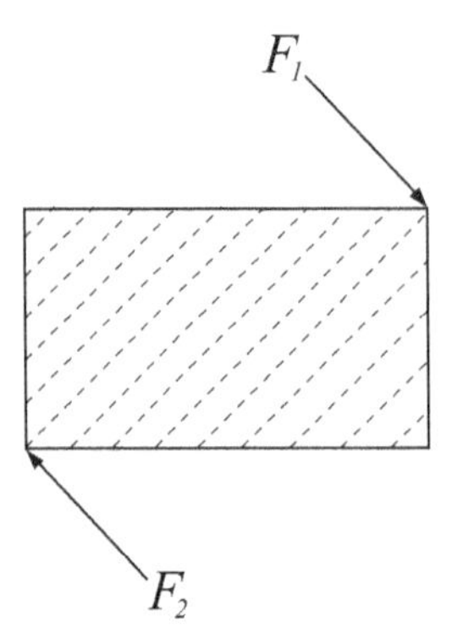

Figure 5.8: This example illustrates why forces must be mapped to the configuration space before they are combined. The two forces illustrated in the figure are vectors of equal magnitude in opposite directions. Vector addition of these two forces produces zero net force, but there is a net torque induced by these forces.

Example 5.7 Two-link planar arm

Consider again the two-link planar arm of Example 5.3, with joint torques as determined in Example 5.5. In this case the total joint torque induced by the attractive and repuslive workspace potential fields is given by

$$\begin{aligned} \tau(q_s) &= \tau_{att,1}(q_s) + \tau_{att,2}(q_s) + \tau_{rep,2}(q_s) \\ &= \begin{bmatrix} 1 \\ 0 \end{bmatrix} + \begin{bmatrix} 2 \\ 1 \end{bmatrix} + \begin{bmatrix} -8 \\ -4 \end{bmatrix} = \begin{bmatrix} -5 \\ -3 \end{bmatrix} \end{aligned}$$

These joint torques have the effect of causing each joint to rotate in a clockwise direction, away from the goal, due to the close proximity of o_2 to the obstacle. By choosing a smaller value for η_2, this effect can be overcome.
◇

5.2.4 Gradient Descent Planning

Gradient descent is a well known approach for solving optimization problems. The idea is simple. Starting at the initial configuration, take a small step in the direction of the negative gradient (which is the direction that decreases the potential as quickly as possible). This gives a new configuration, and the process is repeated until the final configuration is reached. More formally, we can define a gradient descent algorithm as follows.

Gradient Descent Algorithm

1. $q^0 \leftarrow q_s,\ i \leftarrow 0$
2. IF $||q^i - q_f|| > \epsilon$

$$q^{i+1} \leftarrow q^i + \alpha^i \frac{\tau(q^i)}{||\tau(q^i)||}$$

$$i \leftarrow i + 1$$

   ELSE return $< q^0, q^1, \ldots, q^i >$

3. GO TO 2

In this algorithm the notation q^i is used to denote the value of q at the ith iteration (not the ith componenent of the vector q) and the final path consists of the sequence of iterates $< q^0, q^1, \ldots, q^i >$. The value of the scalar α^i determines the step size at the ith iteration; it is multiplied by the unit vector in the direction of the resultant force. It is important that α^i be small enough that the robot is not allowed to "jump into" obstacles while being large enough that the algorithm does not require excessive computation time. In motion planning problems the choice for α^i is often made on an ad hoc or empirical basis, perhaps based on the distance to the nearest obstacle or to the goal. A number of systematic methods for choosing α^i can be found in the optimization literature.

It is unlikely that we will ever exactly satisfy the condition $q^i = q_f$ and for this reason the algorithm terminates in line 2 when q^i is sufficiently near the goal configuration q_f. We choose ϵ to be a sufficiently small constant, based on the task requirements.

There are a number of design choices that must be made when using this algorithm.

ζ_i in Equation (5.4) controls the relative influence of the attractive potential for control point o_i. It is not necessary that all of the ζ_i be set to the same value. Typically we assign a larger weight to one of the o_i than to the others, producing a "follow the leader" type of motion, in which the leader o_i is quickly attracted to its final position and the robot then reorients itself so that the other o_i reach their final positions.

η_i in Equation (5.6) controls the relative influence of the repulsive potential for o_i. As with the ζ_i it is not necessary that all of the η_i be set to the same value. In particular, we typically set the value of η_i to be much smaller for obstacles that are near the goal position of the robot (to avoid having these obstacles repel the robot from the goal).

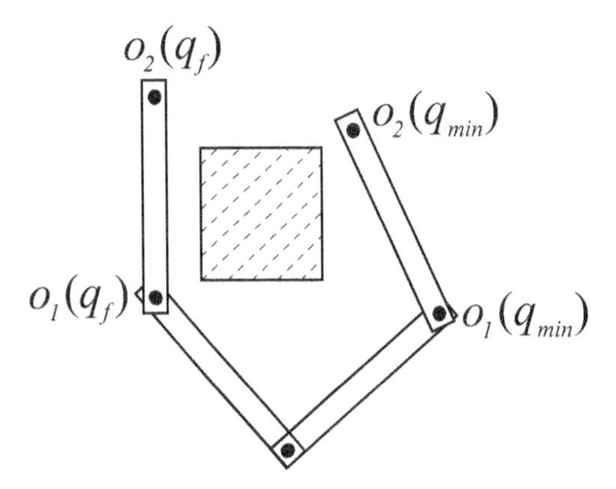

Figure 5.9: The configuration q_{min} is a local minimum in the potential field. At q_{min} the attractive force exactly cancels the repulsive force and the planner fails to make further progress.

ρ_0 in Equation (5.6) defines the distance of influence for obstacles. As with the η_i we can define a distinct value of ρ_0 for each obstacle. In particular, we do not want any obstacle's region of influence to include the goal position of any repulsive control point. We may also wish to assign distinct values of ρ_0 to the obstacles to avoid the possibility of overlapping regions of influence for distinct obstacles.

The problem that plagues all gradient descent algorithms is the possible existence of local minima in the potential field. For appropriate choice of α^i, it can be shown that the gradient descent algorithm is guaranteed to converge to a minimum in the field, but there is no guarantee that this minimum will be the global minimum. In our case this implies that there is no guarantee that this method will find a path to q_{f}. An example of this situation is shown in Figure 5.9. We will discuss ways to deal with this below in Section 5.3.

5.3 ESCAPING LOCAL MINIMA

As noted above, one problem that plagues artificial potential field methods for path planning is the existence of local minima in the potential field. In the case of articulated manipulators the resultant field U is the sum of many attractive and repulsive fields, and this combination is likely to yield many local minima. This problem has long been known in the optimization community, where probabilistic methods such as simulated annealing have been developed to cope with it. Similarly, the robot path planning community has developed what are known as **randomized methods** to deal with this and other problems.

The first method we discuss for escaping local minima combines gradient descent with randomization. This approach uses gradient descent until the planner finds itself stuck in a local minimum, and then uses a random walk to escape the local minimum. The algorithm is a slight modification of the gradient descent algorithm of Section 5.2.4:

1. $q^0 \leftarrow q_s$, $i \leftarrow 0$
2. IF $||q^i - q_f|| > \epsilon$

   $$q^{i+1} \leftarrow q^i + \alpha^i \frac{\tau(q^i)}{||\tau(q^i)||}$$

   $i \leftarrow i+1$

   ELSE return $< q^0, q^1, \ldots, q^i >$
3. IF stuck in a local minimum

   execute a random walk, ending at q'

   $q^{i+1} \leftarrow q'$
4. GO TO 2

The two new problems that must be solved are determining when the planner is stuck in a local minimum and defining the random walk. Typically, a heuristic is used to recognize a local minimum. For example, if several successive q^i lie within a small region of the configuration space, it is likely that there is a nearby local minimum (for example, if for some small positive ϵ_m we have $||q^i - q^{i+1}|| < \epsilon_m$, $||q^i - q^{i+2}|| < \epsilon_m$, and $||q^i - q^{i+3}|| < \epsilon_m$ then assume q^i is near a local minimum).

Defining the random walk requires a bit more care. One approach is to simulate Brownian motion. The random walk consists of t random steps. A random step from $q = (q_1, \ldots, q_n)$ is obtained by randomly adding a small fixed constant to each q_i,

$$q_{\text{random-step}} = (q_1 \pm v_1, \ldots, q_n \pm v_n)$$

with v_i a fixed small constant and the probability of adding $+v_i$ or $-v_i$ equal to 1/2 (that is, a uniform distribution). Without loss of generality, assume that $q = 0$. We can use probability theory to characterize the behavior of the random walk consisting of t random steps. In particular, the probability density function for $q' = (q_1, \ldots, q_n)$ is given by

$$p_i(q_i, t) \approx \frac{1}{v_i\sqrt{2\pi t}} \exp\left(-\frac{q_i^2}{2v_i^2 t}\right) \tag{5.12}$$

which is a zero mean Gaussian density function[2] with variance $v_i^2 t$. This is a result of the central limit theorem, which states that the probability density function for the sum of k independent, identically distributed random variables tends to a Gaussian density function as $k \to \infty$. The variance $v_i^2 t$ essentially determines the range of the random walk. If certain characteristics of local minima (for example, the size of the basin of attraction) are known in advance, these can be used to select the parameters v_i and t. Otherwise, they can be determined empirically or based on weak assumptions about the potential field.

5.4 PROBABILISTIC ROADMAP METHODS

The potential field approaches described above incrementally explore $\mathcal{Q}_{\text{free}}$, searching for a path from q_{s} to q_{f}. At termination, these planners return a single path. Thus, if multiple path planning problems must be solved, such a planner must be applied once for each problem. An alternative approach is to construct a representation of $\mathcal{Q}_{\text{free}}$ that can be used to quickly generate paths when new path planning problems arise. This is useful, for example, when a robot operates for a prolonged period in a single workspace.

One such representation is a known as **configuration space roadmap**. A roadmap is a one-dimensional network of curves that effectively represents $\mathcal{Q}_{\text{free}}$. Using roadmap methods, planning comprises three stages: (1) find a path from q_{s} to a configuration q_a in the roadmap, (2) find a path from q_{f} to a configuration q_b in the roadmap, (3) find a path in the roadmap from q_a to q_b. Steps (1) and (2) are typically much easier than finding a path from q_{s} to q_{f}.

In this section, we will describe **probabilistic roadmaps (PRMs)**. A PRM is a network of simple curve segments, or arcs, that meet at nodes. Each node corresponds to a configuration. Each arc between two nodes corresponds to a collision free path between two configurations. Constructing a PRM is a conceptually straightforward process. First, a set of random configurations is generated to serve as the nodes in the roadmap. Then, a simple, local path planner is used to generate paths that connect pairs of configurations. Finally, if the initial roadmap consists of multiple connected components,[3] it is augmented by an enhancement phase, in which

[2] A Gaussian density function is the classical bell shaped curve. The mean indicates the center of the curve (the peak of the bell) and the variance indicates the width of the bell. The probability density function (pdf) tells how likely it is that the variable q_i will lie in a certain interval. The higher the pdf values, the more likely that q_i will lie in the corresponding interval.

[3] A connected component is a maximal subnetwork of the network such that a path

new nodes and arcs are added in an attempt to connect disjoint components of the roadmap. To solve a path planning problem, the simple, local planner is used to connect q_s and q_f to the roadmap, and the resulting roadmap is searched for a path from q_s to q_f. These four steps are illustrated in Figure 5.10. We now discuss these steps in more detail.

5.4.1 Sampling the Configuration Space

The simplest way to generate sample configurations is with uniform random sampling of the configuration space. Sample configurations that lie in $\mathcal{QO}$ are discarded. A simple collision checking algorithm can determine when this is the case. The disadvantage of this approach is that the number of samples it places in any particular region of $\mathcal{Q}_{\text{free}}$ is proportional to the volume of the region. Therefore, uniform sampling is unlikely to place samples in narrow passages of $\mathcal{Q}_{\text{free}}$. In the PRM literature, this is refered to as the **narrow passage problem.** It can be dealt with either by using more intelligent sampling schemes, or by using an enhancement phase during the construction of the PRM. In this section, we discuss the latter option.

5.4.2 Connecting Pairs of Configurations

Given a set of nodes that correspond to configurations, the next step in building the PRM is to determine which pairs of nodes should be connected by a simple path. The typical approach is to attempt to connect each node to its k nearest neighbors, with k a parameter chosen by the user. Of course, to define the nearest neighbors, a distance function is required. Table 5.1 lists four distance functions that have been popular in the PRM literature. In this table, q and q' are the two configurations corresponding to different nodes in the roadmap, q_i refers to the value of the ith joint variable, $\mathcal{A}$ is a set of reference points on the robot, and $\mathrm{p}(q)$ refers to the workspace reference point p at configuration q. Of these, the simplest, and perhaps most commonly used, is the 2-norm in configuration space.

Once pairs of neighboring nodes have been identified, a simple local planner is used to connect these nodes. Often, a straight line in configuration space is used as the candidate plan, and thus, planning the path between two nodes is reduced to collision checking along a straight line path in the configuration space. If a collision occurs on this path, it can be discarded, or a more sophisticated planner (for example, the planner described in Section 5.3) can be used to attempt to connect the nodes.

---

exists in the subnetwork between any two nodes.

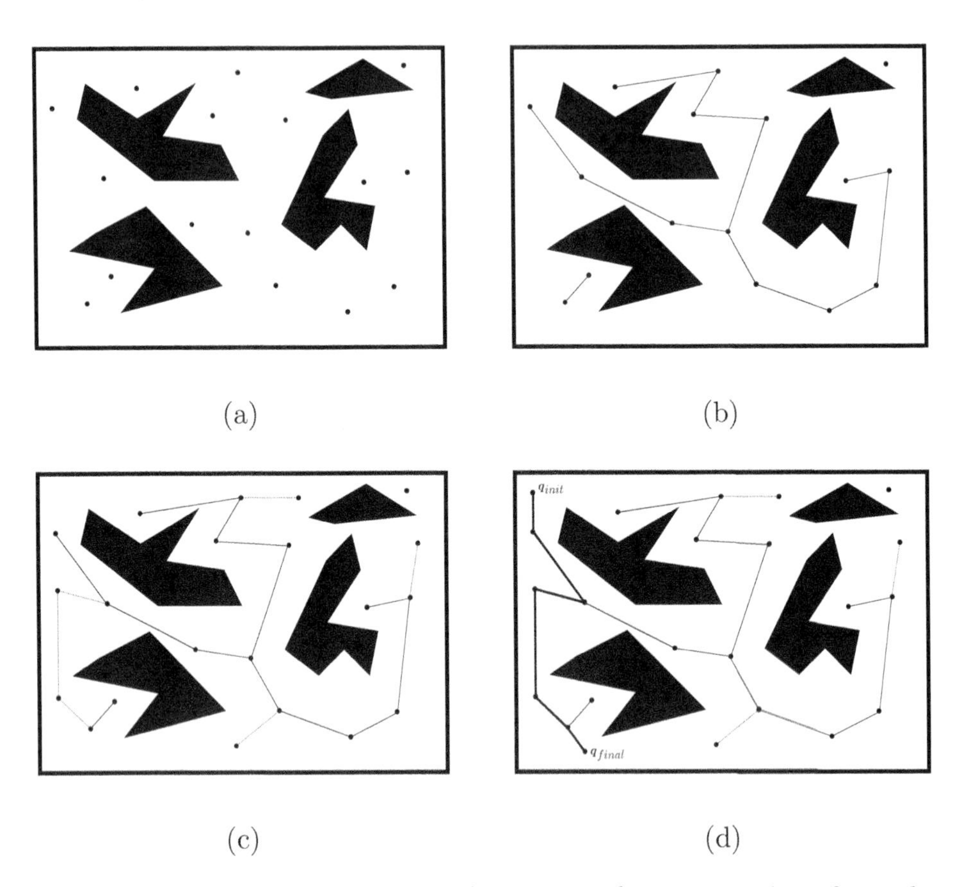

Figure 5.10: This figures illustrates the steps in the construction of a probabilistic roadmap for a two-dimensional configuration space containing polygonal obstacles. (a) First, a set of random samples is generated in the configuration space. Only collision-free samples are retained. (b) Each sample is connected to its nearest neighbors using a simple, straight-line path. If such a path causes a collision, the corresponding samples are not connected in the roadmap. (c) Since the initial roadmap contains multiple connected components, additional samples are generated and connected to the roadmap in an enhancement phase. (d) A path from q_{s} to q_{f} is found by connecting q_{s} and q_{f} to the roadmap and then searching this augmented roadmap for a path from q_{s} to q_{f}.

2-norm in $\mathcal{Q}$:	$\Vert q' - q \Vert = \left[\sum_{i=1}^{n}(q'_i - q_i)^2\right]^{\frac{1}{2}}$
∞-norm in $\mathcal{Q}$:	$\max_n \vert q'_i - q_i \vert$
2-norm in workspace:	$\left[\sum_{\mathrm{p}\in\mathcal{A}} \left\Vert \mathrm{p}(q') - \mathrm{p}(q) \right\Vert^2\right]^{\frac{1}{2}}$
∞-norm in workspace:	$\max_{\mathrm{p}\in\mathcal{A}} \left\Vert \mathrm{p}(q') - \mathrm{p}(q) \right\Vert$

Table 5.1: Four commonly used distance functions.

The simplest approach to collision detection along the straight line path is to sample the path at a sufficiently fine discretization, and to check each sample for collision. This method works, provided the discretization is fine enough, but it is very inefficient. This is because many of the computations required to check for collision at one sample are repeated for the next sample (assuming that the robot has moved only a small amount between the two configurations). For this reason, incremental collision detection approaches have been developed. While these approaches are beyond the scope of this text, a number of collision detection software packages are available in the public domain. Most developers of robot motion planners use one of these packages, rather than implementing their own collision detection routines.

5.4.3 Enhancement

After the initial PRM has been constructed, it is likely that it will consist of multiple connected components. Often these individual components lie in large regions of $\mathcal{Q}_{\text{free}}$ that are connected by narrow passages in $\mathcal{Q}_{\text{free}}$. The goal of the enhancement process is to connect as many of these disjoint components as possible.

One approach to enhancement is merely to attempt to connect pairs of nodes in two disjoint components, perhaps by using a more sophisticated planner such as described in Section 5.3. A common approach is to identify the largest connected component, and to attempt to connect the smaller components to it. The node in the smaller component that is closest to the larger component is typically chosen as the candidate for connection. A second approach is to choose a node randomly as a candidate for connection, and to bias the random choice based on the number of neighbors of the node; a node with fewer neighbors in the roadmap is more likely to be near a narrow passage, and should be a more likely candidate for connection.

Another approach to enhancement is to add more random nodes to the roadmap, in the hope of finding nodes that lie in or near the narrow passages. One approach is to identify nodes that have few neighbors, and to generate sample configurations in regions around these nodes. The local planner is then used to attempt to connect these new configurations to the roadmap.

5.4.4 Path Smoothing

After the PRM has been generated, path planning amounts to connecting q_{s} and q_{f} to the roadmap using the local planner, and then performing path smoothing, since the resulting path will be composed of straight line segments in the configuration space. The simplest path smoothing algorithm is to select two random points on the path and try to connect them with the local planner. This process is repeated until until no significant progress is made.

5.5 TRAJECTORY PLANNING

Recall that a path from q_{s} to q_{f} in configuration space is defined as a continuous map, $\gamma : [0,1] \rightarrow \mathcal{Q}$, with $\gamma(0) = q_{\mathrm{s}}$ and $\gamma(1) = q_{\mathrm{f}}$. A **trajectory** is a function of time $q(t)$ such that $q(t_0) = q_{\mathrm{s}}$ and $q(t_f) = q_{\mathrm{f}}$. In this case, $t_f - t_0$ represents the amount of time taken to execute the trajectory. Since the trajectory is parameterized by time, we can compute velocities and accelerations along the trajectories by differentiation. If we think of the argument to γ as a time variable, then a path is a special case of a trajectory, one that will be executed in one unit of time. In other words, in this case γ gives a complete specification of the robot's trajectory, including the time derivatives (since one need only differentiate γ to obtain these).

As seen above, a path planning algorithm will not typically give the map γ; it will give only a sequence of points (called **via points**) along the path. This is also the case for other ways in which a path could be specified. In some cases, paths are specified by giving a sequence of end-effector poses, $T_6^0(k\Delta t)$. In this case, the inverse kinematic solution must be used to convert this to a sequence of joint configurations. A common way to specify paths for industrial robots is to physically lead the robot through the desired motion with a teach pendant, the so-called **teach and playback mode**. In some

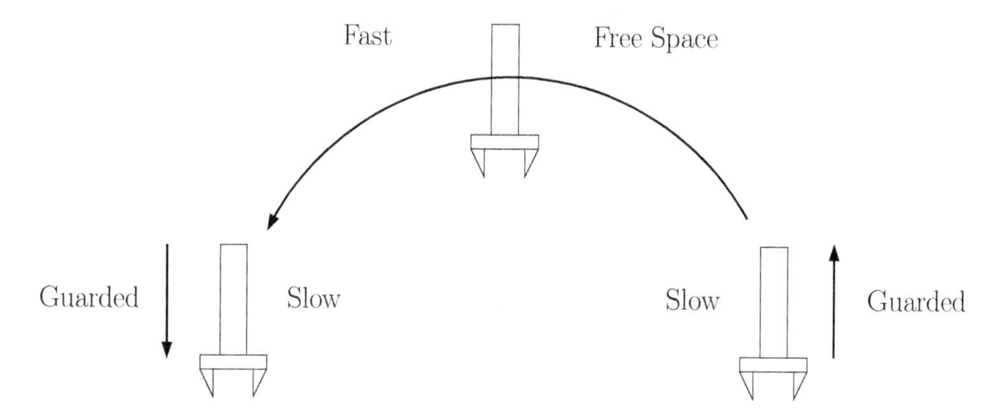

Figure 5.11: Often the end effector trajectory can be decomposed into initial and final guarded motions that are executed at low speeds, and a free motion that is executed at high speed.

cases, this may be more efficient than deploying a path planning system, for example, in static environments when the same path will be executed many times. In this case, there is no need for calculation of the inverse kinematics; the desired motion is simply recorded as a set of joint angles (actually as a set of encoder values).

It is often the case that a manipulator motion can be decomposed into segments consisting of free and **guarded motions**, such as shown in Figure 5.11. During the free motion the manipulator can move very fast, since no obstacles are near by, but at the start and end of the motion, care must be taken to avoid obstacles.

Below, we first consider **point to point** motion. In this case the task is to plan a trajectory from an initial configuration $q(t_0)$ to a final configuration $q(t_f)$. In some cases, there may be constraints on the trajectory (for example, if the robot must start and end with zero velocity). Nevertheless, it is easy to realize that there are infinitely many trajectories that will satisfy a finite number of constraints on the endpoints. It is common practice therefore to choose trajectories from a finitely parameterizable family, for example, polynomials of degree n, where n depends on the number of constraints to be satisfied. This is the approach that we will take in this text. Once we have seen how to construct trajectories between two configurations, it is straightforward to generalize the method to the case of trajectories specified by multiple via points.

5.5.1 Trajectories for Point to Point Motion

As described above, the problem is to find a trajectory that connects the initial and final configurations while satisfying other specified constraints at the endpoints, such as velocity and/or acceleration constraints. Without loss of generality, we will consider planning the trajectory for a single joint, since the trajectories for the remaining joints will be created independently and in exactly the same way. Thus, we will concern ourselves with the problem of determining $q(t)$, where $q(t)$ is a scalar joint variable.

We suppose that at time t_0 the joint variable satisfies

$$q(t_0) = q_0 \tag{5.13}$$
$$\dot{q}(t_0) = v_0 \tag{5.14}$$

and we wish to attain the values at t_f

$$q(t_f) = q_f \tag{5.15}$$
$$\dot{q}(t_f) = v_f \tag{5.16}$$

Figure 5.12 shows a suitable trajectory for this motion. In addition, we may wish to specify the constraints on initial and final accelerations. In this case we have two additional equations

$$\ddot{q}(t_0) = \alpha_0 \tag{5.17}$$
$$\ddot{q}(t_f) = \alpha_f \tag{5.18}$$

Below we will investigate several specific ways to compute trajectories using low order polynomials. We begin with cubic polynomials, which allow specification of initial and final positions and velocities. We then describe quintic polynomial trajectories, which also allow the specification of the initial and final accelerations. After describing these two general polynomial trajectories, we describe trajectories that are pieced together from segments of constant acceleration.

Cubic Polynomial Trajectories

Consider first the case where we wish to generate a polynomial joint trajectory between two configurations, and that we wish to specify the start and end velocities for the trajectory. This gives four constraints that the trajectory must satisfy. Therefore, at a minimum we require a polynomial with four independent coefficients that can be chosen to satisfy these constraints. Thus, we consider a cubic trajectory of the form

$$q(t) = a_0 + a_1 t + a_2 t^2 + a_3 t^3 \tag{5.19}$$

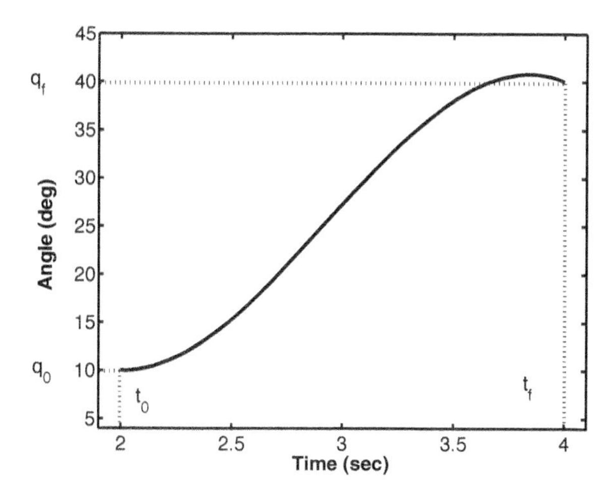

Figure 5.12: A typical joint space trajectory.

Then the desired velocity is given as

$$\dot{q}(t) = a_1 + 2a_2t + 3a_3t^2 \tag{5.20}$$

Combining Equations (5.19) and (5.20) with the four constraints yields four equations in four unknowns

$$\begin{aligned} q_0 &= a_0 + a_1t_0 + a_2t_0^2 + a_3t_0^3 \\ v_0 &= a_1 + 2a_2t_0 + 3a_3t_0^2 \\ q_f &= a_0 + a_1t_f + a_2t_f^2 + a_3t_f^3 \\ v_f &= a_1 + 2a_2t_f + 3a_3t_f^2 \end{aligned}$$

These four equations can be combined into a single matrix equation

$$\begin{bmatrix} 1 & t_0 & t_0^2 & t_0^3 \\ 0 & 1 & 2t_0 & 3t_0^2 \\ 1 & t_f & t_f^2 & t_f^3 \\ 0 & 1 & 2t_f & 3t_f^2 \end{bmatrix} \begin{bmatrix} a_0 \\ a_1 \\ a_2 \\ a_3 \end{bmatrix} = \begin{bmatrix} q_0 \\ v_0 \\ q_f \\ v_f \end{bmatrix} \tag{5.21}$$

It can be shown (Problem 5-17) that the determinant of the coefficient matrix in Equation (5.21) is equal to $(t_f - t_0)^4$ and, hence, Equation (5.21) always has a unique solution provided a nonzero time interval is allowed for the execution of the trajectory.

Example 5.8 Cubic Polynomial Trajectory

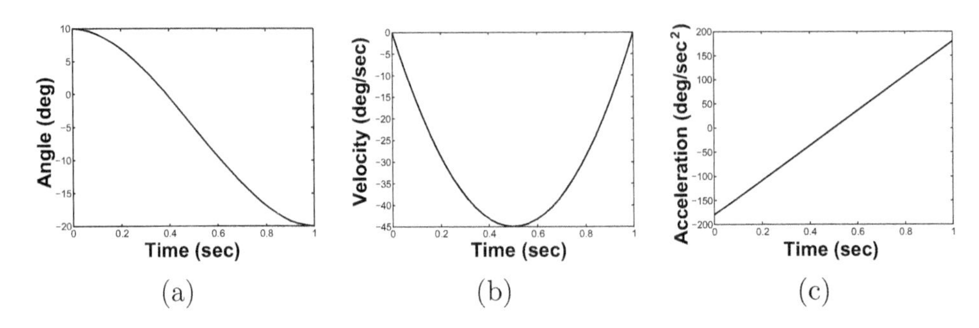

Figure 5.13: (a) Cubic polynomial trajectory. (b) Velocity profile for cubic polynomial trajectory. (c) Acceleration profile for cubic polynomial trajectory.

As an illustrative example, we may consider the special case that the initial and final velocities are zero. Suppose we take $t_0 = 0$ and $t_f = 1$ sec, with

$$v_0 = 0 \qquad v_f = 0$$

Thus, we want to move from the initial position q_0 to the final position q_f in 1 second, starting and ending with zero velocity. From Equation (5.21) we obtain

$$\begin{bmatrix} 1 & 0 & 0 & 0 \\ 0 & 1 & 0 & 0 \\ 1 & 1 & 1 & 1 \\ 0 & 1 & 2 & 3 \end{bmatrix} \begin{bmatrix} a_0 \\ a_1 \\ a_2 \\ a_3 \end{bmatrix} = \begin{bmatrix} q_0 \\ 0 \\ q_f \\ 0 \end{bmatrix}$$

This is then equivalent to the four equations

$$\begin{aligned} a_0 &= q_0 \\ a_1 &= 0 \\ a_2 + a_3 &= q_f - q_0 \\ 2a_2 + 3a_3 &= 0 \end{aligned}$$

These latter two equations can be solved to yield

$$\begin{aligned} a_2 &= 3(q_f - q_0) \\ a_3 &= -2(q_f - q_0) \end{aligned}$$

The required cubic polynomial function is therefore

$$q(t) = q_0 + 3(q_f - q_0)t^2 - 2(q_f - q_0)t^3$$

The corresponding velocity and acceleration curves are given as

$$\begin{aligned}\dot{q}(t) &= 6(q_f - q_0)t - 6(q_f - q_0)t^2 \\ \ddot{q}(t) &= 6(q_f - q_0) - 12(q_f - q_0)t\end{aligned}$$

Figure 5.13 shows these trajectories with $q_0 = 10°$, $q_f = -20°$.
◇

Quintic Polynomial Trajectories

As can be seen in in Figure 5.13, a cubic trajectory gives continuous positions and velocities at the start and finish points times but discontinuities in the acceleration. The derivative of acceleration is called the **jerk**. A discontinuity in acceleration leads to an impulsive jerk, which may excite vibrational modes in the manipulator and reduce tracking accuracy. For this reason, one may wish to specify constraints on the acceleration as well as on the position and velocity. In this case, we have six constraints (one each for initial and final configurations, initial and final velocities, and initial and final accelerations). Therefore we require a fifth order polynomial

$$q(t) = a_0 + a_1 t + a_2 t^2 + a_3 t^3 + a_4 t^4 + a_5 t^5 \tag{5.22}$$

Using Equations (5.13) – (5.18) and taking the appropriate number of derivatives we obtain the following equations,

$$\begin{aligned}q_0 &= a_0 + a_1 t_0 + a_2 t_0^2 + a_3 t_0^3 + a_4 t_0^4 + a_5 t_0^5 \\ v_0 &= a_1 + 2a_2 t_0 + 3a_3 t_0^2 + 4a_4 t_0^3 + 5a_5 t_0^4 \\ \alpha_0 &= 2a_2 + 6a_3 t_0 + 12a_4 t_0^2 + 20a_5 t_0^3 \\ q_f &= a_0 + a_1 t_f + a_2 t_f^2 + a_3 t_f^3 + a_4 t_f^4 + a_5 t_f^5 \\ v_f &= a_1 + 2a_2 t_f + 3a_3 t_f^2 + 4a_4 t_f^3 + 5a_5 t_f^4 \\ \alpha_f &= 2a_2 + 6a_3 t_f + 12a_4 t_f^2 + 20a_5 t_f^3\end{aligned}$$

which can be written as

$$\begin{bmatrix} 1 & t_0 & t_0^2 & t_0^3 & t_0^4 & t_0^5 \\ 0 & 1 & 2t_0 & 3t_0^2 & 4t_0^3 & 5t_0^4 \\ 0 & 0 & 2 & 6t_0 & 12t_0^2 & 20t_0^3 \\ 1 & t_f & t_f^2 & t_f^3 & t_f^4 & t_f^5 \\ 0 & 1 & 2t_f & 3t_f^2 & 4t_f^3 & 5t_f^4 \\ 0 & 0 & 2 & 6t_f & 12t_f^2 & 20t_f^3 \end{bmatrix} \begin{bmatrix} a_0 \\ a_1 \\ a_2 \\ a_3 \\ a_4 \\ a_5 \end{bmatrix} = \begin{bmatrix} q_0 \\ v_0 \\ \alpha_0 \\ q_f \\ v_f \\ \alpha_f \end{bmatrix} \tag{5.23}$$

Example 5.9 Quintic Polynomial Trajectory

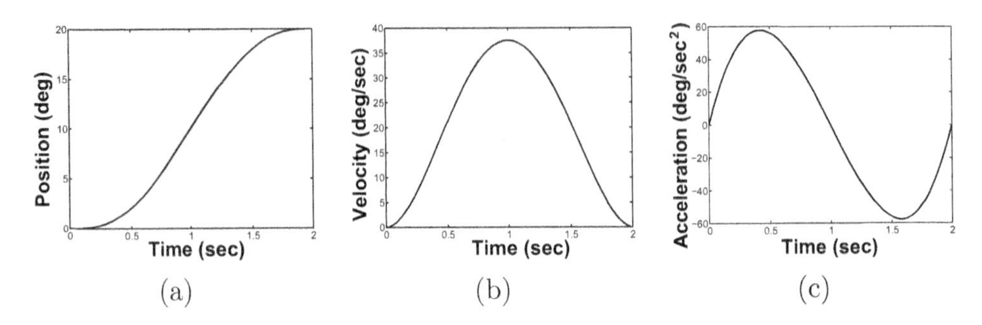

Figure 5.14: (a) Quintic polynomial trajectory, (b) its velocity profile, and (c) its acceleration profile.

Figure 5.14 shows a quintic polynomial trajectory with $q(0) = 0$, $q(2) = 40$ *with zero initial and final velocities and accelerations.*
◇

Linear Segments with Parabolic Blends (LSPB)

Another way to generate suitable joint space trajectories is by using so-called **Linear Segments with Parabolic Blends** (**LSPB**). This type of trajectory has a **Trapezoidal Velocity Profile** and is appropriate when a constant velocity is desired along a portion of the path. The LSPB trajectory is such that the velocity is initially "ramped up" to its desired value and then "ramped down" when it approaches the goal position. To achieve this we specify the desired trajectory in three parts. The first part from time t_0 to time t_b is a quadratic polynomial. This results in a linear "ramp" velocity. At time t_b, called the **blend time**, the trajectory switches to a linear function. This corresponds to a constant velocity. Finally, at $t_f - t_b$ the trajectory switches once again, this time to a quadratic polynomial so that the velocity is linear.

We choose the blend time t_b so that the position curve is symmetric as shown in Figure 5.15. For convenience suppose that $t_0 = 0$ and $\dot{q}(t_f) = 0 = \dot{q}(0)$. Then between times 0 and t_b we have

$$q(t) = a_0 + a_1 t + a_2 t^2$$

so that the velocity is

$$\dot{q}(t) \;=\; a_1 + 2a_2 t$$

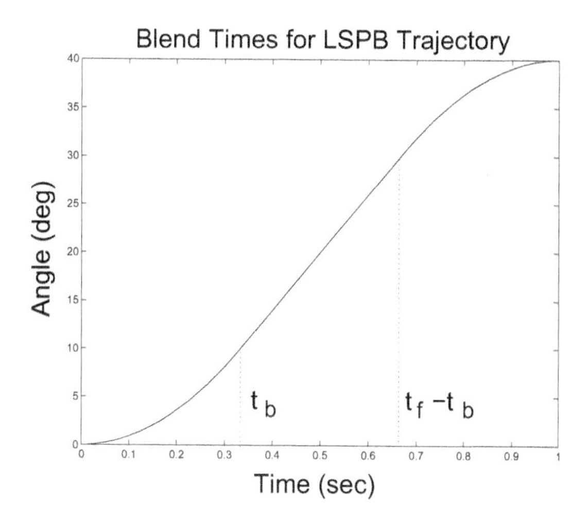

Figure 5.15: Blend times for LSPB trajectory.

The constraints $q_0 = 0$ and $\dot{q}(0) = 0$ imply that

$$\begin{aligned} a_0 &= q_0 \\ a_1 &= 0 \end{aligned}$$

At time t_b we want the velocity to equal a given constant, say V. Thus, we have

$$\dot{q}(t_b) = 2a_2 t_b = V$$

which implies that

$$a_2 = \frac{V}{2t_b}$$

Therefore the required trajectory between 0 and t_b is given as

$$\begin{aligned} q(t) &= q_0 + \frac{V}{2t_b}t^2 = q_0 + \frac{\alpha}{2}t^2 \\ \dot{q}(t) &= \frac{V}{t_b}t = \alpha t \\ \ddot{q} &= \frac{V}{t_b} = \alpha \end{aligned}$$

where α denotes the acceleration.

Now, between time t_b and $t_f - t_b$, the trajectory is a linear segment with velocity V

$$q(t) = q(t_b) + V(t - t_b)$$

Since, by symmetry,

$$q\left(\frac{t_f}{2}\right) = \frac{q_0 + q_f}{2}$$

we have

$$\frac{q_0 + q_f}{2} = q(t_b) + V(\frac{t_f}{2} - t_b)$$

which implies that

$$q(t_b) = \frac{q_0 + q_f}{2} - V(\frac{t_f}{2} - t_b)$$

Since the two segments must "blend" at time t_b we require

$$q_0 + \frac{V}{2}t_b = \frac{q_0 + q_f - Vt_f}{2} + Vt_b$$

which, upon solving for the blend time t_b, gives

$$t_b = \frac{q_0 - q_f + Vt_f}{V} \tag{5.24}$$

Note that we have the constraint $0 < t_b \leq \frac{t_f}{2}$. This leads to the inequality

$$\frac{q_f - q_0}{V} < t_f \leq \frac{2(q_f - q_0)}{V}$$

To put it another way we have the inequality

$$\frac{q_f - q_0}{t_f} < V \leq \frac{2(q_f - q_0)}{t_f}$$

Thus, the specified velocity must be between these limits or the motion is not possible.

The portion of the trajectory between $t_f - t_b$ and t_f is now found by symmetry considerations (Problem 5-21). The complete LSPB trajectory is given by

$$q(t) = \begin{cases} q_0 + \frac{\alpha}{2}t^2 & 0 \leq t \leq t_b \\ \\ \frac{q_f + q_0 - Vt_f}{2} + Vt & t_b < t \leq t_f - t_b \\ \\ q_f - \frac{\alpha t_f^2}{2} + \alpha t_f t - \frac{\alpha}{2}t^2 & t_f - t_b < t \leq t_f \end{cases} \tag{5.25}$$

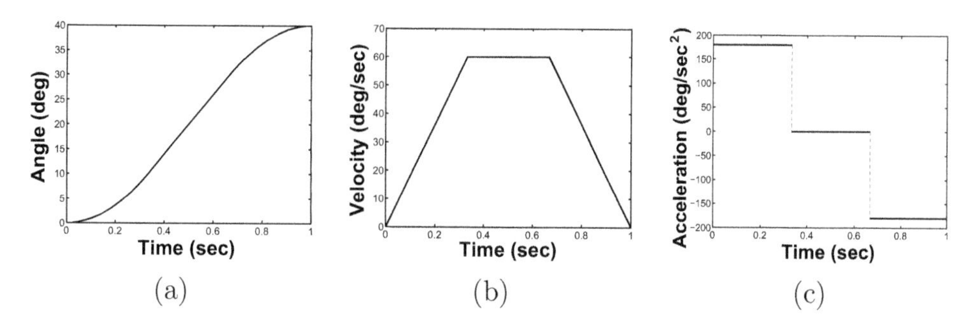

Figure 5.16: (a) LSPB trajectory. (b) Velocity profile for LSPB trajectory. (c) Acceleration profile for LSPB trajectory.

Figure 5.16(a) shows such an LSPB trajectory, where the maximum velocity $V = 60$. In this case $t_b = \frac{1}{3}$. The velocity and acceleration curves are given in Figures 5.16(b) and 5.16(c), respectively.

Minimum Time Trajectories

An important variation of the LSPB trajectory is obtained by leaving the final time t_f unspecified and seeking the "fastest" trajectory between q_0 and q_f with a given constant acceleration α, that is, the trajectory with the final time t_f a minimum. This is sometimes called a **bang-bang** trajectory since the optimal solution is achieved with the acceleration at its maximum value $+\alpha$ until an appropriate **switching time** t_s at which time it abruptly switches to its minimum value $-\alpha$ (maximum deceleration) from t_s to t_f.

Returning to our simple example in which we assume that the trajectory begins and ends at rest, that is, with zero initial and final velocities, symmetry considerations would suggest that the switching time t_s is just $\frac{t_f}{2}$. This is indeed the case. For nonzero initial and/or final velocities, the situation is more complicated and we will not discuss it here. If we let V_s denote the velocity at time t_s then we have $V_s = \alpha t_s$ and using Equation (5.24) with $t_b = t_s$ we obtain

$$t_s = \frac{q_0 - q_f + V_s t_f}{V_s}$$

The symmetry condition $t_s = \frac{t_f}{2}$ implies that

$$V_s = \frac{q_f - q_0}{t_s}$$

Combining these two we have the conditions

$$\frac{q_f - q_0}{t_s} = \alpha t_s$$

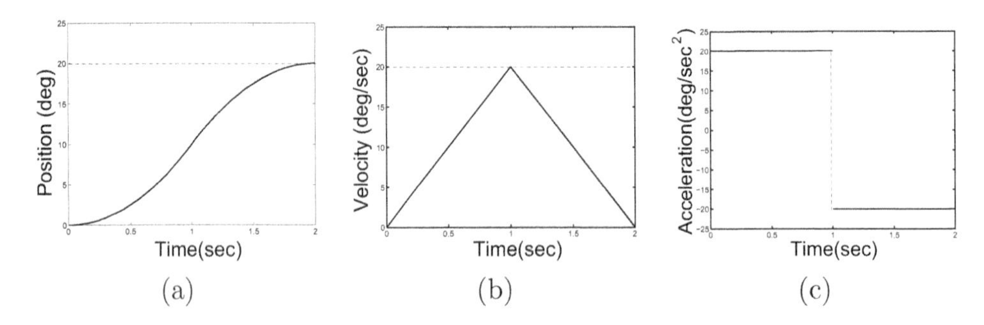

Figure 5.17: (a) Minimum-time trajectory. (b) Velocity profile for minimum-time trajectory. (c) Acceleration profile for minimum-time trajectory.

which implies that

$$t_s = \sqrt{\frac{q_f - q_0}{\alpha}}$$

Figure 5.17 shows the position, velocity, and acceleration for such a minimum time trajectory.

5.5.2 Trajectories for Paths Specified by Via Points

Now that we have examined the problem of planning a trajectory between two configurations, we generalize our approach to the case of planning a trajectory that passes through a sequence of configurations, called **via points**. Consider the simple example of a path specified by three points, q_0, q_1, and q_2, such that the via points are reached at times t_0, t_1, and t_2, respectively. If in addition to these three constraints we impose constraints on the initial and final velocities and accelerations, we obtain the following set of constraints,

$$\begin{aligned}
q(t_0) &= q_0 \\
\dot{q}(t_0) &= v_0 \\
\ddot{q}(t_0) &= \alpha_0 \\
q(t_1) &= q_1 \\
q(t_2) &= q_2 \\
\dot{q}(t_2) &= v_2 \\
\ddot{q}(t_2) &= \alpha_2
\end{aligned}$$

which could be satisfied by generating a trajectory using the sixth order polynomial

$$q(t) = a_0 + a_1 t + a_2 t^2 + a_3 t^3 + a_4 t^4 + a_5 t^5 + a_6 t^6 \tag{5.26}$$

One advantage to this approach is that, since $q(t)$ is continuously differentiable, we need not worry about discontinuities in either velocity or acceleration at the via point, q_1. However, to determine the coefficients for this polynomial, we must solve a linear system of dimension seven. The clear disadvantage to this approach is that as the number of via points increases, the dimension of the corresponding linear system also increases, making the method intractable when many via points are used.

An alternative to using a single high order polynomial for the entire trajectory is to use low order polynomials for trajectory segments between adjacent via points. These polynomials are sometimes referred to as interpolating polynomials or blending polynomials. With this approach, we must take care that velocity and acceleration constraints are satisfied at the via points, where we switch from one polynomial to another.

For the first segment of the trajectory, suppose that the initial and final times are t_0 and t_f, respectively, and the constraints on initial and final velocities are given by

$$\begin{array}{lcl} q(t_0) = q_0 & ; & q(t_f) = q_1 \\ \dot{q}(t_0) = v_0 & ; & \dot{q}(t_f) = v_1 \end{array} \tag{5.27}$$

the required cubic polynomial for this segment of the trajectory can be computed from

$$q(t_0) = a_0 + a_1(t - t_0) + a_2(t - t_0)^2 + a_3(t - t_0)^3 \tag{5.28}$$

where

$$\begin{aligned} a_0 &= q_0 \\ a_1 &= v_0 \\ a_2 &= \frac{3(q_1 - q_0) - (2v_0 + v_1)(t_f - t_0)}{(t_f - t_0)^2} \\ a_3 &= \frac{2(q_0 - q_1) + (v_0 + v_1)(t_f - t_0)}{(t_f - t_0)^3} \end{aligned}$$

A sequence of moves can be planned using the above formula by using the end conditions q_f, v_f of the ith move as initial conditions for the subsequent move.

Example 5.10 Three-segment Cubic Spline Trajectory

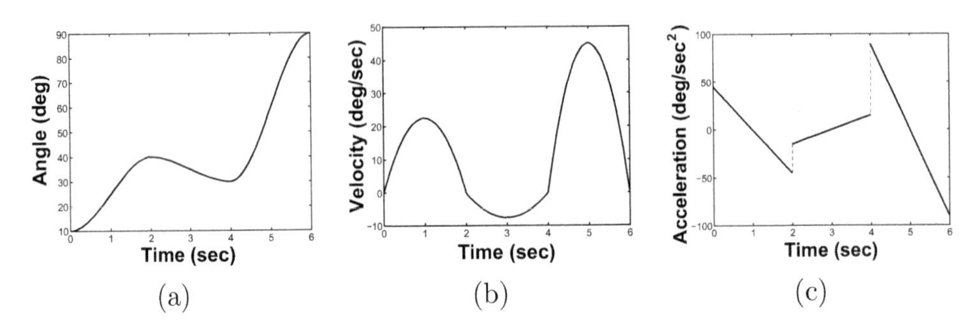

(a) (b) (c)

Figure 5.18: (a) Cubic spline trajectory made from three cubic polynomials. (b) Velocity profile for multiple cubic polynomial trajectory. (c) Acceleration profile for multiple cubic polynomial trajectory.

Figure 5.18 shows a 6-second move, computed in three parts using Equation (5.28), where the trajectory begins at 10° *and is required to reach* 40° *at* 2 *seconds,* 30° *at* 4 *seconds, and* 90° *at* 6 *seconds, with zero velocity at* 0, 2, 4, *and* 6 *seconds.*

◇

Example 5.11 Cubic Spline Trajectory with Blending Constraints

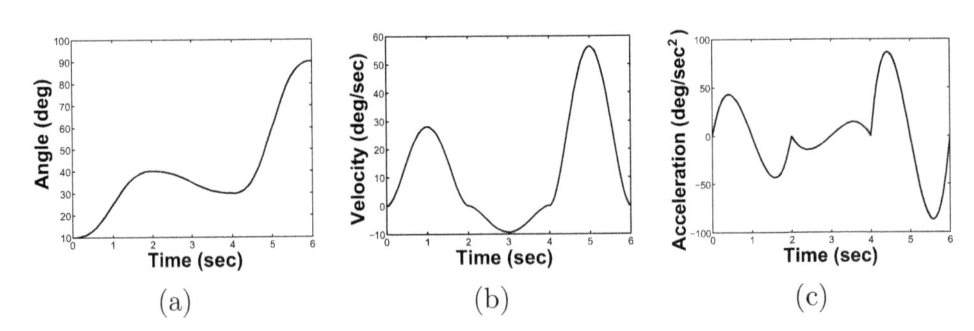

(a) (b) (c)

Figure 5.19: (a) Trajectory with multiple quintic segments. (b) Velocity profile for multiple quintic segments. (c) Acceleration profile for multiple quintic segments.

Figure 5.19 shows the same 6-second trajectory as in Example 5.10 with the added constraints that the accelerations should be zero at the blend times.

◇

5.6 SUMMARY

In this chapter we studied methods for generating collision-free trajectories for robot manipulators. We divided the problem into two parts, first computing a collinion-free path (represented by a sequence of set points), then by using interpolating polynomials to convert these paths into continuous trajectories.

We described two approaches to path planning, incremental search guided by artificial potential fields and probabilistic roadmap methods. In the first, artificial potential fields are constructed in the robot's workspace. These fields attract the origins of the DH frames to their goal positions, while repelling the arm from obstacles in the workspace. The negative gradients of these potentials define artificial forces, and these can be converted from workspace forces to joint torques using the relationship $\tau = J^T F$. Gradient descent methods can then be used to incrementally explore the configuration space in search of a collision-free path. Since gradient descent methods are prone to become trapped by local minima, we introduced randomized motion as a method for escaping local minima.

The second method of path planning constructs a roadmap in the configuration space using a random sampling scheme. A set of random samples are generated, and each of these is connected to a set of its nearest neighbors using a simple local motion planner (often a simple straight-line planner in the configuration space). Such a roadmap is called a probabilistic roadmap (PRM). Once the roadmap has been constructed, planning amounts to connecting the initial and goal configurations to the roadmap (again, using the simple, local planner), then searching the roadmap for a connecting path.

Given a sequence of set points, a trajectory can be constructed using a low-order polynomial defined in terms of initial and final conditions for joint variables and their derivatives. We described cubic and quintic trajectories, along with trajectories that are pieced together from segments of constant acceleration (including minimum time, or bang-bang trajectories).

PROBLEMS

5-1 Describe the configuration space for a mobile robot that can translate and rotate in the plane.

5-2 Describe the configuration space for the three-link manipulator shown in Figure 3.23.

5-3 Describe the configuration space for the two-link manipulator shown in Figure 3.24.

5-4 Describe the configuration space for the two-link manipulator shown in Figure 3.25.

5-5 Describe the configuration space for the three-link manipulator shown in Figure 3.26.

5-6 Describe the configuration space for a six-link anthropomorphic arm equipped with a spherical wrist.

5-7 Verify Equation (5.2).

5-8 Derive the equations needed to compute the shortest distance from a point p to the line segment in the plane with vertices a_1 and a_2.

5-9 Derive the equations needed to compute the shortest distance from a point p to the polygon in the plane with vertices a_i, $i = 1, \ldots, n$.

5-10 Derive the equations needed to compute the shortest distance from a point p to the polygon in three dimensions with vertices a_i, $i = 1, \ldots, n$.

5-11 Verify Equation (5.6).

5-12 Consider a simple polygonal robot with four vertices, such that at $q = (0, 0, 0)$ the vertices are located at $a_1(0) = (0, 0)$, $a_2(0) = (1, 0)$, $a_3(0) = (1, 1)$, and $a_4(0) = (0, 1)$. If two point obstacles are located at $o_1 = (3, 3)$ and $o_2 = (-3, -3)$, determine the artificial workspace and configuration space forces that act on the robot.

5-13 Write a computer program to implement the path planner described in Section 5.2 for a three-link planar arm moving among polygonal obstacles.

5-14 Write a simple computer program to perform collision checking for the case of a polygonal robot moving in the plane among polygonal obstacles. Your program should accept a configuration q as input, and should return a value that indicates whether q is a collision-free configuration.

5-15 Give a procedure for generating random samples of orientations in $SO(n)$ given that you have access to a random number generator that can generate samples from the uniform distribution on the unit interval. Your samples need not be uniformly distributed on $SO(n)$.

5-16 Write a computer program to implement the PRM planner described in Section 5.4 for a three-link planar arm moving among polygonal obstacles.

5-17 Show by direct calculation that the determinant of the coefficient matrix in Equation (5.21) is $(t_f - t_0)^4$.

5-18 Suppose we wish a manipulator to start from an initial configuration at time t_0 and track a conveyor. Discuss the steps needed in planning a suitable trajectory for this problem.

5-19 Suppose we desire a joint space trajectory $\dot{q}_i^d(t)$ for the ith joint (assumed to be revolute) that begins at rest at position q_0 at time t_0 and reaches position q_1 in 2 seconds with a final velocity of 1 radian/sec. Compute a cubic polynomial satisfying these constraints. Sketch the trajectory as a function of time.

5-20 Compute a LSPB trajectory to satisfy the same requirements as in Problem 5-19. Sketch the resulting position, velocity, and acceleration profiles.

5-21 Fill in the details of the computation of the LSPB trajectory. In other words compute the portion of the trajectory between times $t_f - t_b$ and t_f and hence verify Equations (5.25).

5-22 Write a Matlab m-file, lspb.m, to generate an LSPB trajectory, given appropriate initial data.

5-23 Rewrite the Matlab m-files, cubic.m, quintic.m, and lspb.m to turn them into Matlab functions. Document them appropriately.

NOTES AND REFERENCES

The earliest work on robot planning was done in the late sixties and early seventies in a few university-based Artificial Intelligence (AI) labs [32], [36], and [98]. This research dealt with high level planning using symbolic reasoning that was much in vogue at the time in the AI community. Geometry was not often explicitly considered in early robot planners, in part because it was not clear how to represent geometric constraints in a computationally feasible manner. The configuration space and its application to path planning were introduced in [79]. This was the first rigorous, formal treatment of the geometric path planning problem, and it initiated a surge in path planning research.

The earliest work in geometric path planning developed methods to construct volumetric representations of the free configuration space. These included exact methods [112], and approximate methods [15],[60], and [79]. In the former case, the best known algorithms have exponential complexity and require exact descriptions of both the robot and its environment, while in the latter case, the size of the representation of configuration space grows exponentially in the dimension of the configuration space. The best known algorithm for the path planning problem, giving an upper bound on the amount of computation time required to solve the problem, appeared in [16]. That real robots rarely have an exact description of the environment, and a drive for faster planning systems led to the development of potential fields approaches [64], and [66].

By the early nineties, a great deal of research had been done on the geometric path planning problem, and this work is nicely summarized in the textbook [73]. This textbook helped to generate a renewed interest in the path planning problem, and it provided a common framework in which to analyze and express path planning algorithms.

In the early nineties, randomization was introduced in the robot planning community [9], originally to circumvent the problems with local minima in potential fields. Early randomized motion planners proved effective for a large range of problems, but sometimes required extensive computation time for some robots in certain environments [62]. This limitation, together with the idea that a robot will operate in the same environment for a long period of time led to the development of the probabilistic roadmap planners [61, 100, 62].

A comprehensive review of motion planning research, including sensor-based approaches, can be found in [18].

Much work has been done in the area of collision detection in recent years [78], [91], [135], and [136]. This work is primarily focused on finding efficient, incremental methods for detecting collisions between objects when one or both are moving. A number of public domain collision detection software packages are currently available on the internet.

The artificial potential field method for motion planning has much in common with nonlinear optimization (for example, choosing step size α). A thorough discussion of nonlinear optimization can be found in [11].

Chapter 6

INDEPENDENT JOINT CONTROL

The control problem for robot manipulators is the problem of determining the time history of joint inputs required to cause the end effector to execute a commanded motion. The joint inputs may be joint forces and torques, or they may be inputs to the actuators, for example voltage inputs to the motors, depending on the model used for controller design. The commanded motion is typically specified either as a sequence of end-effector positions and orientations, or as a continuous path.

There are many control techniques and methodologies that can be applied to the control of manipulators. The particular control method used can have a significant impact on the performance of the manipulator and consequently on the range of its possible applications. For example, continuous path tracking requires a different control architecture than does point-to-point motion.

In addition, the mechanical design of the manipulator itself will influence the type of control scheme needed. For example, the control problems encountered with a cartesian manipulator are fundamentally different from those encountered with an elbow manipulator. This creates a so-called **hardware/software trade-off** between the mechanical structure of the system and the architecture/programming of the controller.

Technological improvements are continually being made in the mechanical design of robots, which in turn improves their performance potential and broadens their range of applications. Realizing this increased performance, however, requires more sophisticated approaches to control. One can draw an analogy to the aerospace industry. Early aircraft were relatively easy to fly but possessed limited performance capabilities. As performance increased with technological advances, so did the problems of control to the extent that the latest vehicles, such as the space shuttle or forward swept wing fighter aircraft, cannot be flown without sophisticated computer control.

As an illustration of the effect of the mechanical design on the control problem, we may compare a robot actuated by permanent magnet DC motors with gear reduction to a direct-drive robot using high-torque motors with no gear reduction. In the first case, the motor dynamics are linear and well understood and the effect of the gear reduction is largely to decouple the system by reducing the inertia coupling among the joints. However, the presence of the gears introduces friction, drive train compliance, and backlash.

In the case of a direct-drive robot, the problems of backlash, friction, and compliance due to the gears are eliminated. However, the coupling among the links is now significant, and the dynamics of the motors themselves may be much more complex. The result is that in order to achieve high performance from this type of manipulator, a different set of control problems must be addressed.

In this chapter we consider the simplest type of control strategy, namely, independent joint control. In this type of control each axis of the manipulator is controlled as a single-input/single-output (SISO) system. Any coupling effects due to the motion of the other links are treated as disturbances.

The basic structure of a single-input/single-output feedback control system is shown in Figure 6.1.

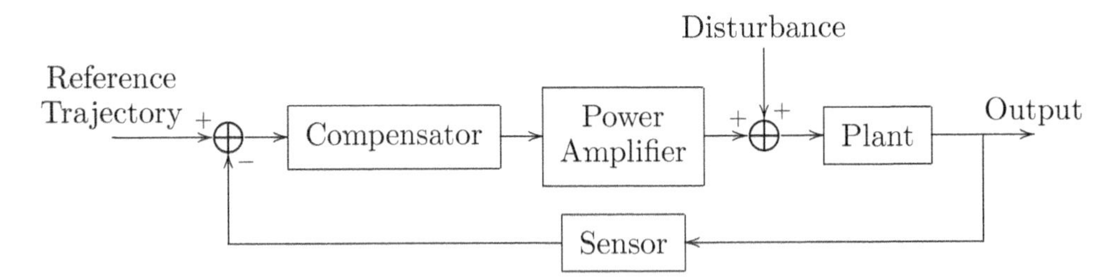

Figure 6.1: Basic structure of a feedback control system. The compensator measures the "error" between a "reference" and a measured "output" and produces signals to the plant that are designed to drive the error to zero despite the presence of disturbances.

The design objective is to choose the compensator in such a way that the plant output "tracks" or follows a desired output, given by the reference signal. The control signal, however, is not the only input acting on the system. Disturbances, which are really inputs that we do not control, also influence the behavior of the output. Therefore, the controller must be designed, in addition, so that the effects of the disturbances on the plant output are reduced. If this is accomplished, the plant is said to "reject" the disturbances. The twin objectives of **tracking** and **disturbance rejection** are central to any control methodology.

6.1 ACTUATOR DYNAMICS

We will analyze in some detail the dynamics of **permanent magnet DC-motors**, as these are the simplest actuators to analyze and are commonly used in robot manipulators. Other types of motors, in particular **AC-motors** and so-called **brushless DC-motors** are also used as actuators for robots but we will not discuss them here.

A DC-motor works on the principle that a current-carrying conductor in a magnetic field experiences a force $F = i \times \phi$, where ϕ is the magnetic flux and i is the current in the conductor. The motor itself consists of a fixed **stator** and a movable **rotor** that rotates inside the stator as shown in Figure 6.2.

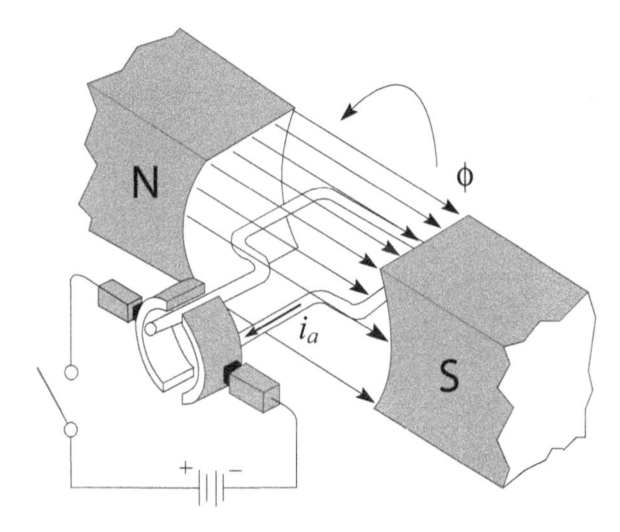

Figure 6.2: Principle of operation of a permanent magnet DC motor. The magnitude of the force (or torque) on the armature is proportional to the product of the current and magnetic flux. A **commutator** is required to periodically switch the direction of the current through the armature to keep it rotating in the same direction.

If the stator produces a radial magnetic flux ϕ and the current in the rotor (also called the **armature**) is i, then there will be a torque on the rotor causing it to rotate. The magnitude of this torque is

$$\tau_m = K_1 \phi i_a \tag{6.1}$$

where τ_m is the motor torque (N-m), ϕ is the magnetic flux (webers), i_a is the armature current (amperes), and K_1 is a physical constant.

In addition, whenever a conductor moves in a magnetic field, a voltage V_b is generated across its terminals that is proportional to the velocity of the conductor in the field. This voltage, called the **back emf**, will tend to oppose the current flow in the conductor. Thus, in addition to the torque τ_m in Equation (6.1), we have the back emf relation

$$V_b = K_2 \phi \omega_m \tag{6.2}$$

where V_b denotes the back emf (volts), ω_m is the angular velocity of the rotor (rad/sec), and K_2 is a proportionality constant.

DC-motors can be classified according to the way in which the magnetic field is produced and the armature is designed. Here we discuss only the so-called **permanent magnet** motors whose stator consists of a permanent magnet. In this case we can take the flux ϕ to be a constant. The torque on the rotor is then controlled by controlling the armature current i_a.

Consider the schematic diagram of Figure 6.3, where

$$\begin{aligned}
V &= \text{armature voltage} \\
L &= \text{armature inductance} \\
R &= \text{armature resistance} \\
V_b &= \text{back emf} \\
i_a &= \text{armature current} \\
\theta_m &= \text{rotor position} \\
\tau_m &= \text{generated torque} \\
\tau_\ell &= \text{load torque} \\
\phi &= \text{magnetic flux due to stator}
\end{aligned}$$

The differential equation for the armature current is then

$$L\frac{di_a}{dt} + Ri_a = V - V_b \tag{6.3}$$

Since the flux ϕ is constant, the torque developed by the motor is

$$\tau_m = K_1 \phi i_a = K_m i_a \tag{6.4}$$

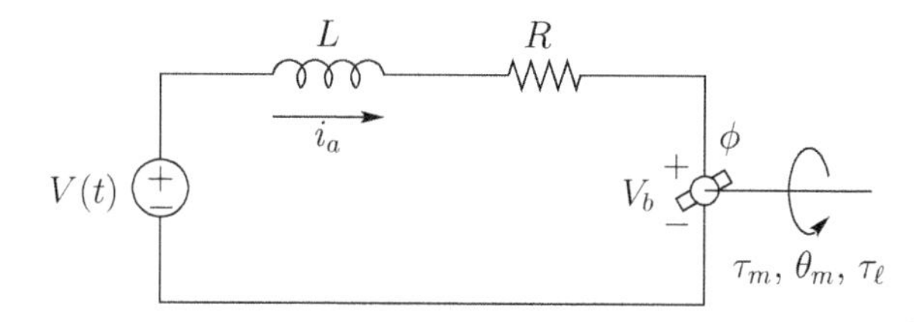

Figure 6.3: Circuit diagram for an armature controlled DC motor. The rotor windings have an effective inductance L and resistance R. The applied voltage V is the control input.

where K_m is the torque constant in N-m/amp. Also, from Equation (6.2) we have

$$V_b = K_2\phi\omega_m = K_b\omega_m = K_b\frac{d\theta_m}{dt} \tag{6.5}$$

where K_b is the back emf constant. It can be shown that the numerical values of K_m and K_b are the same provided MKS units[1] are used.

The torque constant can be determined from a set of torque-speed curves as shown in Figure 6.4 for various values of the applied voltage V.

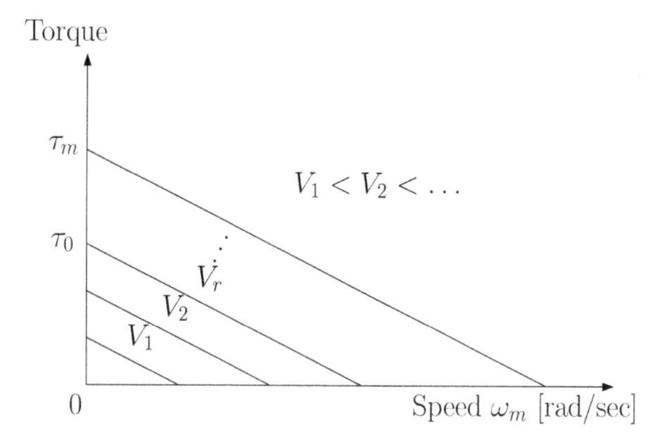

Figure 6.4: Typical torque-speed curves of a DC motor. Each line represents the torque versus speed for a given value of the applied voltage.

When the motor is stalled, the blocked-rotor torque at the rated voltage V_r is denoted τ_0. Using Equations (6.3) and (6.4) with $V_b = 0$ and $di_a/dt = 0$ we have

$$V_r = Ri_a = \frac{R\tau_0}{K_m} \tag{6.6}$$

[1]MKS units are based on the meter, kilogram, and second.

Therefore the torque constant is

$$K_m = \frac{R\tau_0}{V_r} \tag{6.7}$$

6.2 INDEPENDENT JOINT MODEL

In this section we use the DC motor model from the previous section to derive differential equation and transfer function models treating each link of a manipulator as an independent single-input/single-output (SISO) system. Dynamic coupling among the joints is modeled as a disturbance to the SISO system. The SISO model is adequate for applications not involving very fast motion, especially in robots with large gear reduction between the actuators and the links. A large gear reduction reduces the nonlinear coupling among the links. In the following chapters we will derive more detailed models of the dynamics of n-link robots and treat the nonlinear control problem.

The remainder of this section refers to Figure 6.5 consisting of the DC-motor in series with a gear train with gear ratio $r : 1$ and connected to a link of the manipulator. The gear ratio r typically has values in the range 20 to 200 or more. Referring to Figure 6.5, we set $J_m = J_a + J_g$, the sum of the actuator and gear inertias.

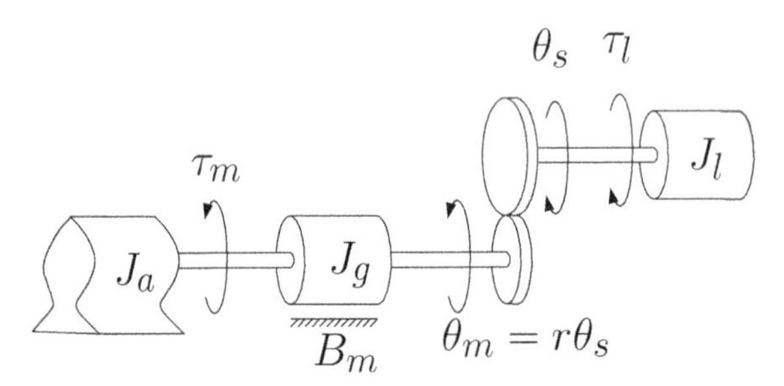

Figure 6.5: Lumped model of a single link with actuator/gear train. J_a, J_g, and J_ℓ are, respectively, the actuator, gear, and load inertias. B_m is the coefficient of motor friction and includes friction in the brushes and gears.

In terms of the motor angle θ_m, the equation of motion of this system is then

$$\begin{aligned} J_m \frac{d^2\theta_m}{dt^2} + B_m \frac{d\theta_m}{dt} &= \tau_m - \tau_\ell / r \\ &= K_m i_a - \tau_\ell / r \end{aligned} \tag{6.8}$$

the latter equality coming from Equation (6.4). In the Laplace domain the three Equations (6.3), (6.5), and (6.8) may be combined and written as

$$(Ls+R)I_a(s) = V(s) - K_b s\Theta_m(s) \tag{6.9}$$

$$(J_m s^2 + B_m s)\Theta_m(s) = K_m I_a(s) - \tau_\ell(s)/r \tag{6.10}$$

The block diagram of the above system is shown in Figure 6.6.

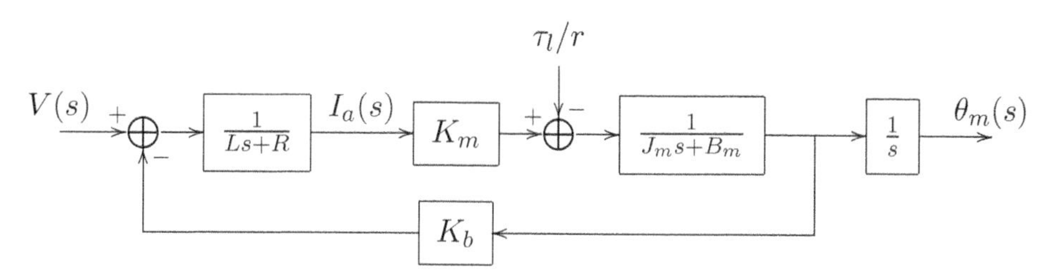

Figure 6.6: Block diagram for a DC motor system. The block diagram represents a third order system from input voltage $V(s)$ to output position $\theta_m(s)$.

The transfer function from $V(s)$ to $\Theta_m(s)$ is given, with $\tau_\ell = 0$, by (Problem 6-1)

$$\frac{\Theta_m(s)}{V(s)} = \frac{K_m}{s\left[(Ls+R)(J_m s+B_m)+K_b K_m\right]} \tag{6.11}$$

The transfer function from the load torque $\tau_\ell(s)$ to $\Theta_m(s)$ is given, with $V = 0$, by (Problem 6-1)

$$\frac{\Theta_m(s)}{\tau_\ell(s)} = \frac{-(Ls+R)/r}{s\left[(Ls+R)(J_m s+B_m)+K_b K_m\right]} \tag{6.12}$$

Notice that the magnitude of this latter transfer function, and hence the effect of the load torque on the motor angle, is reduced by the gear reduction r.

Frequently it is assumed that the "electrical time constant" L/R is much smaller than the "mechanical time constant" J_m/B_m. This is a reasonable assumption for many electro-mechanical systems and leads to a reduced order model of the actuator dynamics. If we divide numerator and denominator of Equations (6.11) and (6.12) by R and neglect the electrical time constant by setting L/R equal to zero, the transfer functions in Equations (6.11) and (6.12) become, respectively, (Problem 6-2)

$$\frac{\Theta_m(s)}{V(s)} = \frac{K_m/R}{s(J_m s+B_m+K_b K_m/R)} \tag{6.13}$$

and

$$\frac{\Theta_m(s)}{\tau_\ell(s)} = \frac{-1/r}{s(J_m(s) + B_m + K_b K_m/R)} \tag{6.14}$$

In the time domain Equations (6.13) and (6.14) represent, by superposition, the second order differential equation

$$J_m \ddot{\theta}_m(t) + (B_m + K_b K_m/R)\dot{\theta}_m(t) = (K_m/R)V(t) - \tau_\ell(t)/r \tag{6.15}$$

Henceforth, we will drop the subscripts and write Equation (6.15) as

$$J\ddot{\theta}(t) + B\dot{\theta}(t) = u(t) - d(t) \tag{6.16}$$

where $B = B_m + K_b K_m/R$ represents the effective damping, $u = (K_m/R)V$ is the control input, and $d = \tau_\ell(t)/r$ represents a disturbance input. The block diagram corresponding to the reduced order system (6.16) is shown in Figure 6.7.

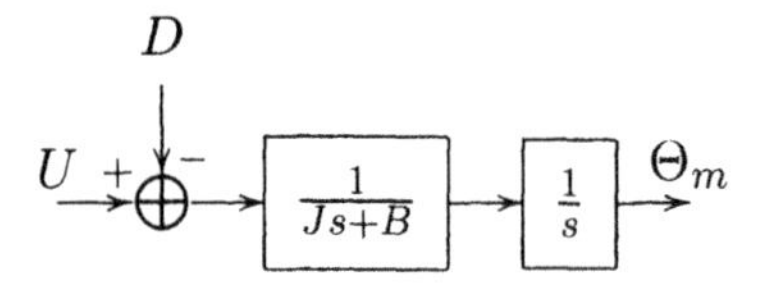

Figure 6.7: Block diagram of the simplified, open-loop system. The disturbance D represents all of the nonlinearities and coupling from the other links.

6.3 SET-POINT TRACKING

In this section we discuss the problem of set-point tracking. The **set-point tracking problem** is the problem of tracking a constant or step reference command θ^d and arises in point-to-point motion. We first consider the simplest types of compensators, so-called PD and PID compensators.

6.3.1 PD Compensator

Using a PD compensator, the control input $U(s)$ is given in the Laplace domain as

$$U(s) = K_P(\Theta^d(s) - \Theta(s)) - K_D s\Theta(s) \tag{6.17}$$

where K_P, K_D are the proportional and derivative gains, respectively. The resulting closed-loop block diagram is shown in Figure 6.8.

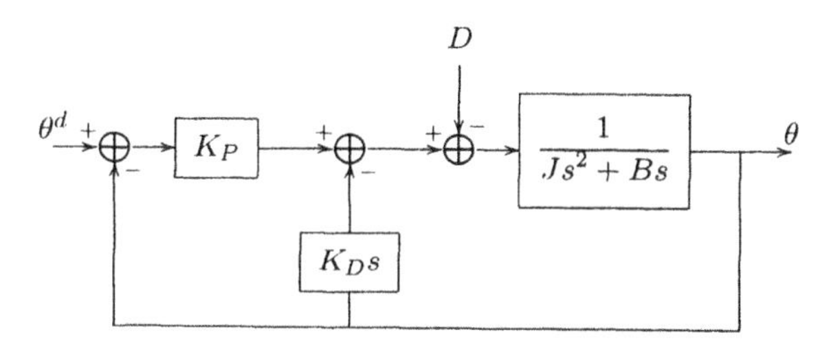

Figure 6.8: The system with PD control.

The resulting closed-loop system is given by

$$\Theta(s) = \frac{K_P}{\Omega(s)}\Theta^d(s) - \frac{1}{\Omega(s)}D(s) \tag{6.18}$$

where $\Omega(s)$ is the closed-loop characteristic polynomial

$$\Omega(s) = Js^2 + (B + K_D)s + K_P \tag{6.19}$$

The closed-loop system will be stable for all positive values of K_P and K_D and bounded disturbances, and the tracking error is given by

$$\begin{aligned} E(s) &= \Theta^d(s) - \Theta(s) \\ &= \frac{Js^2 + (B + K_D)s}{\Omega(s)}\Theta^d(s) + \frac{1}{\Omega(s)}D(s) \end{aligned} \tag{6.20}$$

For a step reference input

$$\Theta^d(s) = \frac{\Omega^d}{s} \tag{6.21}$$

and a constant disturbance

$$D(s) = \frac{D}{s} \tag{6.22}$$

it follows directly from the final value theorem that the steady state error e_{ss} satisfies

$$e_{ss} = \lim_{s \to 0} sE(s) = -\frac{D}{K_P} \tag{6.23}$$

Since the magnitude D of the disturbance is proportional to the gear reduction $1/r$, we see that the steady state error is smaller for larger gear reduction and can be made arbitrarily small by making the position gain K_P large, which is to be expected since the system is type 1.

For the PD compensator given by Equation (6.17) the closed-loop system is second order and hence the step response is determined by the closed-loop natural frequency ω and damping ratio ζ. Given a desired value for these quantities, the gains K_D and K_P can be found from the expression

$$s^2 + \frac{(B + K_D)}{J}s + \frac{K_p P}{J} \quad = \quad s^2 + 2\zeta\omega s + \omega^2 \qquad (6.24)$$

as

$$K_P = \omega^2 J, \qquad K_D = 2\zeta\omega J - B \qquad (6.25)$$

It is customary in robotics applications to take the damping ratio $\zeta = 1$ so that the response is critically damped. This produces the fastest non-oscillatory response. In this context ω determines the speed of response.

Example 6.1

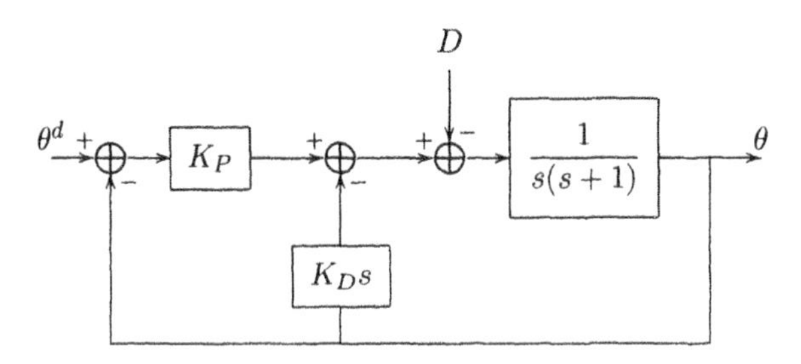

Figure 6.9: Second order system of Example 6.1 with PD Compensator.

Consider the second order system of Figure 6.9 with a PD Compensator. The closed-loop characteristic polynomial is easily computed as

$$p(s) \quad = \quad s^2 + (1 + K_D)s + K_P \qquad (6.26)$$

Suppose $\theta^d = 10$ *and there is no disturbance* ($d = 0$). *With* $\zeta = 1$, *the required PD gains for various values of* ω *are shown in Table 6.1. The corresponding step responses are shown in Figure 6.10.*

◇

Example 6.2

Now, suppose that there is a constant disturbance $d = 40$ *acting on the system. The response of the system with the PD gains of Table 6.1 are shown in Figure 6.11. We see that the steady state error due to the disturbance is smaller for large gains as expected.*

◇

Table 6.1: Proportional and derivative gains for the system of Figure 6.9 for various values of natural frequency ω and damping ratio $\zeta = 1$.

Natural Frequency (ω)	Proportional Gain K_P	Derivative Gain K_D
4	16	7
8	64	15
12	144	23

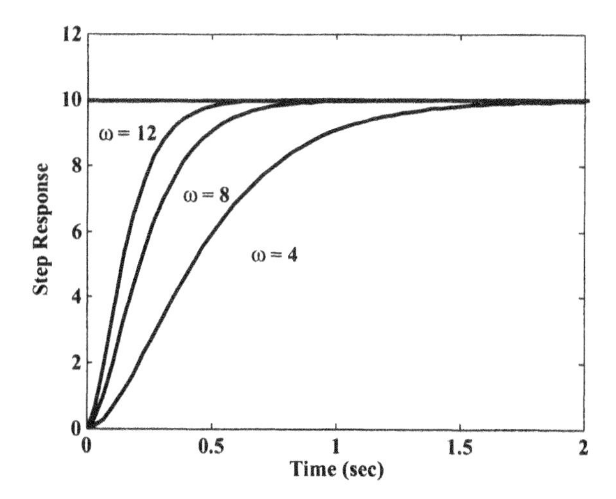

Figure 6.10: Critically damped second order step responses. The rise time decreases for increasing values of ω.

6.3.2 PID Compensator

In order to reject a constant disturbance using PD control, we have seen from Equation (6.23) that large gains may be required. By using integral control we can achieve zero steady state error while keeping the gains small. Thus, let us add an integral term K_I/s to the above PD compensator. This leads to the so-called PID control law, as shown in Figure 6.12. The system is now type 2 and the PID control achieves exact steady tracking of step inputs while rejecting step disturbances, provided of course that the closed-loop system is stable.

With the PID compensator

$$U(s) = (K_P + \frac{K_I}{s})(\Theta^d(s) - \Theta(s)) - K_D s\Theta(s) \tag{6.27}$$

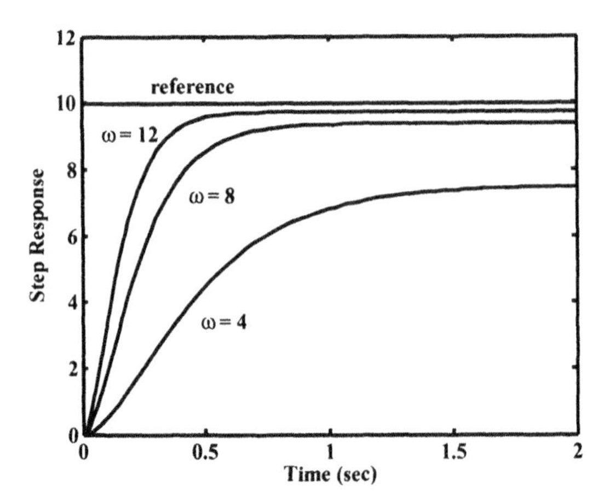

Figure 6.11: Second order system response with PD control and disturbance added. The steady state error decreases for increasing ω.

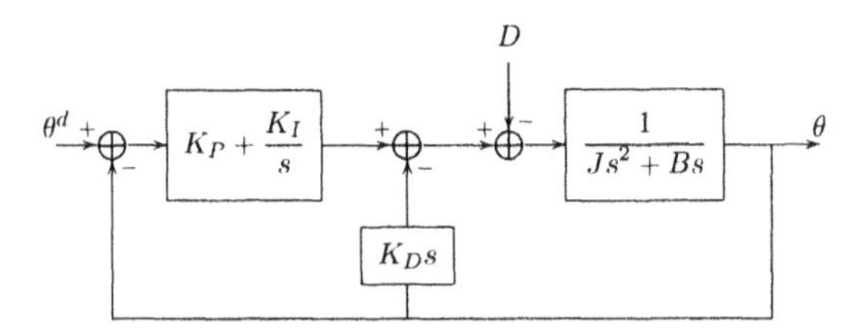

Figure 6.12: Closed-loop system with PID control. The integrator added to the compensator increases the system order from two to three and increases the system type number from 1 to 2.

the closed-loop system is now the third order system

$$\Theta(s) = \frac{(K_P s + K_I)}{\Omega_2(s)}\Theta^d(s) - \frac{s}{\Omega_2(s)}D(s) \tag{6.28}$$

where

$$\Omega_2 = Js^3 + (B + K_D)s^2 + K_P s + K_I \tag{6.29}$$

Applying the Routh-Hurwitz criterion to this polynomial, it follows that the closed-loop system is stable if the gains are positive, and in addition,

$$K_I < \frac{(B + K_D)K_P}{J} \tag{6.30}$$

Example 6.3

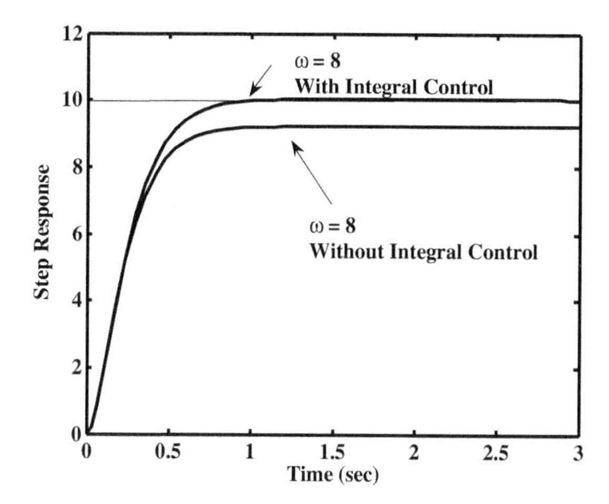

Figure 6.13: Response with integral control action showing that the steady state error to a constant disturbance has been removed.

To the previous system we have added a disturbance and an integral control term in the compensator. The step responses are shown in Figure 6.13. We see that the steady state error due to the disturbance is removed.
◇

PID control is, by far, the most common type of control used in industry due to its simplicity. The main problem in implementing PID control is in the "tuning," that is, in the choice of the proportional, derivative, and integral gains. As we see from the inequality (6.30) the magnitude of the integral gain K_I is limited by the stability constraint. Therefore one common design rule-of-thumb is to first set $K_I = 0$ and design the proportional and derivative gains, K_P and K_D, to achieve the desired transient behavior (rise time, settling time, and so forth) and then to choose K_I within the limits imposed by (6.30) to remove the steady state error.

6.3.3 The Effect of Saturation and Flexibility

In theory, an arbitrarily fast response and arbitrarily small steady state error to a constant disturbance could be achieved by simply increasing the gains in the PD or PID compensator. In practice, however, there is a maximum speed of response achievable from the system. Two major factors, heretofore neglected, limit the achievable performance of the system. The first factor, **saturation**, is due to limits on the maximum torque (or current) input.

Many manipulators, in fact, incorporate current limiters in the servo-system to prevent damage that might result from overdrawing current. The second effect is flexibility in the motor shaft and/or drive train.

Example 6.4

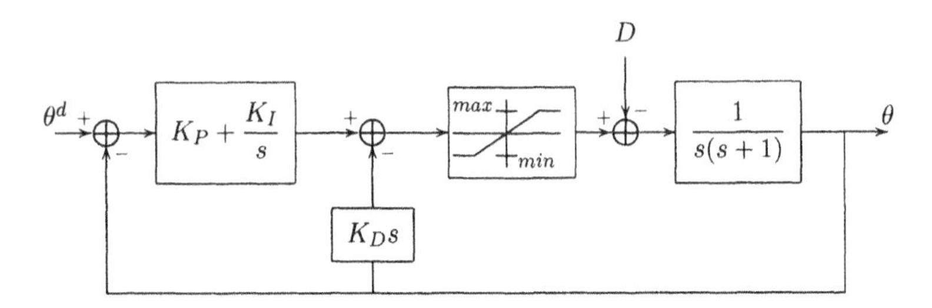

Figure 6.14: Second order system with input saturation limiting the magnitude of the input signal. Increasing the magnitude of the compensator output signal beyond the saturation limit will not increase the input to the plant.

Consider the block diagram of Figure 6.14, where the saturation function represents the maximum allowable input. With PID control and saturation the response is shown in Figure 6.15.

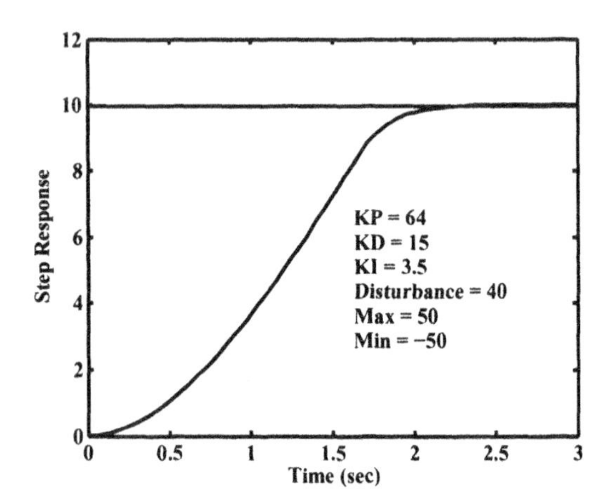

Figure 6.15: Response of the second order system with saturation, disturbance, and PID control. The effect of the saturation is seen in the much slower rise time.

◇

The second effect to consider is the joint flexibility. Let k_r be the effective stiffness at the joint. The joint resonant frequency is then $\omega_r = \sqrt{k_r/J}$. It is common engineering practice to limit ω in Equation (6.25) to no more than half of ω_r to avoid excitation of the joint resonance. We will discuss the effects of the joint flexibility in more detail in Section 6.5. These examples clearly show the limitations of PD control when additional effects such as input saturation, disturbances, and unmodeled dynamics must be considered.

6.4 FEEDFORWARD CONTROL

The analysis in the previous section was carried out under the assumption that the reference signal and disturbance are constant and is not valid for tracking more general time varying trajectories such as a cubic polynomial trajectory of the type generated in Chapter 5. In this section we introduce the notion of **feedforward control** as a method to track time varying trajectories.

Suppose that $\theta^d(t)$ is an arbitrary joint space reference trajectory and consider the block diagram of Figure 6.16, where $G(s)$ represents the forward transfer function of a given system and $H(s)$ is the compensator transfer function.

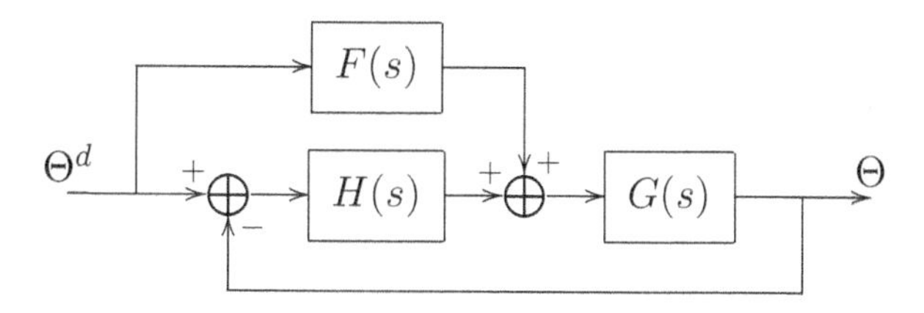

Figure 6.16: Feedforward control scheme. $F(s)$ is the feedforward transfer function which has the reference signal Θ^d as input. The output of the feedforward block is superimposed on the output of the compensator $H(s)$.

A feedforward control scheme consists of adding a feedforward path with transfer function $F(s)$ as shown. Let each of the three transfer functions be represented as ratios of polynomials

$$G(s) = \frac{q(s)}{p(s)}, H(s) = \frac{c(s)}{d(s)}, F(s) = \frac{a(s)}{b(s)} \tag{6.31}$$

We assume that $G(s)$ is strictly proper and $H(s)$ is proper. Simple block diagram manipulation shows that the closed-loop transfer function $T(s) =$

$\frac{Y(s)}{R(s)}$ is given by (Problem 6-8)

$$T(s) = \frac{q(s)(c(s)b(s) + a(s)d(s))}{b(s)(p(s)d(s) + q(s)c(s))} \tag{6.32}$$

The closed-loop characteristic polynomial is $b(s)(p(s)d(s) + q(s)c(s))$. Therefore, for stability of the closed-loop system, we require that the compensator $H(s)$ and the feedforward transfer function $F(s)$ be chosen so that the polynomials $p(s)d(s) + q(s)c(s)$ and $b(s)$ are Hurwitz. This says that, in addition to stability of the closed-loop system, the feedforward transfer function $F(s)$ must itself be stable.

If we choose the feedforward transfer function $F(s)$ equal to $1/G(s)$, the inverse of the forward plant, that is, $a(s) = p(s)$ and $b(s) = q(s)$, then the closed-loop system becomes

$$q(s)(p(s)d(s) + q(s)c(s))Y(s) = q(s)(p(s)d(s) + q(s)c(s))R(s) \tag{6.33}$$

or, in terms of the tracking error $E(s) = R(s) - Y(s)$,

$$q(s)(p(s)d(s) + q(s)c(s))E(s) = 0 \tag{6.34}$$

Thus, assuming stability, the output $\theta(t)$ will track any reference trajectory $\theta^d(t)$. Note that we can only choose $F(s)$ in this manner provided that the numerator polynomial $q(s)$ of the forward plant is Hurwitz, that is, as long as all zeros of the forward plant are in the left half plane. Such systems are called **minimum phase**.

If there is a disturbance $D(s)$ entering the system as shown in Figure 6.17, then it is easily shown (Problem 6-9) that the tracking error $E(s)$ is given by

$$E(s) = \frac{q(s)d(s)}{p(s)d(s) + q(s)c(s)}D(s) \tag{6.35}$$

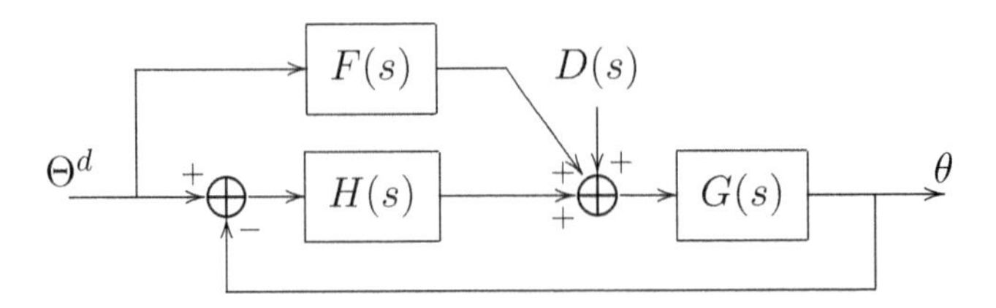

Figure 6.17: Feedforward control with disturbance $D(s)$.

We have thus shown that, in the absence of disturbances, the closed-loop system will track *any* desired trajectory $\theta^d(t)$ provided that the closed-loop system is stable. The steady state error is thus due only to the disturbance.

Let us apply this idea to the robot model of Section 6.3. Suppose that $\theta^d(t)$ is an arbitrary trajectory that we wish the system to track. In this case we have $G(s) = 1/(Js^2 + Bs)$ together with a PD compensator $H(s) = K_P + K_D s$.

We see that the plant transfer function $G(s)$ has no finite zeros and hence is minimum phase. Thus, we can choose the feedforward transfer $F(s)$ as $F(s) = Js^2 + Bs$. The resulting system is shown in Figure 6.18.

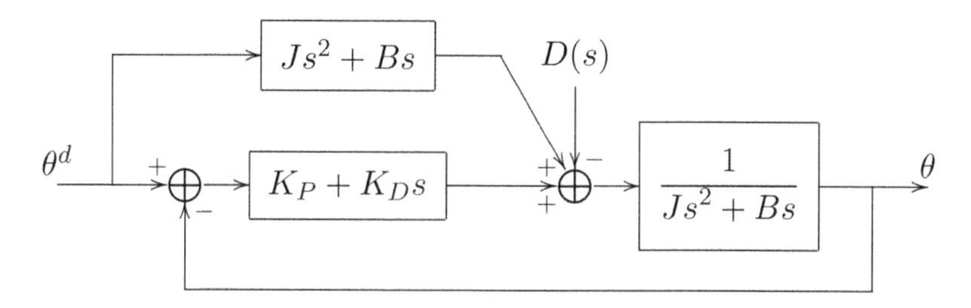

Figure 6.18: Feedforward compensator for the second order system of Section 6.3.

Note that $1/G(s)$ is not a proper rational function. However, since the derivatives of the reference trajectory $\theta^d(t)$ are known and precomputed, the implementation of the above scheme does not require differentiation of an actual signal. It is easy to see from Equation (6.35) that the steady state error to a step disturbance is now given by the same expression as in Equation (6.23) independent of the reference trajectory. As before, a PID compensator would result in zero steady state error to a step disturbance. In the time domain the control law of Figure 6.18 can be written as

$$\begin{aligned} V(t) &= J\ddot{\theta}^d + B\dot{\theta}^d + K_D(\dot{\theta}^d - \dot{\theta}) + K_P(\theta^d - \theta) \\ &= f(t) + K_D\dot{e}(t) + K_P e(t) \end{aligned} \tag{6.36}$$

where $f(t)$ is the feedforward signal

$$f(t) = J\ddot{\theta}^d + B\dot{\theta}^d \tag{6.37}$$

and $e(t)$ is the tracking error $\theta^d(t) - \theta(t)$. Since the forward plant equation is

$$J\ddot{\theta} + B\dot{\theta} = V(t) - rd(t)$$

the closed-loop error $e(t) = \theta(t) - \theta^d(t)$ satisfies the second order differential equation

$$J\ddot{e} + (B + K_D)\dot{e} + K_P e(t) \quad = \quad -rd(t) \qquad (6.38)$$

We note from Equation (6.38) that the characteristic polynomial of the closed-loop system is identical to Equation (6.19). However, the system (6.38) is now written in terms of the tracking error $e(t)$. Therefore, assuming that the closed-loop system is stable, the tracking error will approach zero asymptotically for any desired joint space trajectory in the absence of disturbances, that is, if $d = 0$.

6.5 DRIVE TRAIN DYNAMICS

In this section we discuss in more detail the problem of joint flexibility. For many manipulators, particularly those using harmonic drives for torque transmission, the joint flexibility is significant. In addition to torsional flexibility in the gears, joint flexibility is caused by effects such as shaft windup, bearing deformation, and compressibility of the hydraulic fluid in hydraulic robots.

The Harmonic Drive

Harmonic drives are a type of gear mechanism that are very popular for use in robots due to their low backlash, high torque transmission, and compact size.

Figure 6.19: The harmonic drive. The rotation of the elliptical wave generator meshes the teeth of the flexspline and circular spline resulting in low backlash and high torque transmission. (Courtesy of of HD Systems, www.hdsi.net.)

A typical harmonic drive is shown in Figure 6.19 and consists of a rigid **circular spline**, a flexible **flexspline**, and an elliptical **wave generator**. The wave generator is attached to the actuator and hence is turned at high speed by the motor. The circular spline is attached to the load. As the wave generator rotates it deforms the flexspline causing a number of teeth of the flexspline to mesh with the teeth of the circular spline. The effective gear ratio is determined by the difference in the number of teeth of the flexspline and circular spline.

The low backlash and high torque throughput of the harmonic drive results from the relative large number of teeth that are meshed at any given time. However, the principle of the harmonic drive relies on the flexibility of the flexspline. This flexibility is the limiting factor to the achievable performance in many cases.

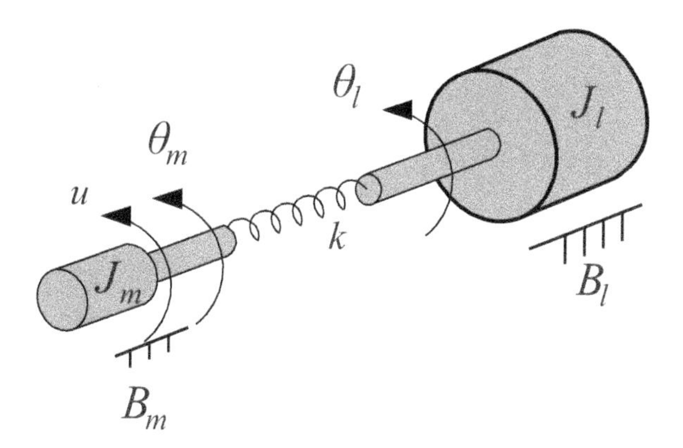

Figure 6.20: Idealized model to represent joint flexibility. The stiffness constant k represents the effective torsional stiffness of the harmonic drive.

Consider the idealized situation of Figure 6.20 consisting of an actuator connected to a load through a torsional spring representing the joint flexibility. For simplicity we take the motor torque u, rather than the armature voltage, as input. The equations of motion are

$$J_\ell\ddot{\theta}_\ell + B_\ell\dot{\theta}_\ell + k(\theta_\ell - \theta_m) = 0 \qquad (6.39)$$

$$J_m\ddot{\theta}_m + B_m\dot{\theta}_m - k(\theta_\ell - \theta_m) = u \qquad (6.40)$$

where J_ℓ, J_m are the load and motor inertias, B_ℓ and B_m are the load and motor damping constants, and u is the input torque applied to the motor shaft. The joint stiffness constant k represents the torsional stiffness of the harmonic drive gears. In the Laplace domain we can write the above system as

$$p_\ell(s)\Theta_\ell(s) = k\Theta_m(s) \quad (6.41)$$
$$p_m(s)\Theta_m(s) = k\Theta_\ell(s) + U(s) \quad (6.42)$$

where

$$p_\ell(s) = J_\ell s^2 + B_\ell s + k \quad (6.43)$$
$$p_m(s) = J_m s^2 + B_m s + k \quad (6.44)$$

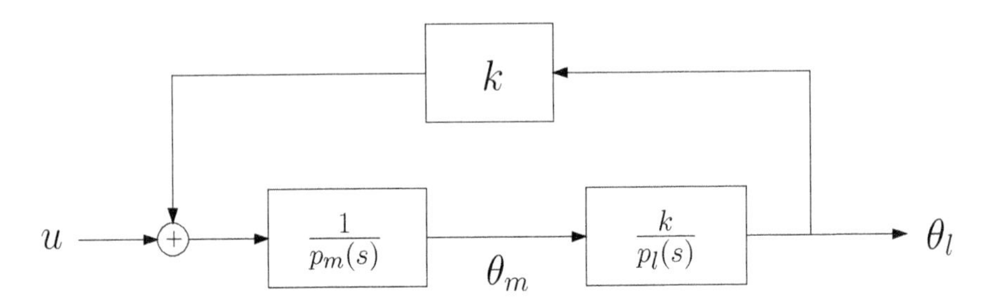

Figure 6.21: Block diagram for the system (6.41) and (6.42).

This system is represented by the block diagram of Figure 6.21. The output to be controlled is, of course, the load angle θ_ℓ. The open-loop transfer function between U and Θ_ℓ is given by

$$\frac{\Theta_\ell(s)}{U(s)} = \frac{k}{p_\ell(s)p_m(s) - k^2} \quad (6.45)$$

The open-loop characteristic polynomial $p_\ell p_m - k^2$ is

$$J_\ell J_m s^4 + (J_\ell B_m + J_m B_\ell)s^3 + (k(J_\ell + J_m) + B_\ell B_m)s^2 + k(B_\ell + B_m)s \quad (6.46)$$

We can obtain some insight into the behavior of the system by first neglecting the damping coefficients B_ℓ and B_m. In this case the open-loop characteristic polynomial would be

$$J_\ell J_m s^4 + k(J_\ell + J_m)s^2 \quad (6.47)$$

which has a double pole at the origin and a pair of complex conjugate poles on the $j\omega$-axis at $s = \pm j\omega$ where $\omega^2 = k\left(\frac{1}{J_\ell} + \frac{1}{J_m}\right)$. Note that the frequency of the imaginary poles increases with increasing joint stiffness k.

In practice the stiffness of the harmonic drive is large and the damping is small, which results in a difficult system to control. Assuming that the open-loop damping constants B_ℓ and B_m are small, the open-loop poles of the system (6.41) and (6.42) will be in the left half plane near the poles of the undamped system.

Suppose we implement a PD compensator $C(s) = K_P + K_D s$. At this point the analysis depends on whether the position/velocity sensors are placed on the motor shaft or on the load shaft, that is, whether the PD compensator is a function of the motor variables or the load variables. If the motor variables are measured then the closed-loop system is given by the block diagram of Figure 6.22.

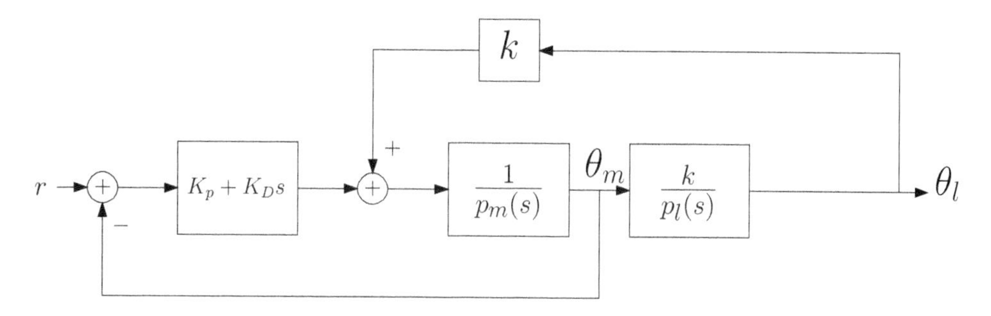

Figure 6.22: PD control with motor angle feedback.

In order to perform a root locus we set $K_P + K_D s = K_D(s + a)$ with $a = K_P/K_D$. The root locus for the closed-loop system in terms of K_D is shown in Figure 6.23.

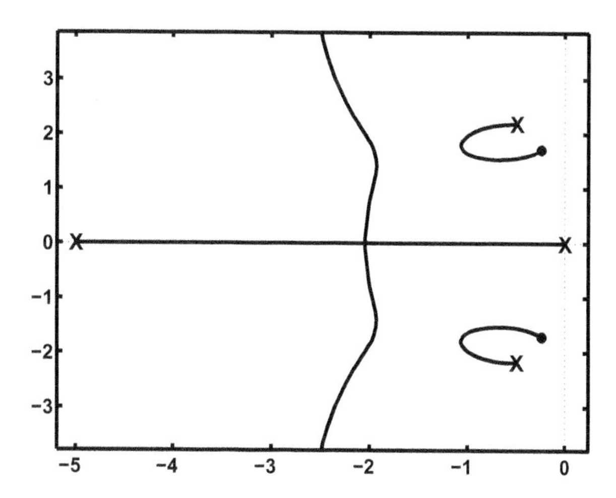

Figure 6.23: Root locus for the system of Figure 6.22.

We see that the system is stable for all values of the gain K_D but that the presence of the open-loop zeros near the $j\omega$ axis may result in undesirable oscillations. Also the poor relative stability means that disturbances and other unmodeled dynamics could render the system unstable.

If we measure the load angle θ_ℓ instead, the system with PD control is represented by the block diagram of Figure 6.24. The corresponding root locus is shown in Figure 6.25.

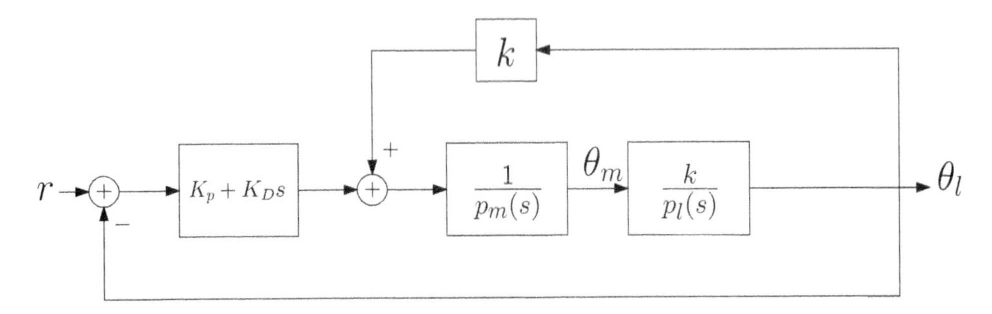

Figure 6.24: PD control with load angle feedback.

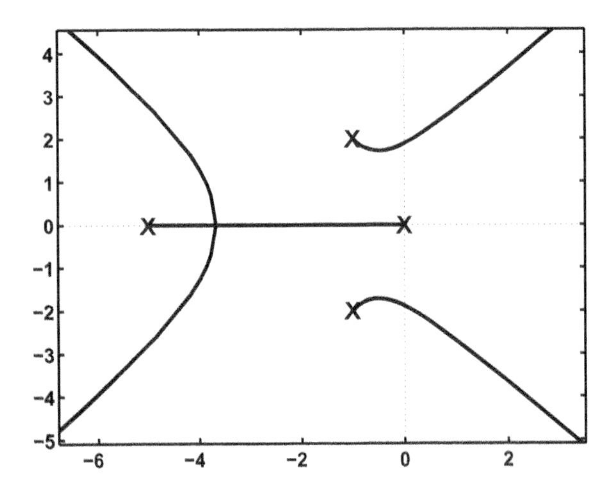

Figure 6.25: Root locus for the system of Figure 6.24.

In this case the system is unstable for large K_D. The critical value of K_D, that is, the value of K_D for which the system becomes unstable, can be found from the Routh-Hurwitz criterion. The best that one can do in this case is to limit the gain K_D so that the closed-loop poles remain within the left half plane with a reasonable stability margin.

Figure 6.26 (respectively, Figure 6.27) shows the response of the system with motor (respectively, load) feedback using the PD controller $K_D(s+a)$.

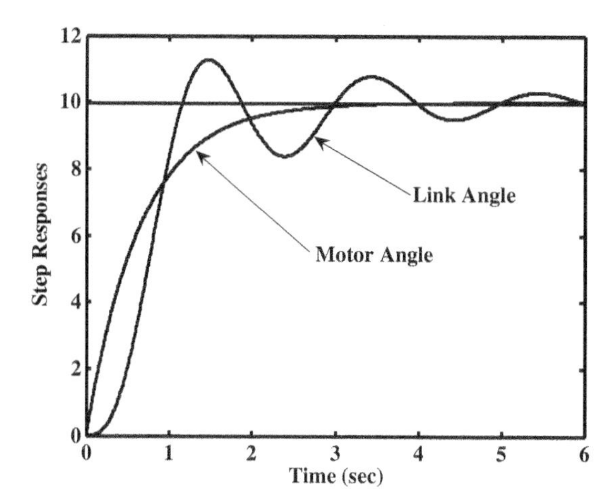

Figure 6.26: Step response — PD control with motor angle feedback. The motor shaft angle, collocated with the motor torque, shows the desired response without overshoot. The motion of the motor shaft excites an oscillation in the load angle, which is effectively outside the feedback loop.

6.6 STATE SPACE DESIGN

In this section we consider the application of state space methods for the control of the flexible joint system above.[2] The previous analysis has shown that PD control is inadequate for robot control unless the joint flexibility is negligible or unless one is content with relatively slow response of the manipulator. Not only does the joint flexibility limit the magnitude of the gain for stability reasons, it also introduces lightly damped poles into the closed-loop system that may result in oscillation of the transient response. We can write the system given by Equations (6.39) and (6.40) in state space by choosing state variables

$$\begin{array}{ll} x_1 = \theta_\ell & x_2 = \dot{\theta}_\ell \\ x_3 = \theta_m & x_4 = \dot{\theta}_m \end{array} \tag{6.48}$$

[2]This section assumes more knowledge of control theory than previous sections.

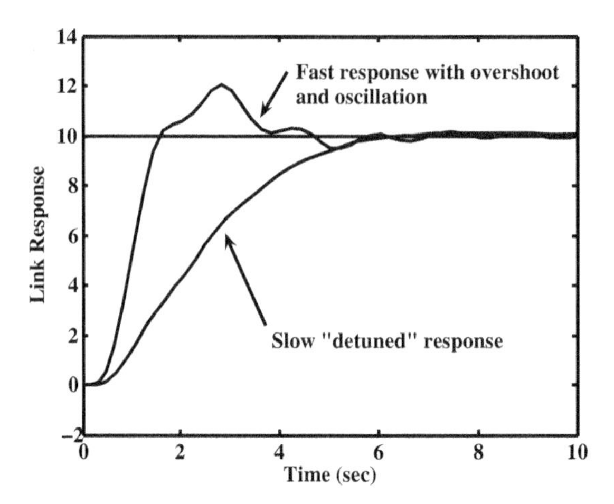

Figure 6.27: Step response — PD control with load angle feedback. The load angle is shown for two different sets of gain parameters. As we know that the system is unstable for large gain, we must effectively "detune" the system for stability, which results in a slower than desired response.

In terms of these state variables the system given by Equations (6.39) and (6.40) becomes

$$\begin{aligned} \dot{x}_1 &= x_2 \\ \dot{x}_2 &= -\frac{k}{J_\ell}x_1 - \frac{B_\ell}{J_\ell}x_2 + \frac{k}{J_\ell}x_3 \\ \dot{x}_3 &= x_4 \\ \dot{x}_4 &= \frac{k}{J_m}x_1 - \frac{B_\ell}{J_m}x_4 - \frac{k}{J_m}x_3 + \frac{1}{J_m}u \end{aligned} \tag{6.49}$$

which, in matrix form, can be written as

$$\dot{x} = Ax + bu \tag{6.50}$$

where

$$A = \begin{bmatrix} 0 & 1 & 0 & 0 \\ -\frac{k}{J_\ell} & -\frac{B_\ell}{J_\ell} & \frac{k}{J_\ell} & 0 \\ 0 & 0 & 0 & 1 \\ \frac{k}{J_m} & 0 & -\frac{k}{J_m} & \frac{B_m}{J_m} \end{bmatrix}, \quad b = \begin{bmatrix} 0 \\ 0 \\ 0 \\ \frac{1}{J_m} \end{bmatrix} \tag{6.51}$$

If we choose an output $y(t)$, say the measured load angle $\theta_\ell(t)$, then we have an output equation

$$y = x_1 = c^T x \tag{6.52}$$

where

$$c^T = [1, 0, 0, 0] \tag{6.53}$$

The relationship between the state space form given by Equations (6.50)–(6.53) and the transfer function (6.45) is found by taking Laplace transforms of Equations (6.50)–(6.52) with initial conditions set to zero. This yields

$$G(s) = \frac{\Theta_\ell(s)}{U(s)} = \frac{Y(s)}{U(s)} = c^T(sI - A)^{-1}b \tag{6.54}$$

where I is the $n \times n$ identity matrix. The poles of $G(s)$ are eigenvalues of the matrix A. For the system (6.50)–(6.53), the converse holds as well, that is, all of the eigenvalues of A are poles of $G(s)$. This is always true if the state space system is defined using a minimal number of state variables.

6.6.1 State Feedback Control

Given a linear system in state space form, such as Equation (6.50), a **linear state feedback control law** is an input u of the form

$$u(t) = -k^T x + r = -\sum_{i=1}^{4} k_i x_i + r \tag{6.55}$$

where k_i are constant gains to be determined and r is a reference input. In other words, the control is determined as a linear combination of the system states which, in this case, are the motor and load positions and velocities. Compare this to the previous PD/PID control, which was a function either of the motor position and velocity or of the load position and velocity, but not both. If we substitute the control law given by Equation (6.55) into Equation (6.50) we obtain

$$\dot{x} = (A - bk^T)x + br \tag{6.56}$$

Thus, we see that the linear feedback control has the effect of changing the poles of the system from those determined by A to those determined by $A - bk^T$.

In the previous PD design the closed-loop pole locations were restricted to lie on the root locus shown in Figure 6.23 or 6.25. Since there are more free parameters in Equation (6.55) than in the PD controller, it may be possible to achieve a much larger range of closed-loop poles. This turns out to be the case if the system given by Equation (6.50) satisfies a property known as **controllability**.

Definition 6.1 *A linear system is said to be* **completely controllable**, *or* **controllable** *for short, if for each initial state $x(t_0)$ and each final state $x(t_f)$ there is a control input $t \rightarrow u(t)$ that transfers the system from $x(t_0)$ at time t_0 to $x(t_f)$ at time t_f.*

The above definition says, in essence, that if a system is controllable we can achieve any state whatsoever in finite time starting from an arbitrary initial state. To check whether a system is controllable we have the following simple test.

Lemma 6.1 *A linear system of the form (6.50) is controllable if and only if*

$$\det[b, Ab, A^2b, \ldots, A^{n-1}b] \neq 0 \tag{6.57}$$

The $n \times n$ matrix $[b, Ab, \ldots, A^{n-1}b]$ is called the **controllability matrix** for the linear system defined by the pair (A, b). The fundamental importance of controllability of a linear system is shown by the following

Theorem 1 *Let $\alpha(x) = s^n + \alpha_n s^{n-1} + \cdots + \alpha_2 s + \alpha_1$ be an arbitrary polynomial of degree n with real coefficients. Then there exists a state feedback control law of the form Equation (6.55) such that*

$$\det(sI - A + bk^T) = \alpha(s) \tag{6.58}$$

if and only if the system (6.50) is controllable.

This fundamental result says that, for a controllable linear system, we may achieve **arbitrary**[3] closed-loop poles using state feedback. Returning to the specific fourth-order system given by Equation (6.51), we see that the system is indeed controllable since

$$\det[b, Ab, A^2b, A^3b] = \frac{k^2}{J_m^4 J_\ell^2} \tag{6.59}$$

[3]Since the coefficients of the polynomial $a(s)$ are real, the only restriction on the pole locations is that they occur in complex conjugate pairs.

which is never zero since $k > 0$. Thus, we can achieve any desired set of closed-loop poles that we wish, which is much more than was possible using the previous PD compensator.

There are many algorithms that can be used to determine the feedback gains in Equation (6.55) to achieve a desired set of closed-loop poles. This is known as the **pole assignment** problem. In this case most of the difficulty lies in choosing an appropriate set of closed-loop poles based on the desired performance, the limits on the available torque, etc. We would like to achieve a fast response from the system without requiring too much torque from the motor. One way to design the feedback gains is through an optimization procedure. This takes us into the realm of optimal control theory. For example, we may choose as our goal the minimization of the performance criterion

$$J = \int_0^\infty \{x^T(t)Qx(t) + Ru^2(t)\}dt \tag{6.60}$$

where Q is a given symmetric, positive definite matrix and $R > 0$.

Choosing a control law to minimize Equation (6.60) frees us from having to decide beforehand what the closed-loop poles should be as they are automatically dictated by the weighting matrices Q and R in Equation (6.60). It is shown in optimal control texts that the optimum linear control law that minimizes Equation (6.60) is given as

$$u = -k_*^T x \tag{6.61}$$

where

$$k_* = \frac{1}{R} b^T P \tag{6.62}$$

and P is the (unique) symmetric, positive definite $n \times n$ matrix satisfying the so-called **matrix algebraic Riccati equation**

$$A^T P + PA - \frac{1}{R} Pbb^T P + Q = 0 \tag{6.63}$$

The control law (6.61) is referred to as a **linear quadratic (LQ) optimal control**, since the performance index is quadratic and the control system is linear.

6.6.2 Observers

The above result that any set of closed-loop poles may be achieved for a controllable linear system is remarkable. In effect, this result says that we may achieve any closed-loop response that we desire. However, to achieve it we have had to pay a price, namely, that the control law must be a function of *all* of the states. In order to build a compensator that requires only the measured output, in this case θ_ℓ, we need to introduce the concept of an **observer**. An observer is a state estimator. It is a dynamical system (constructed in software) that attempts to estimate the full state $x(t)$ using only the system model, Equations (6.50)–6.53), and the measured output $y(t)$. A complete discussion of observers is beyond the scope of the present text. We give here only a brief introduction to the main idea of observers for linear systems.

Assuming that we know the parameters of the system (6.50) we could simulate the response of the system in software and recover the value of the state $x(t)$ at time t from the simulation and we could use this simulated or estimated state, call it $\hat{x}(t)$, in place of the true state in Equation (6.61). However, since the true initial condition $x(t_0)$ for Equation (6.50) will generally be unknown, this idea is not feasible. However the idea of using the model of the system given by Equation (6.50) is a good starting point to construct a state estimator in software. Let us, therefore, consider an estimate $\hat{x}(t)$ satisfying the system

$$\dot{\hat{x}} = A\hat{x} + bu + \ell(y - c^T\hat{x}) \tag{6.64}$$

Equation (6.64) is called an **observer** for Equation (6.50) and represents a model of the system (6.50) with an additional term $\ell(y - c^T\hat{x})$. This additional term is a measure of the error between the output $y(t) = c^T x(t)$ of the plant and the estimate of the output, $c^T\hat{x}(t)$. Since we know the coefficient matrices in Equation (6.64) and can measure y directly, we can solve the above system for $\hat{x}(t)$ starting from any initial condition, and use this $\hat{x}$ in place of the true state x in the feedback law (6.61). The additional term ℓ in Equation (6.64) is to be designed so that $\hat{x} \to x$ as $t \to \infty$, that is, so that the estimated state converges to the true (unknown) state, independent of the initial condition $x(t_0)$. Let us see how this is done.

Define $e(t) = x - \hat{x}$ as the **estimation error**. Combining Equations (6.50) and (6.64), since $y = c^T x$, we see that the estimation error satisfies the system

$$\dot{e} = (A - \ell c^T)e \tag{6.65}$$

From Equation (6.65) we see that the dynamics of the estimation error are determined by the eigenvalues of $A-\ell c^T$. Since ℓ is a design quantity we can attempt to choose it so that $e(t) \to 0$ as $t \to \infty$, in which case the estimate $\hat{x}$ converges to the true state x. In order to do this we obviously want to choose ℓ so that the eigenvalues of $A - \ell c^T$ are in the left half plane. This is similar to the pole assignment problem considered previously. In fact it is dual, in a mathematical sense, to the pole assignment problem. It turns out that the eigenvalues of $A - \ell c^T$ can be assigned arbitrarily if and only if the pair (A, c) satisfies the property known as **observability**. Observability is defined by the following:

Definition 6.2 *A linear system is* **completely observable**, *or* **observable** *for short, if every initial state $x(t_0)$ can be exactly determined from measurements of the output $y(t)$ and the input $u(t)$ in a finite time interval $t_0 \leq t \leq t_f$.*

To check whether a system is observable we have the following

Theorem 2 *The pair (A, c) is observable if and only if*

$$\det\left[c, A^T c, \ldots, A^{T^{n-1}} c\right] \neq 0 \tag{6.66}$$

The $n \times n$ matrix $[c^T, c^T A^T, \ldots, c^T A^{T^{n-1}}]$ is called the **observability matrix** for the pair (A, c^T). In the system given by Equations (6.50)–(6.53) above we have that

$$\det\left[c, A^T c, A^{T^2} c, A^{T^3} c\right] = \frac{k^2}{J_\ell^2} \tag{6.67}$$

and hence the system is observable.

If we use the estimated state $\hat{x}$ in place of the true state, we have the system (with $r = 0$)

$$\begin{aligned} \dot{x} &= Ax + bu \\ u &= -k^T \hat{x} \end{aligned}$$

It is easy to show from the above that the state x and estimation error e jointly satisfy the equation

$$\begin{bmatrix} \dot{x} \\ \dot{e} \end{bmatrix} = \begin{bmatrix} A - bk^T & bk^T \\ 0 & A - \ell c^T \end{bmatrix} \begin{bmatrix} x \\ e \end{bmatrix} \tag{6.68}$$

and therefore the set of closed-loop poles of the system will consist of the union of the eigenvalues of $A - \ell c^T$ and the eigenvalues of $A - bk^T$.

This result is known as the **separation principle**. As the name suggests, the separation principle allows us to separate the design of the state feedback control law (6.61) from the design of the state estimator (6.64). A typical procedure is to place the observer poles to the left of the desired pole locations of $A - bk^T$. This results in rapid convergence of the estimated state to the true state, after which the response of the system is nearly the same as if the true state were being used in Equation (6.61).

The result that the closed-loop poles of the system may be placed arbitrarily, under the assumption of controllability and observability, is a powerful theoretical result. There are always practical considerations to be taken into account, however. The most serious factor to be considered in observer design is noise in the measurement of the output. To place the poles of the observer very far to the left of the imaginary axis in the complex plane requires that the observer gains be large. Large gains can amplify noise in the output measurement and result in poor overall performance. Large gains in the state feedback control law (6.61) can result in saturation of the input, again resulting in poor performance. Also uncertainties in the system parameters, or nonlinearities such as a nonlinear spring characteristic and backlash, will reduce the achievable performance from the above design. Therefore, the above ideas are intended only to illustrate what may be possible by using more advanced concepts from control theory. In Chapter 8 we will develop more advanced, nonlinear control methods to control systems with uncertainties in the parameters.

6.7 SUMMARY

This chapter is a basic introduction to robot control treating each joint of the manipulator as an independent single-input/single-output (SISO) system. In this approach one is primarily concerned with the actuator and drive-train dynamics. We first derived a reduced-order linear model for the dynamics of a permanent-magnet DC-motor and showed that the transfer function from the motor voltage $V(s)$ to the motor shaft angle $\Theta_m(s)$ can be expressed as

$$\frac{\Theta_m(s)}{V(s)} = \frac{K_m/R}{s(J_m s + B_m + K_b K_m/R)}$$

while the transfer function from a load disturbance $D(s)$ to $\Theta_m(s)$ is

$$\frac{\Theta_m(s)}{D(s)} = \frac{-1/r}{s(J_m(s) + B_m + K_b K_m/R)}$$

We then considered the set-point tracking problem using PD and PID compensators. A PD compensator is of the form

$$U(s) = K_P(\Theta^d(s) - \Theta(s)) - K_D s\Theta(s)$$

which results in a closed-loop system

$$\Theta(s) = \frac{K_P}{\Omega(s)}\Theta^d(s) - \frac{1}{\Omega(s)}D(s)$$

where

$$\Omega(s) = Js^2 + (B + K_D)s + K_P$$

is the closed-loop characteristic polynomial whose roots determine the closed-loop poles and, hence, the performance of the system.

A PID compensator is of the form

$$U(s) = (K_P + \frac{K_I}{s})(\Theta^d(s) - \Theta(s)) - K_D s\Theta(s)$$

The closed-loop system is now the third order system

$$\Theta(s) = \frac{(K_P s + K_I)}{\Omega_2(s)}\Theta^d(s) - \frac{rs}{\Omega_2(s)}D(s)$$

where

$$\Omega_2 = Js^3 + (B + K_D)s^2 + K_P s + K_I \qquad (6.69)$$

We discussed methods to design the PD and PID gains for a desired transient and steady state response. We then discussed the effects of saturation and flexibility on the performance of the system. Both of these effects limit the achievable performance of the closed-loop system.

We next discussed the use of feedforward control as a method to track time varying reference trajectories such as the cubic polynomial trajectories that we derived in Chapter 5. A feedforward control scheme consists of adding a feedforward path from the reference signal to the control signal with transfer function $F(s)$. We showed that choosing $F(s)$ as the inverse of the forward plant allows tracking of arbitrary reference trajectories provided the forward plant is minimum phase.

Next, we considered the effect of drive train dynamics in more detail. We derived a simple model a single link system that included the joint elasticity and showed the limitations of PD control for this case. We then introduced state space control methods, which are much more powerful than the simple PD and PID control methods.

We introduced the fundamental notions of controllability and observability and showed that, if the state space model is both controllable and observable, we could design a linear control law to achieve any set of desired closed-loop poles. Specifically, given the linear system

$$\begin{aligned} \dot{x} &= Ax + bu \\ y &= c^T x \end{aligned}$$

then the state feedback control law $u = -k^T \hat{x}$ where $\hat{x}$ is the estimate of the state x computed from a linear observer

$$\dot{\hat{x}} = A\hat{x} + bu + \ell(y - c^T \hat{x})$$

results in the closed-loop system (in terms of the state x and estimation error $e = x - \hat{x}$)

$$\begin{bmatrix} \dot{x} \\ \dot{e} \end{bmatrix} = \begin{bmatrix} A - bk^T & bk^T \\ 0 & A - \ell c^T \end{bmatrix} \begin{bmatrix} x \\ e \end{bmatrix}$$

The set of closed-loop poles of the system will therefore consist of the union of the eigenvalues of $A - \ell c^T$ and the eigenvalues of $A - bk^T$, a result known as the separation principle.

We also introduced the notion of linear quadratic optimal control and showed that the control

$$u = -k_*^T x$$

where

$$k_* = \frac{1}{R} b^T P$$

and P is the (unique) symmetric, positive definite $n \times n$ matrix satisfying the so-called matrix algebraic Riccati equation

$$A^T P + PA - \frac{1}{R} P b b^T P + Q = 0$$

not only stabilizes the system but minimizes the quadratic performance measure

$$J = \int_0^\infty \{x^T(t) Q x(t) + R u^2(t)\} dt$$

PROBLEMS

6-1 Using block diagram reduction techniques derive the transfer functions given by Equations (6.11) and (6.12).

6-2 Derive the transfer functions for the reduced order model given by Equations (6.13) and (6.14).

6-3 Derive Equations (6.18) and (6.19).

6-4 Verify the expression given by Equation (6.20) for the tracking error for the system in Figure 6.8. State the Final Value Theorem and use it to show that the steady state error e_{ss} is indeed given by Equation (6.23).

6-5 Derive Equations (6.28) and (6.29).

6-6 Derive the inequality (6.30) using the Routh-Hurwitz criterion.

6-7 For the system of Figure 6.14 investigate the effect of saturation with various values of the PID gains and disturbance magnitude.

6-8 Verify Equation (6.32)

6-9 Verify Equation (6.35)

6-10 Derive Equations (6.45), (6.46), and (6.47).

6-11 Given the state space model defined by Equation (6.50) show that the transfer function

$$G(s) = c^T(sI - A)^{-1}b$$

is identical to Equation (6.45).

6-12 Search the control literature (for example, [58]) and find two or more algorithms for the pole assignment problem for linear systems.

6-13 Derive Equations (6.59) and (6.67).

6-14 Search the control literature to find out what is meant by **integrator windup**. Find out what is meant by **anti-windup** (or anti-reset windup). Simulate a PID control with anti-reset windup for the system of Figure 6.14. Compare the response with and without anti-reset windup.

6-15 Include the dynamics of a permanent magnet DC-motor for the system given by Equations (6.39) and (6.40). What can you say now about controllability and observability of the system?

6-16 Choose appropriate state variables and write the system Equations (6.9) and (6.10) in state space form. What is the dimension of the state space?

6-17 Suppose in the flexible joint system represented by Equations (6.39) and (6.40) the following parameters are given

$$J_\ell = 10 \quad B_\ell = 1 \quad k = 100$$
$$J_m = 2 \quad B_m = 0.5$$

(a) Sketch the open-loop poles of the transfer functions given by Equation (6.45).

(b) Apply a PD compensator to the system (6.45). Sketch the root locus for the system. Choose a reasonable location for the compensator zero. Using the Routh criterion find the value of the compensator gain when the root locus crosses the imaginary axis.

6-18 One of the problems encountered in space applications of robots is the fact that the base of the robot cannot be anchored, that is, cannot be fixed in an inertial coordinate frame. Consider the idealized situation shown in Figure 6.28, consisting of an inertia J_1 connected to the rotor of a motor whose stator is connected to an inertia J_2.

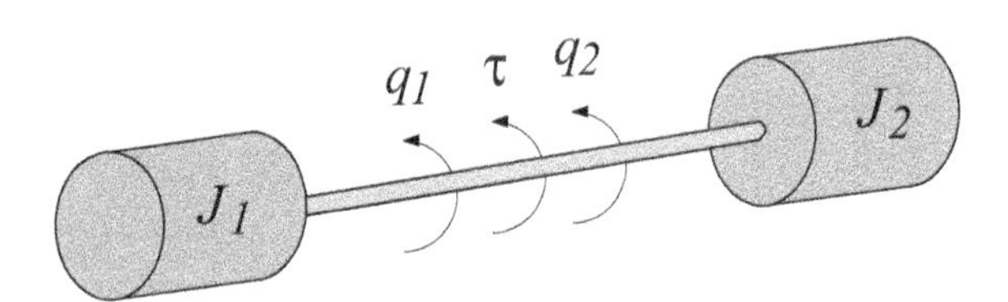

Figure 6.28: Coupled Inertias in Free Space.

For example, J_1 could represent the space shuttle robot arm and J_2 the inertia of the shuttle itself. The simplified equations of motion are thus

$$\begin{aligned} J_1\ddot{q}_1 &= \tau \\ J_2\ddot{q}_2 &= \tau \end{aligned}$$

Write this system in state space form and show that it is uncontrollable. Discuss the implications of this and suggest possible solutions.

6-19 Given the linear second order system

$$\begin{bmatrix} \dot{x}_1 \\ \dot{x}_2 \end{bmatrix} = \begin{bmatrix} 1 & -3 \\ 1 & -2 \end{bmatrix} \begin{bmatrix} x_1 \\ x_2 \end{bmatrix} + \begin{bmatrix} 1 \\ -2 \end{bmatrix} u$$

find a linear state feedback control $u = k_1 x_1 + k_2 x_2$ so that the closed-loop system has poles at $s = -2, 2$.

6-20 Repeat the above if possible for the system

$$\begin{bmatrix} \dot{x}_1 \\ \dot{x}_2 \end{bmatrix} = \begin{bmatrix} -1 & 0 \\ 0 & 2 \end{bmatrix} \begin{bmatrix} x_1 \\ x_2 \end{bmatrix} + \begin{bmatrix} 0 \\ 1 \end{bmatrix} u$$

Can the closed-loop poles be placed at -2?
Can this system be stabilized? Explain.

The above system is said to be **stabilizable**, which is a weaker notion than controllability.

6-21 Repeat the above for the system

$$\begin{bmatrix} \dot{x}_1 \\ \dot{x}_2 \end{bmatrix} = \begin{bmatrix} +1 & 0 \\ 0 & 2 \end{bmatrix} \begin{bmatrix} x_1 \\ x_2 \end{bmatrix} + \begin{bmatrix} 0 \\ 1 \end{bmatrix} u$$

6-22 Consider the block diagram of Figure 6.16. Suppose that $G(s) = 1/(2s^2 + s)$ and suppose that it is desired to track a reference signal $\theta^d(t) = \sin(t) + \cos(2t)$. If we further specify that the closed-loop system should have a natural frequency less than 10 radians with a damping ratio greater than 0.707, compute an appropriate compensator $C(s)$ and feedforward transfer function $F(s)$.

NOTES AND REFERENCES

Although we treated only the dynamics of permanent-magnet DC motors, the use of AC motors is increasing in robotics and other types of motion control applications. AC motors do not require commutators and brushes and so are inherently more maintenance free and reliable. However, they are more difficult to control and require more sophisticated power electronics. With recent advances in power electronics together with their decreasing cost, AC motors may soon replace DC motors as the dominant actuation method for robot manipulators. A reference that treats different types of motors is [48].

A good background text on linear control systems is [70]. For a text that treats PID control in depth, consult [7].

The pole assignment theorem is due to Wonham [141]. The notion of controllability and observability, introduced by Kalman in [59], arises in several fundamental ways in addition to those discussed here. The interested reader should consult various references on the Kalman filter, which is a linear state estimator for systems whose output measurements are corrupted by (stochastic, white) noise. The linear observer that we discuss here was introduced by Luenberger [81] and is often referred to as the deterministic Kalman filter. Luenberger's main contribution in [81] was in the so-called reduced-order observer, which allows one to reduce the dimension of the observer.

Linear quadratic optimal control, in its present form, was introduced by Kalman in [59], where the importance of the Riccati equation was emphasized. The field of linear control theory is broad and there are many other techniques available to design state-feedback and output-feedback control laws in addition to the basic optimal control approach considered here. More recent control system design methods include the H_∞ approach [28] as well as approaches based on fuzzy logic [101] and neural networks [77] and [44].

The problem of drive-train dynamics in robotics was first pointed out by Good and Sweet, who studied the dynamics of the General Electric P-50 robot (see [130]). For this and other early robots the limiting factors to performance were current limiters that limited how much current could be drawn by the motors and elasticity in the joints due to gear flexibility. Both effects limit the maximum attainable safe speed at which the robot can operate. This work stimulated considerable research into the control of robots with input constraints (see [126]) and the control of robots with flexible joints (see [122]).

Finally, virtually all robot control systems today are implemented digitally. A treatment of digital control requires consideration of issues of sampling, quantization, resolution, as well as computer architecture, real-time programming and other issues that are not considered in this chapter. The interested reader should consult, for example, [39] for these latter subjects.

Chapter 7

DYNAMICS

This chapter deals with the dynamics of robot manipulators. Whereas the kinematic equations describe the motion of the robot without consideration of the forces and torques producing the motion, the dynamic equations explicitly describe the relationship between force and motion. The equations of motion are important to consider in the design of robots, in simulation and animation of robot motion, and in the design of control algorithms. We introduce the so-called **Euler-Lagrange equations**, which describe the evolution of a mechanical system subject to **holonomic constraints** (this term is defined later on). To motivate the Euler-Lagrange approach we begin with a simple derivation of these equations from Newton's second law for a one-degree-of-freedom system. We then derive the Euler-Lagrange equations from the **principle of virtual work** in the general case.

In order to determine the Euler-Lagrange equations in a specific situation, one has to form the **Lagrangian** of the system, which is the difference between the **kinetic energy** and the **potential energy**; we show how to do this in several commonly encountered situations. We then derive the dynamic equations of several example robotic manipulators, including a two-link cartesian robot, a two-link planar robot, and a two-link robot with remotely driven joints.

We also discuss several important properties of the Euler-Lagrange equations that can be exploited to design and analyze feedback control algorithms. Among these are explicit bounds on the inertia matrix, linearity in the inertia parameters, and the skew symmetry and passivity properties.

This chapter is concluded with a derivation of an alternate formulation of the dynamical equations of a robot, known as the **Newton-Euler formulation**, which is a recursive formulation of the dynamic equations that is often used for numerical calculation.

7.1 THE EULER-LAGRANGE EQUATIONS

In this section we derive a general set of differential equations that describe the time evolution of mechanical systems subjected to holonomic constraints when the constraint forces satisfy the principle of virtual work. These are called the **Euler-Lagrange equations** of motion. Note that there are at least two distinct ways of deriving these equations. The method presented here is based on the method of virtual work, but it is also possible to derive the same equations using Hamilton's principle of least action.

7.1.1 Motivation

To motivate the subsequent derivation, we show first how the Euler-Lagrange equations can be derived from Newton's second law for the one-degree-of-freedom system shown in Figure 7.1.

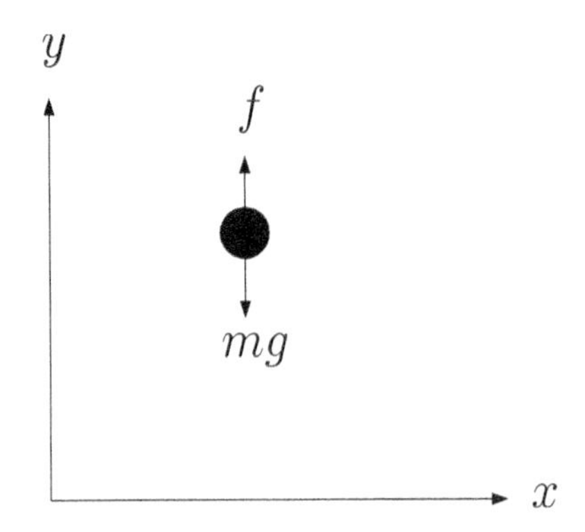

Figure 7.1: A particle of mass m constrained to move vertically constitutes a one-degree-of-freedom system. The gravitational force mg acts downward and an external force f acts upward.

By Newton's second law, the equation of motion of the particle is

$$m\ddot{y} = f - mg \tag{7.1}$$

Notice that the left-hand side of Equation (7.1) can be written as

$$m\ddot{y} = \frac{d}{dt}(m\dot{y}) = \frac{d}{dt}\frac{\partial}{\partial \dot{y}}\left(\frac{1}{2}m\dot{y}^2\right) = \frac{d}{dt}\frac{\partial \mathcal{K}}{\partial \dot{y}} \tag{7.2}$$

where $\mathcal{K} = \frac{1}{2}m\dot{y}^2$ is the **kinetic energy**. We use the partial derivative notation in the above expression to be consistent with systems considered later when the kinetic energy will be a function of several variables. Likewise we can express the gravitational force in Equation (7.1) as

$$mg = \frac{\partial}{\partial y}(mgy) = \frac{\partial \mathcal{P}}{\partial y} \tag{7.3}$$

where $\mathcal{P} = mgy$ is the **potential energy due to gravity**. If we define

$$\mathcal{L} = \mathcal{K} - \mathcal{P} = \frac{1}{2}m\dot{y}^2 - mgy \tag{7.4}$$

and note that

$$\frac{\partial \mathcal{L}}{\partial \dot{y}} = \frac{\partial \mathcal{K}}{\partial \dot{y}} \quad \text{and} \quad \frac{\partial \mathcal{L}}{\partial y} = -\frac{\partial \mathcal{P}}{\partial y}$$

then we can write Equation (7.1) as

$$\frac{d}{dt}\frac{\partial \mathcal{L}}{\partial \dot{y}} - \frac{\partial \mathcal{L}}{\partial y} = f \tag{7.5}$$

The function $\mathcal{L}$, which is the difference of the kinetic and potential energy, is called the **Lagrangian** of the system, and Equation (7.5) is called the **Euler-Lagrange Equation**.

The general procedure that we discuss below is, of course, the reverse of the above; namely, one first writes the kinetic and potential energies of a system in terms of a set of so-called **generalized coordinates** $(q_1, \ldots, q_n)$, where n is the number of degrees of freedom of the system and then computes the equations of motion of the n-DOF system according to

$$\frac{d}{dt}\frac{\partial \mathcal{L}}{\partial \dot{q}_k} - \frac{\partial \mathcal{L}}{\partial q_k} = \tau_k \; ; \; k = 1, \ldots, n \tag{7.6}$$

where τ_k is the (generalized) force associated with q_k. In the above single DOF example, the variable y serves as the generalized coordinate. Application of the Euler-Lagrange equations leads to a set of coupled second-order ordinary differential equations and provides a formulation of the dynamic equations of motion equivalent to those derived using Newton's second law. However, as we shall see, the Lagrangian approach is advantageous for complex systems such as multi-link robots.

Example 7.1 Single-Link Manipulator

Consider the single-link robot arm shown in Figure 7.2, consisting of a rigid link coupled through a gear train to a DC motor. Let θ_ℓ and θ_m denote

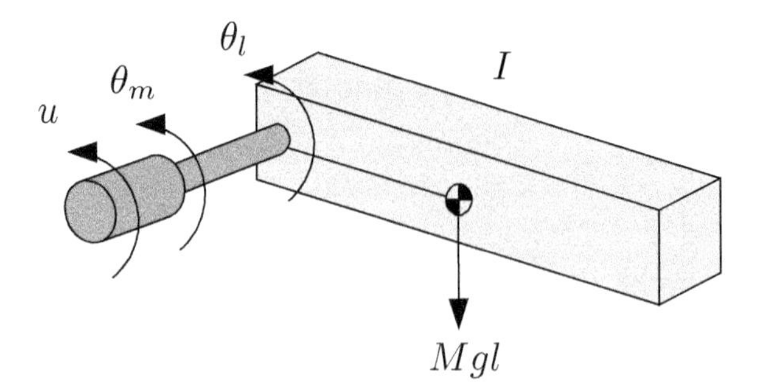

Figure 7.2: Single-link robot. The motor shaft is coupled to the axis of rotation of the link through a gear train which amplifies the motor torque and reduces the motor speed.

the angles of the link and motor shaft, respectively. Then $\theta_m = r\theta_\ell$ *where* $r : 1$ *is the gear ratio. The algebraic relation between the link and motor shaft angles means that the system has only one degree of freedom and we can therefore use as generalized coordinate either* θ_m *or* θ_ℓ*.*

In terms of θ_ℓ*, the kinetic energy of the system is given by*

$$\begin{aligned} K &= \frac{1}{2}J_m\dot{\theta}_m^2 + \frac{1}{2}J_\ell\dot{\theta}_\ell^2 \\ &= \frac{1}{2}(r^2 J_m + J_\ell)\dot{\theta}_\ell^2 \end{aligned} \tag{7.7}$$

where J_m, J_ℓ *are the rotational inertias of the motor and link, respectively. The potential energy is given as*

$$P = Mg\ell(1 - \cos\theta_\ell) \tag{7.8}$$

where M *is the total mass of the link and* ℓ *is the distance from the joint axis to the link center of mass. Defining* $J = r^2 J_m + J_\ell$*, the Lagrangian* $\mathcal{L}$ *is given by*

$$\mathcal{L} = \frac{1}{2}J\dot{\theta}_\ell^2 - Mg\ell(1 - \cos\theta_\ell) \tag{7.9}$$

Substituting this expression into the Equation (7.6) with $n = 1$ *and generalized coordinate* θ_ℓ *yields the equation of motion*

$$J\ddot{\theta}_\ell + Mg\ell\sin\theta_\ell = \tau_\ell \tag{7.10}$$

The generalized force τ_ℓ *represents those external forces and torques that are not derivable from a potential function. For this example,* τ_ℓ *consists of the input motor torque* $u = r\tau_m$, *reflected to the link, and (nonconservative) damping torques* $B_m\dot{\theta}_m$, *and* $B_\ell, \dot{\theta}_\ell$. *Reflecting the motor damping to the link yields*

$$\tau_\ell = u - B\dot{\theta}_\ell$$

where $B = rB_m + B_\ell$. *Therefore, the complete expression for the dynamics of this system is*

$$J\ddot{\theta}_\ell + B\dot{\theta}_\ell + Mg\ell\sin\theta_\ell = u \tag{7.11}$$

◇

7.1.2 Holonomic Constraints and Virtual Work

Now, consider a system of k particles with corresponding position vectors $r_1, \ldots, r_k$ as shown in Figure 7.3.

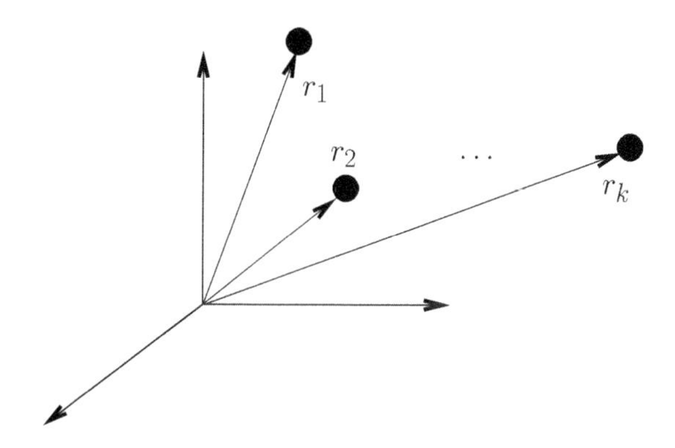

Figure 7.3: An unconstrained system of k particles has $3k$ degrees of freedom. If the particles are constrained, the number of degrees of freedom is reduced.

If these particles are free to move about without any restrictions, then it is quite an easy matter to describe their motion, by noting that the mass times acceleration of each particle equals the external force applied to it. However, if the motion of the particles is constrained in some fashion, then one must take into account not only the externally applied forces, but also the so-called **constraint forces**, that is, the forces needed to make the constraints hold. As a simple illustration of this, suppose the system consists of two particles joined by a massless rigid wire of length ℓ. Then the two coordinates r_1 and r_2 must satisfy the constraint

$$\|r_1 - r_2\| = \ell \quad \text{or} \quad (r_1 - r_2)^T(r_1 - r_2) = \ell^2 \tag{7.12}$$

If one applies some external forces to each particle, then the particles experience not only these external forces but also the force exerted by the wire, which is along the direction $r_2 - r_1$ and of appropriate magnitude. Therefore, in order to analyze the motion of the two particles, we could follow one of two options. We could compute, under each set of external forces, what the corresponding constraint force must be in order that the equation above continues to hold. Alternatively, we can search for a method of analysis that does not require us to know the constraint force. Clearly, the second alternative is preferable, since it is generally quite an involved task to compute the constraint forces. This section is aimed at achieving this latter objective.

First, it is necessary to introduce some terminology. A constraint on the k coordinates $r_1, \ldots, r_k$ is called **holonomic** if it is an equality constraint of the form

$$g_i(r_1, \ldots, r_k) = 0, \qquad i = 1, \ldots, \ell \tag{7.13}$$

The constraint given in Equation (7.12) imposed by connecting two particles by a massless rigid wire is an example of a holonomic constraint. By differentiating Equation (7.13) we have an expression of the form

$$\sum_{j=1}^{k} \frac{\partial g_i}{\partial r_j} \cdot dr_j = 0 \tag{7.14}$$

A constraint of the form

$$\sum_{j=1}^{k} \omega_j \cdot dr_j = 0 \tag{7.15}$$

is called **nonholonomic** if it cannot be integrated to an equality constraint of the form (7.13). It is interesting to note that, while the method of deriving the equations of motion using the principle of virtual work remains valid for nonholonomic systems, methods based on variational principles, such as Hamilton's principle, can no longer be applied to derive the equations of motion. We will discuss systems subject to nonholonomic constraints in Chapter 10.

If a system is subjected to ℓ holonomic constraints, then one can think in terms of the constrained system having ℓ fewer degrees of freedom than the unconstrained system. In this case, it may be possible to express the coordinates of the k particles in terms of n **generalized coordinates** $q_1, \ldots, q_n$. In other words, we assume that the coordinates of the various particles, subjected to the set of constraints given by Equation (7.13), can be expressed in the form

$$r_i = r_i(q_1, \ldots, q_n), \qquad i = 1, \ldots, k \tag{7.16}$$

where $q_1, \ldots, q_n$ are all independent. In fact, the idea of generalized coordinates can be used even when there are infinitely many particles. For example, a physical rigid object such as a bar contains an infinity of particles; but since the distance between each pair of particles is fixed throughout the motion of the bar, six coordinates are sufficient to specify completely the coordinates of any particle in the bar. In particular, one could use three position coordinates to specify the location of the center of mass of the bar, and three Euler angles to specify the orientation of the body. Typically, generalized coordinates are positions, angles, etc. In fact, in Chapter 3 we chose to denote the joint variables by the symbols $q_1, \ldots, q_n$ precisely because these joint variables form a set of generalized coordinates for an n-link robot manipulator.

One can now speak of **virtual displacements**, which are any set of infinitesimal displacements, $\delta r_1, \ldots, \delta r_k$, that are consistent with the constraints. For example, consider once again the constraint (7.12) and suppose r_1, r_2 are perturbed to $r_1 + \delta r_1$, $r_2 + \delta r_2$, respectively. Then, in order that the perturbed coordinates continue to satisfy the constraint, we must have

$$(r_1 + \delta r_1 - r_2 - \delta r_2)^T(r_1 + \delta r_1 - r_2 - \delta r_2) \quad = \quad \ell^2 \tag{7.17}$$

Now, let us expand the above product and take advantage of the fact that the original coordinates r_1, r_2 satisfy the constraint given by Equation (7.12). If we neglect quadratic terms in $\delta r_1, \delta r_2$ we obtain after some algebra (Problem 7-1)

$$(r_1 - r_2)^T(\delta r_1 - \delta r_2) \quad = \quad 0 \tag{7.18}$$

Thus, any infinitesimal perturbations in the positions of the two particles must satisfy the above equation in order that the perturbed positions continue to satisfy the constraint Equation (7.12). Any pair of infinitesimal vectors $\delta r_1, \delta r_2$ that satisfy Equation (7.18) constitutes a set of virtual displacements for this problem. Figure 7.4 shows some representative virtual displacements for a rigid bar.

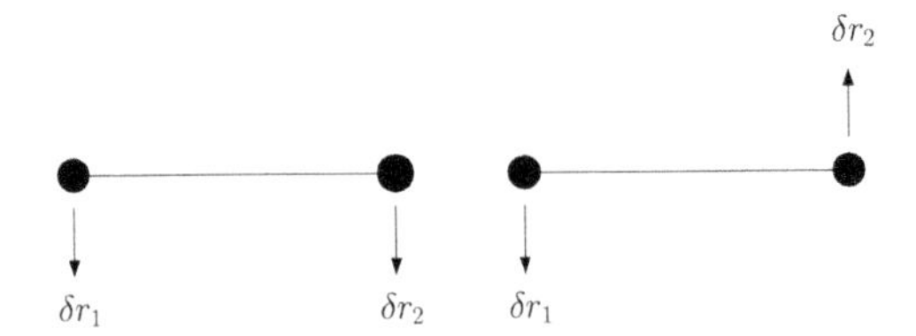

Figure 7.4: Examples of virtual displacements for a rigid bar. These infinitesimal motions do not change the distance between the endpoints and are thus compatible with the assumption that the bar is rigid.

Now, the reason for using generalized coordinates is to avoid dealing with complicated relationships such as Equation (7.18) above. If Equation (7.16) holds, then one can see that the set of all virtual displacements is precisely

$$\delta r_i = \sum_{j=1}^{n} \frac{\partial r_i}{\partial q_j} \delta q_j, \quad i = 1, \ldots, k \tag{7.19}$$

where the virtual displacements $\delta q_1, \ldots, \delta q_n$ of the generalized coordinates are unconstrained (that is what makes them generalized coordinates).

Next, we begin a discussion of constrained systems in equilibrium. Suppose each particle is in equilibrium. Then the net force on each particle is zero, which in turn implies that the work done by each set of virtual displacements is zero. Hence, the sum of the work done by any set of virtual displacements is also zero; that is,

$$\sum_{i=1}^{k} F_i^T \delta r_i = 0 \tag{7.20}$$

where F_i is the total force on particle i. As mentioned earlier, the force F_i is the sum of two quantities, namely (i) the externally applied force f_i, and (ii) the constraint force f_i^a. Now, suppose that the total work done by the constraint forces corresponding to any set of virtual displacements is zero, that is,

$$\sum_{i=1}^{k} f_i^{aT} \delta r_i = 0 \tag{7.21}$$

This will be true whenever the constraint force between a pair of particles is directed along the radial vector connecting the two particles (see the discussion in the next paragraph). Substituting Equation (7.21) into

Equation (7.20) results in

$$\sum_{i=1}^{k} f_i^T \delta r_i = 0 \tag{7.22}$$

The beauty of this equation is that it does not involve the unknown constraint forces, but only the known external forces. This equation expresses the **principle of virtual work**, which can be stated in words as follows:
Principle of Virtual Work: The work done by external forces corresponding to any set of virtual displacements is zero.

Note that the principle is not universally applicable; it requires that Equation (7.21) hold, that is, that the constraint forces do no work. Thus, if the principle of virtual work applies, one can analyze the dynamics of a system *without* having to evaluate the constraint forces.

It is easy to verify that the principle of virtual work applies whenever the constraint force between a pair of particles acts along the vector connecting the position coordinates of the two particles. In particular, when the constraints are of the form (7.12), the principle applies. To see this, consider once again a single constraint of the form (7.12). In this case, the constraint force, if any, must be exerted by the rigid massless wire, and therefore must be directed along the radial vector connecting the two particles. In other words, the force exerted on the first particle by the wire must be of the form

$$f_1^a = c(r_1 - r_2) \tag{7.23}$$

for some constant c (which could change as the particles move about). By the law of action and reaction, the force exerted on the second particle by the wire must be just the negative of the above, that is,

$$f_2^a = -c(r_1 - r_2) \tag{7.24}$$

Now, the work done by the constraint forces corresponding to a set of virtual displacements is

$$f_1^{aT} \delta r_1 + f_2^{aT} \delta r_2 = c(r_1 - r_2)^T (\delta r_1 - \delta r_2) \tag{7.25}$$

But, Equation (7.18) shows that the above expression must be zero for any set of virtual displacements. The same reasoning can be applied if the system consists of several particles that are pairwise connected by rigid massless wires of fixed lengths, in which case the system is subjected to several constraints of the form (7.12). Now, the requirement that the motion of a body be rigid can be equivalently expressed as the requirement that the distance between any pair of points on the body remain constant as the body moves, that is, as an infinity of constraints of the form (7.12). Thus, the principle of virtual work applies whenever rigidity is the only constraint on the motion. There are indeed situations when this principle does not apply, such as in the presence of magnetic fields. However, in all situations encountered in this book, we can safely assume that the principle of virtual work is valid.

7.1.3 D'Alembert's Principle

In Equation (7.22), the virtual displacements δr_i are not independent, so we cannot conclude from this equation that each coefficient F_i *individually* equals zero. In order to apply such reasoning, we must transform to generalized coordinates. Before doing this, we consider systems that are not necessarily in equilibrium. For such systems, **D'Alembert's principle** states that, if one introduces a fictitious additional force $-\dot{p}_i$ on each particle, where p_i is the momentum of particle i, then each particle will be in equilibrium. Thus, if one modifies Equation (7.20) by replacing F_i by $F_i - \dot{p}_i$, then the resulting equation is valid for arbitrary systems. One can then remove the constraint forces as before using the principle of virtual work. This results in the equation

$$\sum_{i=1}^{k} f_i^T \delta r_i - \sum_{i=1}^{k} \dot{p}_i^T \delta r_i \quad = \quad 0 \tag{7.26}$$

The above equation does not mean that each coefficient of δr_i is zero since the virtual constraints δr_i are not independent. The remainder of this derivation is aimed at expressing the above equation in terms of the generalized coordinates which are independent. For this purpose, we express each δr_i in terms of the corresponding virtual displacements of generalized coordinates, as is done in Equation (7.19). Then, the virtual work done by the forces f_i is given by

$$\sum_{i=1}^{k} f_i^T \delta r_i \quad = \quad \sum_{i=1}^{k} \sum_{j=1}^{n} f_i^T \frac{\partial r_i}{\partial q_j} \delta q_j = \sum_{j=1}^{n} \psi_j \delta q_j \tag{7.27}$$

where

$$\psi_j \quad = \quad \sum_{i=1}^{k} f_i^T \frac{\partial r_i}{\partial q_j} \tag{7.28}$$

is called the jth **generalized force**. Note that ψ_j need not have dimensions of force, just as q_j need not have dimensions of length; however, $\psi_j \delta q_j$ must always have dimensions of work.

Now, let us study the second summation in Equation (7.26). Since $p_i = m_i \dot{r}_i$, it follows that

$$\sum_{i=1}^{k} \dot{p}_i^T \delta r_i = \sum_{i=1}^{k} m_i \ddot{r}_i^T \delta r_i = \sum_{i=1}^{k} \sum_{j=1}^{n} m_i \ddot{r}_i^T \frac{\partial r_i}{\partial q_j} \delta q_j \tag{7.29}$$

Next, using the product rule of differentiation, we have

$$\frac{d}{dt}\left[m_i \dot{r}_i^T \frac{\partial r_i}{\partial q_j} \right] = m_i \ddot{r}_i^T \frac{\partial r_i}{\partial q_j} + m_i \dot{r}_i^T \frac{d}{dt}\left[\frac{\partial r_i}{\partial q_j} \right] \tag{7.30}$$

Rearranging the above and summing over all $i = 1, \ldots, n$ yields

$$\sum_{i=1}^{k} m_i \ddot{r}_i^T \frac{\partial r_i}{\partial q_j} = \sum_{i=1}^{k} \left\{ \frac{d}{dt}\left[m_i \dot{r}_i^T \frac{\partial r_i}{\partial q_j} \right] - m_i \dot{r}_i^T \frac{d}{dt}\left[\frac{\partial r_i}{\partial q_j} \right] \right\} \tag{7.31}$$

Now, differentiating Equation (7.16) using the chain rule gives

$$v_i = \dot{r}_i = \sum_{j=1}^{n} \frac{\partial r_i}{\partial q_j} \dot{q}_j \tag{7.32}$$

Observe from the above equation that

$$\frac{\partial v_i}{\partial \dot{q}_j} = \frac{\partial r_i}{\partial q_j} \tag{7.33}$$

Next,

$$\frac{d}{dt}\left[\frac{\partial r_i}{\partial q_j} \right] = \sum_{\ell=1}^{n} \frac{\partial^2 r_i}{\partial q_j \partial q_\ell} \dot{q}_\ell = \frac{\partial}{\partial q_j} \sum_{\ell=1}^{n} \frac{\partial r_i}{\partial q_\ell} \dot{q}_\ell = \frac{\partial v_i}{\partial q_j} \tag{7.34}$$

where the last equality follows from Equation (7.32).

Substituting from Equation (7.33) and Equation (7.34) into Equation (7.31) and noting that $\dot{r}_i = v_i$ gives

$$\sum_{i=1}^{k} m_i \ddot{r}_i^T \frac{\partial r_i}{\partial q_j} = \sum_{i=1}^{k} \left\{ \frac{d}{dt}\left[m_i v_i^T \frac{\partial v_i}{\partial \dot{q}_j} \right] - m_i v_i^T \frac{\partial v_i}{\partial q_j} \right\} \tag{7.35}$$

If we define the *kinetic energy* K to be the quantity

$$K = \sum_{i=1}^{k} \frac{1}{2} m_i v_i^T v_i \tag{7.36}$$

then Equation (7.35) can be compactly expressed as

$$\sum_{i=1}^{k} m_i \ddot{r}_i^T \frac{\partial r_i}{\partial q_j} = \frac{d}{dt}\frac{\partial K}{\partial \dot{q}_j} - \frac{\partial K}{\partial q_j} \tag{7.37}$$

Now, substituting from Equation (7.37) into Equation (7.29) shows that the second summation in Equation (7.26) is

$$\sum_{i=1}^{k} \dot{p}_i^T \delta r_i = \sum_{j=1}^{n} \left\{ \frac{d}{dt}\frac{\partial K}{\partial \dot{q}_j} - \frac{\partial K}{\partial q_j} \right\} \delta q_j \tag{7.38}$$

Finally, combining Equations (7.26), (7.27), and(7.38) gives

$$\sum_{j=1}^{n} \left\{ \frac{d}{dt}\frac{\partial K}{\partial \dot{q}_j} - \frac{\partial K}{\partial q_j} - \psi_j \right\} \delta q_j = 0 \tag{7.39}$$

Now, since the virtual displacements δq_j are independent, we can conclude that each coefficient in Equation (7.39) is zero, that is,

$$\frac{d}{dt}\frac{\partial K}{\partial \dot{q}_j} - \frac{\partial K}{\partial q_j} = \psi_j, \quad j = 1, \ldots, n \tag{7.40}$$

If the generalized force ψ_j is the sum of an externally applied generalized force and another one due to a potential field, then a further modification is possible. Suppose there exist functions τ_j and a potential energy function $P(q)$ such that

$$\psi_j = -\frac{\partial P}{\partial q_j} + \tau_j \tag{7.41}$$

Then Equation (7.40) can be written in the form

$$\frac{d}{dt}\frac{\partial \mathcal{L}}{\partial \dot{q}_j} - \frac{\partial \mathcal{L}}{\partial q_j} = \tau_j \tag{7.42}$$

where $\mathcal{L} = K - P$ is the Lagrangian and we have recovered the **Euler-Lagrange equations of motion** as in Equation (7.6).

7.2 KINETIC AND POTENTIAL ENERGY

In the previous section, we showed that the Euler-Lagrange equations can be used to derive the dynamical equations in a straightforward manner, provided one is able to express the kinetic and potential energy of the system in terms of a set of generalized coordinates. In order for this result to be useful in a practical context, it is important that one be able to compute these terms readily for an n-link robotic manipulator. In this section we derive formulas for the kinetic energy and potential energy of a robot with rigid links using the Denavit-Hartenberg joint variables as generalized coordinates.

To begin we note that the kinetic energy of a rigid object is the sum of two terms, the translational kinetic energy obtained by concentrating the entire mass of the object at the center of mass, and the rotational kinetic energy of the body about the center of mass. Referring to Figure 7.5 we attach a coordinate frame at the center of mass (called the **body attached frame**) as shown.

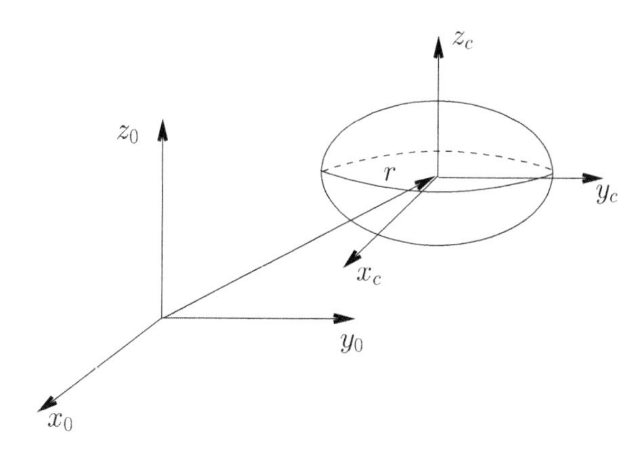

Figure 7.5: A general rigid body has six degrees of freedom. The kinetic energy consists of kinetic energy of rotation and kinetic energy of translation.

The kinetic energy of the rigid body is then given as

$$\mathcal{K} = \frac{1}{2}mv^Tv + \frac{1}{2}\omega^T\mathcal{I}\omega \tag{7.43}$$

where m is the total mass of the object, v and ω are the linear and angular velocity vectors, respectively, and $\mathcal{I}$ is a symmetric 3×3 matrix called the **inertia tensor**.

7.2.1 The Inertia Tensor

It is understood that the above linear and angular velocity vectors, v and ω, respectively, are expressed in the inertial frame. In this case, we know that ω is found from the skew symmetric matrix

$$S(\omega) = \dot{R}R^T \tag{7.44}$$

where R is the orientation transformation from the body attached frame and the inertial frame. It is therefore necessary to express the inertia tensor, $\mathcal{I}$, also in the inertial frame in order to compute the triple product $\omega^T\mathcal{I}\omega$. The inertia tensor relative to the inertial reference frame will depend on the configuration of the object. If we denote as I the inertia tensor expressed instead in the body attached frame, then the two matrices are related via a similarity transformation according to

$$\mathcal{I} = RIR^T \tag{7.45}$$

This is an important observation because the inertia matrix expressed in the body attached frame is a constant matrix independent of the motion of the object and easily computed.

We next show how to compute this matrix explicitly. Let the mass density of the object be represented as a function of position, $\rho(x, y, z)$. Then the inertia tensor in the body attached frame is computed as

$$I = \begin{bmatrix} I_{xx} & I_{xy} & I_{xz} \\ I_{yx} & I_{yy} & I_{yz} \\ I_{zx} & I_{zy} & I_{zz} \end{bmatrix} \tag{7.46}$$

where

$$\begin{aligned} I_{xx} &= \int\int\int (y^2 + z^2)\rho(x, y, z)dx\ dy\ dz \\ I_{yy} &= \int\int\int (x^2 + z^2)\rho(x, y, z)dx\ dy\ dz \\ I_{zz} &= \int\int\int (x^2 + y^2)\rho(x, y, z)dx\ dy\ dz \end{aligned}$$

and

$$\begin{aligned} I_{xy} = I_{yx} &= -\int\int\int xy\rho(x, y, z)dx\ dy\ dz \\ I_{xz} = I_{zx} &= -\int\int\int xz\rho(x, y, z)dx\ dy\ dz \\ I_{yz} = I_{zy} &= -\int\int\int yz\rho(x, y, z)dx\ dy\ dz \end{aligned}$$

The integrals in the above expression are computed over the region of space occupied by the rigid body. The diagonal elements of the inertia tensor, I_{xx},

I_{yy}, I_{zz}, are called the **principal moments of inertia** about the x, y, and z axes, respectively. The off-diagonal terms I_{xy}, I_{xz}, etc., are called the **cross products of inertia**. If the mass distribution of the body is symmetric with respect to the body attached frame, then the cross products of inertia are identically zero.

Example 7.2 Uniform Rectangular Solid

Consider the rectangular solid of length a width b and height c shown in Figure 7.6 and suppose that the density is constant, $\rho(x, y, z) = \rho$.

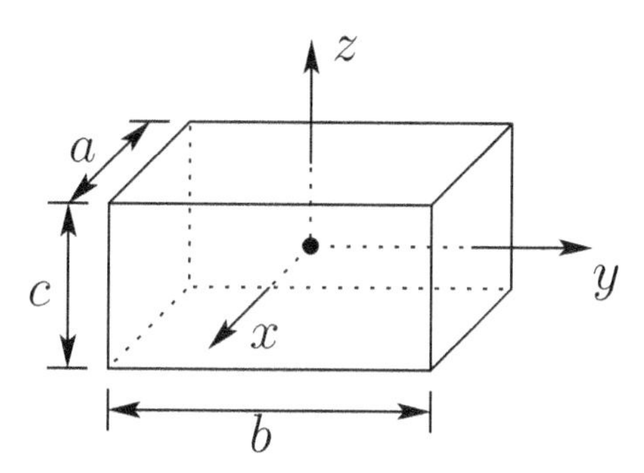

Figure 7.6: A rectangular solid with uniform mass density and coordinate frame attached at the geometric center of the solid.

If the body frame is attached at the geometric center of the object, then by symmetry, the cross products of inertia are all zero and it is a simple exercise to compute

$$\begin{aligned} I_{xx} &= \int_{-c/2}^{c/2} \int_{-b/2}^{b/2} \int_{-a/2}^{a/2} (y^2 + z^2)\rho(x, y, z)dx\ dy\ dz \\ &= \rho \frac{abc}{12}(b^2 + c^2) = \frac{m}{12}(b^2 + c^2) \end{aligned}$$

since $\rho abc = m$, the total mass. Likewise, a similar calculation shows that

$$I_{yy} = \frac{m}{12}(a^2 + c^2) \quad ; \quad I_{zz} = \frac{m}{12}(a^2 + b^2)$$

◇

7.2.2 Kinetic Energy for an n-Link Robot

Now, consider a manipulator with n links. We have seen in Chapter 4 that the linear and angular velocities of any point on any link can be expressed in terms of the Jacobian matrix and the derivatives of the joint variables. Since, in our case, the joint variables are indeed generalized coordinates, it follows that, for appropriate Jacobian matrices J_{v_i} and J_{ω_i}, we have

$$v_i = J_{v_i}(q)\dot{q}, \qquad \omega_i = J_{\omega_i}(q)\dot{q} \tag{7.47}$$

Now, suppose the mass of link i is m_i and that the inertia matrix of link i, evaluated around a coordinate frame parallel to frame i but whose origin is at the center of mass, equals I_i. Then from Equations (7.43) and (7.47) it follows that the overall kinetic energy of the manipulator equals

$$K = \frac{1}{2}\dot{q}^T\left[\sum_{i=1}^{n}\{m_i J_{v_i}(q)^T J_{v_i}(q) + J_{\omega_i}(q)^T R_i(q) I_i R_i(q)^T J_{\omega_i}(q)\}\right]\dot{q} \tag{7.48}$$

$$= \frac{1}{2}\dot{q}^T D(q)\dot{q} \tag{7.49}$$

where

$$D(q) = \left[\sum_{i=1}^{n}\{m_i J_{v_i}(q)^T J_{v_i}(q) + J_{\omega_i}(q)^T R_i(q) I_i R_i(q)^T J_{\omega_i}(q)\}\right] \tag{7.50}$$

is an $n \times n$ configuration dependent matrix called the **inertia matrix**. In Section 7.4 we will compute this matrix for several commonly occurring manipulator configurations. The inertia matrix is **symmetric** and **positive definite** for any manipulator. Symmetry of $D(q)$ is easily seen from Equation (7.50). Positive definiteness can be inferred from the fact that the kinetic energy is always nonnegative and is zero if and only if all of the joint velocities are zero. The formal proof is left as an exercise (Problem 7-5).

7.2.3 Potential Energy for an n-Link Robot

Now, consider the potential energy term. In the case of rigid dynamics, the only source of potential energy is gravity. The potential energy of the ith link can be computed by assuming that the mass of the entire object is concentrated at its center of mass and is given by

$$P_i = m_i g^T r_{ci} \tag{7.51}$$

where g is the vector giving the direction of gravity in the inertial frame and the vector r_{ci} gives the coordinates of the center of mass of link i. The total potential energy of the n-link robot is therefore

$$P = \sum_{i=1}^{n} P_i = \sum_{i=1}^{n} m_i g^T r_{ci} \tag{7.52}$$

In the case that the robot contains elasticity, for example if the joints are flexible, then the potential energy will include terms containing the energy stored in the elastic elements. Note that the potential energy is a function only of the generalized coordinates and not their derivatives.

7.3 EQUATIONS OF MOTION

In this section we specialize the Euler-Lagrange equations derived in Section 7.1 to the case when two conditions hold. First, the kinetic energy is a quadratic function of the vector $\dot{q}$ of the form

$$K = \frac{1}{2}\dot{q}^T D(q)\dot{q} = \sum_{i,j} d_{ij}(q)\dot{q}_i\dot{q}_j \tag{7.53}$$

where $d_{i,j}$ are the entries of the $n \times n$ inertia matrix $D(q)$, which is symmetric and positive definite for each $q \in \mathbb{R}^n$, and second, the potential energy $P = P(q)$ is independent of $\dot{q}$. We have already remarked that robotic manipulators satisfy these conditions.

The Euler-Lagrange equations for such a system can be derived as follows. Using Equation (7.53) we can write the Lagrangian as

$$L = K - P = \frac{1}{2}\sum_{i,j} d_{ij}(q)\dot{q}_i\dot{q}_j - P(q) \tag{7.54}$$

The partial derivatives of the Lagrangian with respect to the kth joint velocity is given by

$$\frac{\partial L}{\partial \dot{q}_k} = \sum_j d_{kj}\dot{q}_j \tag{7.55}$$

and therefore

$$\begin{aligned}\frac{d}{dt}\frac{\partial L}{\partial \dot{q}_k} &= \sum_j d_{kj}\ddot{q}_j + \sum_j \frac{d}{dt}d_{kj}\dot{q}_j \\ &= \sum_j d_{kj}\ddot{q}_j + \sum_{i,j}\frac{\partial d_{kj}}{\partial q_i}\dot{q}_i\dot{q}_j \end{aligned} \tag{7.56}$$

Similarly the partial derivative of the Lagrangian with respect to the kth joint position is given by

$$\frac{\partial L}{\partial q_k} = \frac{1}{2}\sum_{i,j}\frac{\partial d_{ij}}{\partial q_k}\dot{q}_i\dot{q}_j - \frac{\partial P}{\partial q_k} \tag{7.57}$$

Thus, for each $k = 1, \ldots, n$, the Euler-Lagrange equations can be written

$$\sum_j d_{kj}\ddot{q}_j + \sum_{i,j}\left\{\frac{\partial d_{kj}}{\partial q_i} - \frac{1}{2}\frac{\partial d_{ij}}{\partial q_k}\right\}\dot{q}_i\dot{q}_j + \frac{\partial P}{\partial q_k} = \tau_k \tag{7.58}$$

By interchanging the order of summation and taking advantage of symmetry, one can show (Problem 7-6) that

$$\sum_{i,j}\left\{\frac{\partial d_{kj}}{\partial q_i}\right\}\dot{q}_i\dot{q}_j = \frac{1}{2}\sum_{i,j}\left\{\frac{\partial d_{kj}}{\partial q_i} + \frac{\partial d_{ki}}{\partial q_j}\right\}\dot{q}_i\dot{q}_j \tag{7.59}$$

Hence

$$\begin{aligned}\sum_{i,j}\left\{\frac{\partial d_{kj}}{\partial q_i} - \frac{1}{2}\frac{\partial d_{ij}}{\partial q_k}\right\}\dot{q}_i\dot{q}_j &= \sum_{i,j}\frac{1}{2}\left\{\frac{\partial d_{kj}}{\partial q_i} + \frac{\partial d_{ki}}{\partial q_j} - \frac{\partial d_{ij}}{\partial q_k}\right\}\dot{q}_i\dot{q}_j \\ &= \sum_{i,j} c_{ijk}\dot{q}_i\dot{q}_j\end{aligned}$$

where we define

$$c_{ijk} := \frac{1}{2}\left\{\frac{\partial d_{kj}}{\partial q_i} + \frac{\partial d_{ki}}{\partial q_j} - \frac{\partial d_{ij}}{\partial q_k}\right\} \tag{7.60}$$

The terms c_{ijk} in Equation (7.60) are known as **Christoffel symbols** (of the first kind). Note that, for a fixed k, we have $c_{ijk} = c_{jik}$, which reduces the effort involved in computing these symbols by a factor of about one half. Finally, if we define

$$g_k = \frac{\partial P}{\partial q_k} \tag{7.61}$$

then we can write the Euler-Lagrange equations as

$$\sum_{j=1}^{n} d_{kj}(q)\ddot{q}_j + \sum_{i=1}^{n}\sum_{j=1}^{n} c_{ijk}(q)\dot{q}_i\dot{q}_j + g_k(q) = \tau_k, \qquad k = 1, \ldots, n \tag{7.62}$$

In the above equations, there are three types of terms. The first type involves the second derivative of the generalized coordinates. The second type involves quadratic terms in the first derivatives of q, where the coefficients may depend on q. These latter terms are further classified into those involving a product of the type $\dot{q}_i^2$ and those involving a product of the type $\dot{q}_i\dot{q}_j$ where $i \neq j$. Terms of the type $\dot{q}_i^2$ are called **centrifugal**, while terms of the type $\dot{q}_i\dot{q}_j$ are called **Coriolis** terms. The third type of terms are those involving only q but not its derivatives. This third type arises from differentiating the potential energy. It is common to write Equation (7.62) in matrix form as

$$D(q)\ddot{q} + C(q,\dot{q})\dot{q} + g(q) \quad = \quad \tau \tag{7.63}$$

where the $(k,j)^{th}$ element of the matrix $C(q,\dot{q})$ is defined as

$$\begin{aligned} c_{kj} &= \sum_{i=1}^{n} c_{ijk}(q)\dot{q}_i \\ &= \sum_{i=1}^{n} \frac{1}{2}\left\{\frac{\partial d_{kj}}{\partial q_i} + \frac{\partial d_{ki}}{\partial q_j} - \frac{\partial d_{ij}}{\partial q_k}\right\}\dot{q}_i \end{aligned} \tag{7.64}$$

and the gravity vector $g(q)$ is given by

$$g(q) = [g_1(q), \ldots, g_n(q)]^T \tag{7.65}$$

In summary, the development in this section is very general and applies to *any* mechanical system whose kinetic energy is of the form (7.53) and whose potential energy is independent of $\dot{q}$. In the next section we apply this discussion to study specific robot configurations.

7.4 SOME COMMON CONFIGURATIONS

In this section we apply the above method of analysis to several manipulator configurations and derive the corresponding equations of motion. The configurations are progressively more complex, beginning with a two-link cartesian manipulator and ending with a five-bar linkage mechanism that has a particularly simple inertia matrix.

Two-Link Cartesian Manipulator

Consider the manipulator shown in Figure 7.7 consisting of two links and two prismatic joints. Denote the masses of the two links by m_1 and m_2, respectively, and denote the displacement of the two prismatic joints by q_1 and q_2, respectively. It is easy to see, as mentioned in Section 7.1, that these two quantities serve as generalized coordinates for the manipulator. Since the generalized coordinates have dimensions of distance, the corresponding generalized forces have units of force. In fact, they are just the forces applied at each joint. Let us denote these by f_i, $i = 1, 2$.

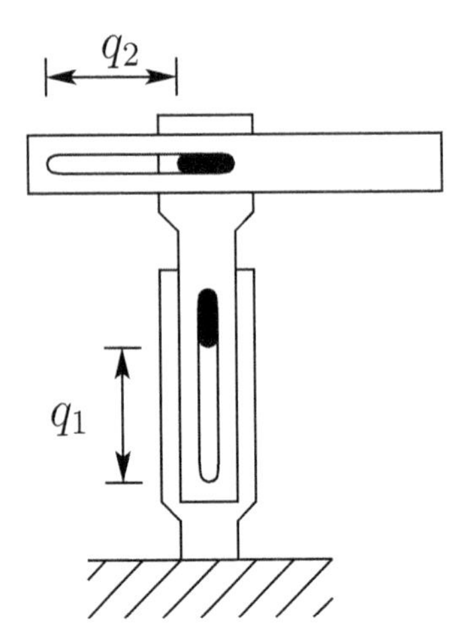

Figure 7.7: Two-link planar Cartesian robot. The orthogonal joint axes and linear joint motion of the Cartesian robot result in simple kinematics and dynamics.

Since we are using the joint variables as the generalized coordinates, we know that the kinetic energy is of the form (7.53) and that the potential energy is only a function of q_1 and q_2. Hence, we can use the formulae in Section 7.3 to obtain the dynamical equations. Also, since both joints are prismatic, the angular velocity Jacobian is zero and the kinetic energy of each link consists solely of the translational term.

It follows that the velocity of the center of mass of link 1 is given by

$$v_{c1} = J_{v_{c1}}\dot{q} \tag{7.66}$$

where

$$J_{v_{c1}} = \begin{bmatrix} 0 & 0 \\ 0 & 0 \\ 1 & 0 \end{bmatrix}, \qquad \dot{q} = \begin{bmatrix} \dot{q}_1 \\ \dot{q}_2 \end{bmatrix} \tag{7.67}$$

Similarly,

$$v_{c2} = J_{v_{c2}}\dot{q} \tag{7.68}$$

where

$$J_{v_{c2}} = \begin{bmatrix} 0 & 0 \\ 0 & 1 \\ 1 & 0 \end{bmatrix} \tag{7.69}$$

Hence, the kinetic energy is given by

$$K = \frac{1}{2}\dot{q}^T \left\{m_1 J_{v_c}^T J_{v_{c1}} + m_2 J_{v_{c2}}^T J_{v_{c2}}\right\} \dot{q} \tag{7.70}$$

Comparing with Equation (7.53), we see that the inertia matrix D is given simply by

$$D = \begin{bmatrix} m_1 + m_2 & 0 \\ 0 & m_2 \end{bmatrix} \tag{7.71}$$

Next, the potential energy of link 1 is $m_1 g q_1$, while that of link 2 is $m_2 g q_1$, where g is the acceleration due to gravity. Hence, the overall potential energy is

$$P = g(m_1 + m_2) q_1 \tag{7.72}$$

Now, we are ready to write down the equations of motion. Since the inertia matrix is constant, all Christoffel symbols are zero. Furthermore, the components g_k of the gravity vector are given by

$$g_1 = \frac{\partial P}{\partial q_1} = g(m_1 + m_2), \qquad g_2 = \frac{\partial P}{\partial q_2} = 0 \tag{7.73}$$

Substituting into Equation (7.62) gives the dynamical equations as

$$\begin{aligned} (m_1 + m_2)\ddot{q}_1 + g(m_1 + m_2) &= f_1 \\ m_2 \ddot{q}_2 &= f_2 \end{aligned} \tag{7.74}$$

Planar Elbow Manipulator

Now, consider the planar manipulator with two revolute joints shown in Figure 7.8. Let us fix notation as follows. For $i = 1, 2$, q_i denotes the joint angle, which also serves as a generalized coordinate; m_i denotes the mass of link i; ℓ_i denotes the length of link i; ℓ_{ci} denotes the distance from the previous joint to the center of mass of link i; and I_i denotes the moment of inertia of link i about an axis coming out of the page, passing through the center of mass of link i.

We will use the Denavit-Hartenberg joint variables as generalized coordinates, which will allow us to make effective use of the Jacobian expressions in Chapter 4 in computing the kinetic energy. First,

$$v_{c1} = J_{v_{c1}} \dot{q} \tag{7.75}$$

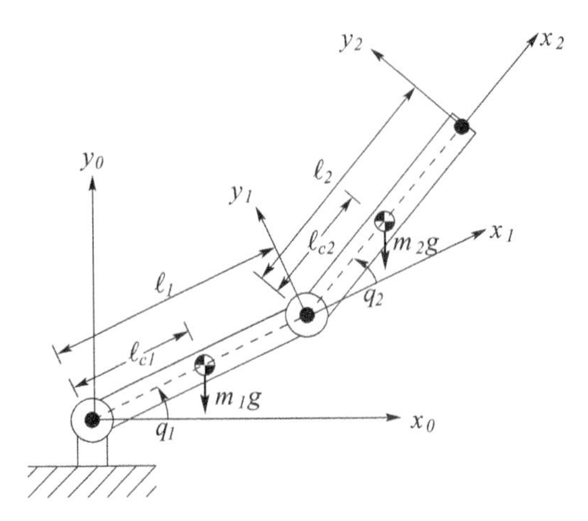

Figure 7.8: Two-link revolute joint arm. The rotational joint motion introduces dynamic coupling between the joints.

where,

$$J_{v_{c1}} = \begin{bmatrix} -\ell_c \sin q_1 & 0 \\ \ell_{c1} \cos q_1 & 0 \\ 0 & 0 \end{bmatrix} \tag{7.76}$$

Similarly,

$$v_{c2} = J_{v_{c2}} \dot{q} \tag{7.77}$$

where

$$J_{v_{c2}} = \begin{bmatrix} -\ell_1 \sin q_1 - \ell_{c2} \sin(q_1 + q_2) & -\ell_{c2} \sin(q_1 + q_2) \\ \ell_1 \cos q_1 + \ell_{c2} \cos(q_1 + q_2) & \ell_{c2} \cos(q_1 + q_2) \\ 0 & 0 \end{bmatrix} \tag{7.78}$$

Hence, the translational part of the kinetic energy is

$$\frac{1}{2} m_1 v_{c1}^T v_{c1} + \frac{1}{2} m_2 v_{c2}^T v_{c2} = \frac{1}{2} \dot{q} \left\{ m_1 J_{v_{c1}}^T J_{v_{c1}} + m_2 J_{v_{c2}}^T J_{v_{c2}} \right\} \dot{q} \tag{7.79}$$

Next, we consider the angular velocity terms. Because of the particularly simple nature of this manipulator, many of the potential difficulties do not arise. First, it is clear that

$$\omega_1 = \dot{q}_1 k, \qquad \omega_2 = (\dot{q}_1 + \dot{q}_2) k \tag{7.80}$$

when expressed in the base inertial frame. Moreover, since ω_i is aligned with the z-axes of each joint coordinate frame, the rotational kinetic energy reduces simply to $I_i\omega_i^2$, where I_i is the moment of inertia about an axis through the center of mass of link i parallel to the z_i-axis. Hence, the rotational kinetic energy of the overall system in terms of the generalized coordinates is

$$\frac{1}{2}\dot{q}^T\left\{I_1\begin{bmatrix}1 & 0\\ 0 & 0\end{bmatrix}+I_2\begin{bmatrix}1 & 1\\ 1 & 1\end{bmatrix}\right\}\dot{q} \tag{7.81}$$

Now, we are ready to form the inertia matrix $D(q)$. For this purpose, we merely have to add the two matrices in Equation (7.79) and Equation (7.81), respectively. Thus

$$D(q) = m_1 J_{v_{c1}}^T J_{v_{c1}} + m_2 J_{v_{c2}}^T J_{v_{c2}} + \begin{bmatrix} I_1+I_2 & I_2\\ I_2 & I_2\end{bmatrix} \tag{7.82}$$

Carrying out the above multiplications and using the standard trigonometric identities $\cos^2\theta+\sin^2\theta=1$, $\cos\alpha\cos\beta+\sin\alpha\sin\beta=\cos(\alpha-\beta)$ leads to

$$\begin{aligned} d_{11} &= m_1\ell_{c1}^2 + m_2(\ell_1^2+\ell_{c2}^2+2\ell_1\ell_{c2}\cos q_2)+I_1+I_2\\ d_{12} &= d_{21} = m_2(\ell_{c2}^2+\ell_1\ell_{c2}\cos q_2)+I_2\\ d_{22} &= m_2\ell_{c2}^2+I_2 \end{aligned} \tag{7.83}$$

Now, we can compute the Christoffel symbols using Equation (7.60). This gives

$$\begin{aligned} c_{111} &= \frac{1}{2}\frac{\partial d_{11}}{\partial q_1} = 0\\ c_{121} &= c_{211} = \frac{1}{2}\frac{\partial d_{11}}{\partial q_2} = -m_2\ell_1\ell_{c2}\sin q_2 = h\\ c_{221} &= \frac{\partial d_{12}}{\partial q_2} - \frac{1}{2}\frac{\partial d_{22}}{\partial q_1} = h\\ c_{112} &= \frac{\partial d_{21}}{\partial q_1} - \frac{1}{2}\frac{\partial d_{11}}{\partial q_2} = -h\\ c_{122} &= c_{212} = \frac{1}{2}\frac{\partial d_{22}}{\partial q_1} = 0\\ c_{222} &= \frac{1}{2}\frac{\partial d_{22}}{\partial q_2} = 0 \end{aligned}$$

Next, the potential energy of the manipulator is just the sum of those of the two links. For each link, the potential energy is just its mass multiplied by the gravitational acceleration and the height of its center of mass. Thus

$$\begin{aligned} P_1 &= m_1 g\ell_{c1}\sin q_1\\ P_2 &= m_2 g(\ell_1\sin q_1+\ell_{c2}\sin(q_1+q_2)) \end{aligned}$$

and so the total potential energy is

$$P = P_1 + P_2 = (m_1\ell_{c1} + m_2\ell_1)g\sin q_1 + m_2\ell_{c2}g\sin(q_1+q_2) \qquad (7.84)$$

Therefore, the functions g_k defined in Equation (7.61) become

$$g_1 = \frac{\partial P}{\partial q_1} = (m_1\ell_{c1} + m_2\ell_1)g\cos q_1 + m_2\ell_{c2}g\cos(q_1+q_2) \qquad (7.85)$$

$$g_2 = \frac{\partial P}{\partial q_2} = m_2\ell_{c2}g\cos(q_1+q_2) \qquad (7.86)$$

Finally, we can write down the dynamical equations of the system as in Equation (7.62). Substituting for the various quantities in this equation and omitting zero terms leads to

$$\begin{array}{lcl} d_{11}\ddot{q}_1 + d_{12}\ddot{q}_2 + c_{121}\dot{q}_1\dot{q}_2 + c_{211}\dot{q}_2\dot{q}_1 + c_{221}\dot{q}_2^2 + g_1 & = & \tau_1 \\ d_{21}\ddot{q}_1 + d_{22}\ddot{q}_2 + c_{112}\dot{q}_1^2 + g_2 & = & \tau_2 \end{array} \qquad (7.87)$$

In this case, the matrix $C(q, \dot{q})$ is given as

$$C = \begin{bmatrix} h\dot{q}_2 & h\dot{q}_2 + h\dot{q}_1 \\ -h\dot{q}_1 & 0 \end{bmatrix} \qquad (7.88)$$

Planar Elbow Manipulator with Remotely Driven Link

Now, we illustrate the use of Lagrangian equations in a situation where the generalized coordinates are not the joint variables defined in earlier chapters. Consider again the planar elbow manipulator, but suppose now that both joints are driven by motors mounted at the base. The first joint is turned directly by one of the motors, while the other is turned via a gearing mechanism or a timing belt (see Figure 7.9).

In this case, one should choose the generalized coordinates as shown in Figure 7.10, because the angle p_2 is determined by driving motor number 2, and is not affected by the angle p_1. We will derive the dynamical equations for this configuration and show that some simplifications will result.

Since p_1 and p_2 are not the joint angles used earlier, we cannot use the velocity Jacobians derived in Chapter 4 in order to find the kinetic energy of each link. Instead, we have to carry out the analysis directly. It is easy to see that

$$v_{c1} = \begin{bmatrix} -\ell_{c1}\sin p_1 & 0 \\ \ell_{c1}\cos p_1 & 0 \\ 0 & 0 \end{bmatrix} \begin{bmatrix} \dot{p}_1 \\ \dot{p}_2 \end{bmatrix} \qquad (7.89)$$

$$v_{c2} = \begin{bmatrix} \ell_1\sin p_1 & -\ell_{c2}\sin p_2 \\ \ell_1\cos p_1 & \ell_{c2}\cos p_2 \\ 0 & 0 \end{bmatrix} \begin{bmatrix} \dot{p}_1 \\ \dot{p}_2 \end{bmatrix} \qquad (7.90)$$

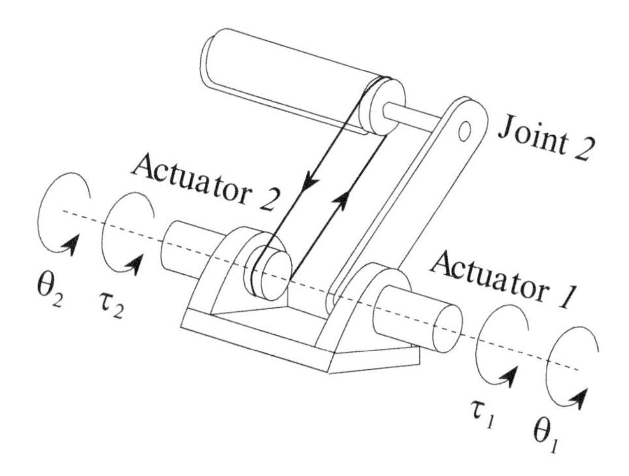

Figure 7.9: Two-link revolute joint arm with remotely driven link. Because of the remote drive the motor shaft angles are not proportional to the joint angles.

$$\omega_1 = \dot{p}_1 k, \qquad \omega_2 = \dot{p}_2 k \tag{7.91}$$

Hence, the kinetic energy of the manipulator equals

$$K = \frac{1}{2}\dot{p}^T D(p)\dot{p} \tag{7.92}$$

where

$$D(p) = \begin{bmatrix} m_1\ell_{c1}^2 + m_2\ell_1^2 + I_1 & m_2\ell_1\ell_{c2}\cos(p_2 - p_1) \\ m_2\ell_1\ell_{c2}\cos(p_2 - p_1) & m_2\ell_{c2}^2 + I_2 \end{bmatrix} \tag{7.93}$$

Computing the Christoffel symbols as in Equation (7.60) gives

$$\begin{aligned}
c_{111} &= \frac{1}{2}\frac{\partial d_{11}}{\partial p_1} = 0 \\
c_{121} &= c_{211} = \frac{1}{2}\frac{\partial d_{11}}{\partial p_2} = 0 \\
c_{221} &= \frac{\partial d_{12}}{\partial p_2} - \frac{1}{2}\frac{\partial d_{22}}{\partial p_1} = -m_2\ell_1\ell_{c2}\sin(p_2 - p_1) \\
c_{112} &= \frac{\partial d_{21}}{\partial p_1} - \frac{1}{2}\frac{\partial d_{11}}{\partial p_2} = m_2\ell_1\ell_{c2}\sin(p_2 - p_1) \\
c_{212} &= = c_{122} = \frac{1}{2}\frac{\partial d_{22}}{\partial p_1} = 0 \\
c_{222} &= \frac{1}{2}\frac{\partial d_{22}}{\partial p_2} = 0
\end{aligned} \tag{7.94}$$

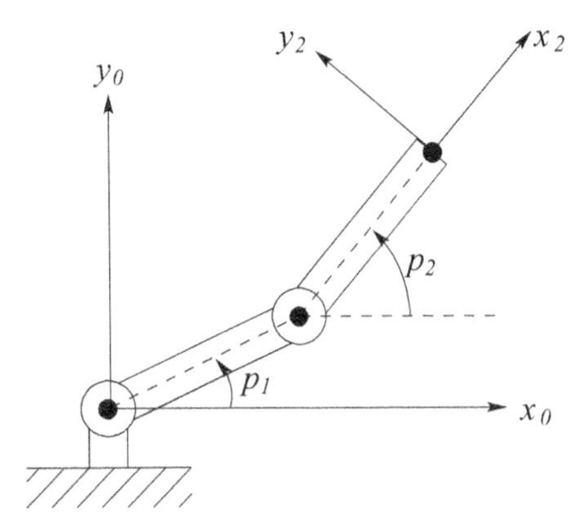

Figure 7.10: Generalized coordinates for the robot of Figure 6.4.

Next, the potential energy of the manipulator, in terms of p_1 and p_2, equals

$$P = m_1 g \ell_{c1} \sin p_1 + m_2 g(\ell_1 \sin p_1 + \ell_{c2} \sin p_2) \tag{7.95}$$

Hence, the gravitational generalized forces are

$$\begin{aligned} g_1 &= (m_1 \ell_{c1} + m_2 \ell_1) g \cos p_1 \\ g_2 &= m_2 \ell_{c2} g \cos p_2 \end{aligned}$$

Finally, the equations of motion are

$$\begin{aligned} d_{11}\ddot{p}_1 + d_{12}\ddot{p}_2 + c_{221}\dot{p}_2^2 + g_1 &= \tau_1 \\ d_{21}\ddot{p}_1 + d_{22}\ddot{p}_2 + c_{112}\dot{p}_1^2 + g_2 &= \tau_2 \end{aligned} \tag{7.96}$$

Comparing Equation (7.96) and Equation (7.87), we see that by driving the second joint remotely from the base we have eliminated the Coriolis forces, but we still have the centrifugal forces coupling the two joints.

Five-Bar Linkage

Now, consider the manipulator shown in Figure 7.11. We will show that, if the parameters of the manipulator satisfy a simple relationship, then the equations of the manipulator are decoupled, so that each quantity q_1 and q_2 can be controlled independently of the other. The mechanism in Figure 7.11 is called a **five-bar linkage**. Clearly, there are only four bars in the figure, but in the theory of mechanisms it is a convention to count the ground as an additional linkage, which explains the terminology. It is assumed that the lengths of links 1 and 3 are the same, and that the two lengths

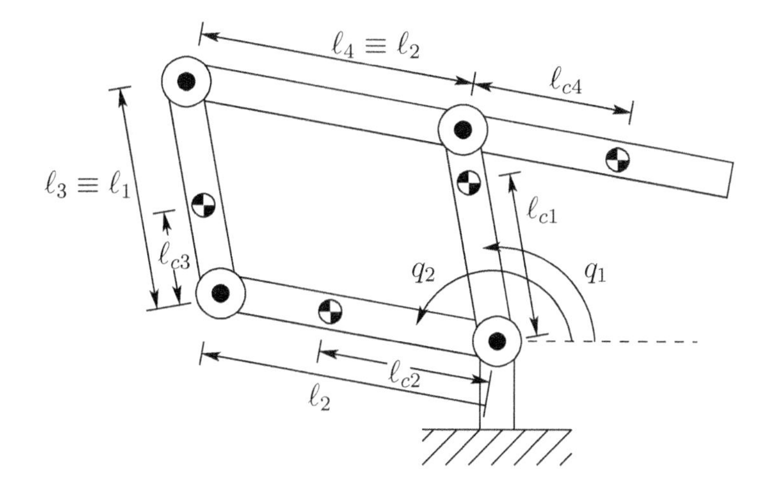

Figure 7.11: Five-bar linkage.

marked ℓ_2 are the same; in this way the closed path in the figure is in fact a parallelogram, which greatly simplifies the computations. Notice, however, that the quantities ℓ_{c1} and ℓ_{c3} need not be equal. For example, even though links 1 and 3 have the same length, they need not have the same mass distribution.

It is clear from the figure that, even though there are four links being moved, there are in fact only two degrees of freedom, identified as q_1 and q_2. Thus, in contrast to the earlier mechanisms studied in this book, this one is a closed kinematic chain (though of a particularly simple kind). As a result, we cannot use the earlier results on Jacobian matrices, and instead have to start from scratch. As a first step we write down the coordinates of the centers of mass of the various links as a function of the generalized coordinates. This gives

$$\begin{bmatrix} x_{c1} \\ y_{c1} \end{bmatrix} = \begin{bmatrix} \ell_{c1}\cos q_1 \\ \ell_{c1}\sin q_1 \end{bmatrix} \tag{7.97}$$

$$\begin{bmatrix} x_{c2} \\ y_{c2} \end{bmatrix} = \begin{bmatrix} \ell_{c2}\cos q_2 \\ \ell_{c2}\sin q_2 \end{bmatrix} \tag{7.98}$$

$$\begin{bmatrix} x_{c3} \\ y_{c3} \end{bmatrix} = \begin{bmatrix} \ell_{c2}\cos q_1 \\ \ell_{c2}\sin q_2 \end{bmatrix} + \begin{bmatrix} \ell_{c3}\cos q_1 \\ \ell_{c3}\sin q_1 \end{bmatrix} \tag{7.99}$$

$$\begin{aligned} \begin{bmatrix} x_{c4} \\ y_{c4} \end{bmatrix} &= \begin{bmatrix} \ell_1\cos q_1 \\ \ell_1\sin q_1 \end{bmatrix} + \begin{bmatrix} \ell_{c4}\cos(q_2 - \pi) \\ \ell_{c4}\sin(q_2 - \pi) \end{bmatrix} \\ &= \begin{bmatrix} \ell_1\cos q_1 \\ \ell_1\sin q_1 \end{bmatrix} - \begin{bmatrix} \ell_{c4}\cos q_2 \\ \ell_{c4}\sin q_2 \end{bmatrix} \end{aligned} \tag{7.100}$$

Next, with the aid of these expressions, we can write down the velocities of the various centers of mass as a function of $\dot{q}_1$ and $\dot{q}_2$. For convenience we drop the third row of each of the following Jacobian matrices as it is always zero. The result is

$$\begin{aligned} v_{c1} &= \begin{bmatrix} -\ell_{c1}\sin q_1 & 0 \\ \ell_{c1}\cos q_1 & 0 \end{bmatrix} \dot{q} \\ v_{c2} &= \begin{bmatrix} 0 & -\ell_{c2}\sin q_2 \\ 0 & \ell_{c2}\cos q_2 \end{bmatrix} \dot{q} \\ v_{c3} &= \begin{bmatrix} -\ell_{c3}\sin q_1 & -\ell_2\sin q_2 \\ \ell_{c3}\cos q_1 & \ell_2\cos q_2 \end{bmatrix} \dot{q} \\ v_{c4} &= \begin{bmatrix} -\ell_1\sin q_1 & \ell_{c4}\sin q_2 \\ \ell_1\cos q_1 & \ell_{c4}\cos q_2 \end{bmatrix} \dot{q} \end{aligned} \tag{7.101}$$

Let us define the velocity Jacobians $J_{v_{ci}}$, $i \in \{1, \ldots, 4\}$ in the obvious fashion, that is, as the four matrices appearing in the above equations. Next, it is clear that the angular velocities of the four links are simply given by

$$\omega_1 = \omega_3 = q_1 k, \omega_2 = \omega_4 = \dot{q}_2 k \tag{7.102}$$

Thus, the inertia matrix is given by

$$D(q) = \sum_{i=1}^{4} m_i J_{vc}^T J_{vc} + \begin{bmatrix} I_1 + I_3 & 0 \\ 0 & I_2 + I_4 \end{bmatrix} \tag{7.103}$$

If we now substitute from Equation (7.101) into the above equation and use the standard trigonometric identities, we are left with

$$\begin{aligned} d_{11}(q) &= m_1\ell_{c1}^2 + m_3\ell_{c3}^2 + m_4\ell_1^2 + I_1 + I_3 \\ d_{12}(q) &= d_{21}(q) = (m_3\ell_2\ell_{c3} - m_4\ell_1\ell_{c4})\cos(q_2 - q_1) \\ d_{22}(q) &= m_2\ell_{c2}^2 + m_3\ell_2^2 + m_4\ell_{c4}^2 + I_2 + I_4 \end{aligned} \tag{7.104}$$

Now, we note from the above expressions that if

$$m_3\ell_2\ell_{c3} = m_4\ell_1\ell_{c4} \tag{7.105}$$

then d_{12} and d_{21} are zero, that is, the inertia matrix is diagonal and constant. As a consequence the dynamical equations will contain neither Coriolis nor centrifugal terms.

Turning now to the potential energy, we have that

$$\begin{aligned} P &= g\textstyle\sum_{i=1}^{4} y_{ci} \\ &= g\sin q_1(m_1\ell_{c1} + m_3\ell_{c3} + m_4\ell_1) \\ &+ g\sin q_2(m_2\ell_{c2} + m_3\ell_2 - m_4\ell_{c4}) \end{aligned} \tag{7.106}$$

Hence,

$$\begin{aligned} g_1 &= g\cos q_1(m_1\ell_{c1} + m_3\ell_{c3} + m_4\ell_1) \\ g_2 &= g\cos q_2(m_2\ell_{c2} + m_3\ell_2 - m_4\ell_{c4}) \end{aligned} \tag{7.107}$$

Notice that g_1 depends only on q_1 but not on q_2 and similarly that g_2 depends only on q_2 but not on q_1. Hence, if the relationship (7.105) is satisfied, then the rather complex-looking manipulator in Figure 7.11 is described by the decoupled set of equations

$$d_{11}\ddot{q}_1 + g_1(q_1) = \tau_1, \qquad d_{22}\ddot{q}_2 + g_2(q_2) = \tau_2 \tag{7.108}$$

This discussion helps to explain the popularity of the parallelogram configuration in industrial robots. If the relationship (7.105) is satisfied, then one can adjust the two angles q_1 and q_2 independently, without worrying about interactions between the two angles. Compare this with the situation in the case of the planar elbow manipulators discussed earlier in this section.

7.5 PROPERTIES OF ROBOT DYNAMIC EQUATIONS

The equations of motion for an n-link robot can be quite formidable especially if the robot contains one or more revolute joints. Fortunately, these equations contain some important structural properties that can be exploited to good advantage for developing control algorithms. We will see this in subsequent chapters. Here we will discuss some of these properties, the most important of which are the so-called **skew symmetry** property and the related **passivity** property, and the **linearity-in-the-parameters** property. For revolute joint robots, the inertia matrix also satisfies global bounds that are useful for control design.

7.5.1 Skew Symmetry and Passivity

The **skew symmetry** property refers to an important relationship between the inertia matrix $D(q)$ and the matrix $C(q, \dot{q})$ appearing in Equation (7.63).

Proposition: 7.1 The Skew Symmetry Property
Let $D(q)$ be the inertia matrix for an n-link robot and define $C(q, \dot{q})$ in terms

of the elements of $D(q)$ according to Equation (7.64). Then the matrix $N(q,\dot{q}) = \dot{D}(q) - 2C(q,\dot{q})$ is skew symmetric, that is, the components n_{jk} of N satisfy $n_{jk} = -n_{kj}$.

Proof: Given the inertia matrix $D(q)$, the $(k,j)^{th}$ component of $\dot{D}(q)$ is given by the chain rule as

$$\dot{d}_{kj} = \sum_{i=1}^{n} \frac{\partial d_{kj}}{\partial q_i}\dot{q}_i \tag{7.109}$$

Therefore, the $(k,j)^{th}$ component of $N = \dot{D} - 2C$ is given by

$$\begin{aligned} n_{kj} &= \dot{d}_{kj} - 2c_{kj} \\ &= \sum_{i=1}^{n}\left[\frac{\partial d_{kj}}{\partial q_i} - \left\{\frac{\partial d_{kj}}{\partial q_i} + \frac{\partial d_{ki}}{\partial q_j} - \frac{\partial d_{ij}}{\partial q_k}\right\}\right]\dot{q}_i \\ &= \sum_{i=1}^{n}\left[\frac{\partial d_{ij}}{\partial q_k} - \frac{\partial d_{ki}}{\partial q_j}\right]\dot{q}_i \end{aligned} \tag{7.110}$$

Since the inertia matrix $D(q)$ is symmetric, that is, $d_{ij} = d_{ji}$, it follows from Equation (7.110) by interchanging the indices k and j that

$$n_{jk} = -n_{kj} \tag{7.111}$$

which completes the proof.

It is important to note that, in order for $N = \dot{D} - 2C$ to be skew-symmetric, one must define C according to Equation (7.64). This will be important in later chapters when we discuss robust and adaptive control algorithms.

Related to the skew symmetry property is the so-called **passivity property** which, in the present context, means that there exists a constant, $\beta \geq 0$, such that

$$\int_0^T \dot{q}^T(\zeta)\tau(\zeta)d\zeta \geq -\beta, \quad \forall\, T > 0 \tag{7.112}$$

The term $\dot{q}^T\tau$ has units of power. Thus, the expression $\int_0^T \dot{q}^T(\zeta)\tau(\zeta)d\zeta$ is the energy produced by the system over the time interval $[0,T]$. Passivity means that the amount of energy dissipated by the system has a lower bound given by $-\beta$. The word passivity comes from circuit theory where a passive system according to the above definition is one that can be built from passive components (resistors, capacitors, inductors). Likewise a passive mechanical system can be built from masses, springs, and dampers.

To prove the passivity property, let H be the total energy of the system, that is, the sum of the kinetic and potential energies,

$$H = \frac{1}{2}\dot{q}^T D(q)\dot{q} + P(q) \tag{7.113}$$

The derivative $\dot{H}$ satisfies

$$\begin{aligned} \dot{H} &= \dot{q}^T D(q)\ddot{q} + \tfrac{1}{2}\dot{q}^T \dot{D}(q)\dot{q} + \dot{q}^T \frac{\partial P}{\partial q} \\ &= \dot{q}^T\{\tau - C(q,\dot{q}) - g(q)\} + \tfrac{1}{2}\dot{q}^T \dot{D}(q)\dot{q} + \dot{q}^T \frac{\partial P}{\partial q} \end{aligned} \tag{7.114}$$

where we have substituted for $D(q)\ddot{q}$ using the equations of motion. Collecting terms and using the fact that $g(q) = \frac{\partial P}{\partial q}$ yields

$$\begin{aligned} \dot{H} &= \dot{q}^T\tau + \frac{1}{2}\dot{q}^T\{\dot{D}(q) - 2C(q,\dot{q})\}\dot{q} \\ &= \dot{q}^T\tau \end{aligned} \tag{7.115}$$

the latter equality following from the skew-symmetry property. Integrating both sides of Equation (7.115) with respect to time gives,

$$\int_0^T \dot{q}^T(\zeta)\tau(\zeta)d\zeta = H(T) - H(0) \geq -H(0) \tag{7.116}$$

since the total energy $H(T)$ is nonnegative, and the passivity property therefore follows with $\beta = H(0)$.

7.5.2 Bounds on the Inertia Matrix

We have remarked previously that the inertia matrix for an n-link rigid robot is symmetric and positive definite. For a fixed value of the generalized coordinate q, let $0 < \lambda_1(q) \leq \cdots \leq \lambda_n(q)$ denote the n eigenvalues of $D(q)$. These eigenvalues are positive as a consequence of the positive definiteness of $D(q)$. As a result, it can easily be shown that

$$\lambda_1(q)I_{n\times n} \leq D(q) \leq \lambda_n(q)I_{n\times n} \tag{7.117}$$

where $I_{n\times n}$ denotes the $n \times n$ identity matrix. The above inequalities are interpreted in the standard sense of matrix inequalities, namely, if A and B are $n \times n$ matrices, then $B < A$ means that the matrix $A - B$ is positive definite and $B \leq A$ means that $A - B$ is positive semi-definite.

If all of the joints are revolute, then the inertia matrix contains only terms involving sine and cosine functions and, hence, is bounded as a function of the generalized coordinates. As a result one can find constants λ_m and λ_M that provide uniform (independent of q) bounds in the inertia matrix

$$\lambda_m I_{n \times n} \leq D(q) \leq \lambda_M I_{n \times n} < \infty \tag{7.118}$$

7.5.3 Linearity in the Parameters

The robot equations of motion are defined in terms of certain parameters, such as link masses, moments of inertia, etc., that must be determined for each particular robot in order, for example, to simulate the equations or to tune controllers. The complexity of the dynamic equations makes the determination of these parameters a difficult task. Fortunately, the equations of motion are linear in these inertia parameters in the following sense. There exists an $n \times \ell$ function, $Y(q, \dot{q}, \ddot{q})$ and an ℓ-dimensional vector Θ such that the Euler-Lagrange equations can be written as

$$D(q)\ddot{q} + C(q, \dot{q})\dot{q} + g(q) = Y(q, \dot{q}, \ddot{q})\Theta \tag{7.119}$$

The function, $Y(q, \dot{q}, \ddot{q})$ is called the **regressor** and $\Theta \in \mathbb{R}^\ell$ is the **parameter vector**. The dimension of the parameter space, that is, the number of parameters needed to write the dynamics in this way, is not unique. In general, a given rigid body is described by ten parameters, namely, the total mass, the six independent entries of the inertia tensor, and the three coordinates of the center of mass. An n-link robot then has a maximum of $10n$ dynamics parameters. However, since the link motions are constrained and coupled by the joint interconnections, there are actually fewer than $10n$ independent parameters. Finding a minimal set of parameters that can parametrize the dynamic equations is, however, difficult in general.

Example 7.3 Two Link Planar Robot

Consider the two link, revolute joint, planar robot from Section 7.4. If we group the inertia terms appearing in Equation (7.83) as

$$\Theta_1 = m_1\ell_{c1}^2 + m_2(\ell_1^2 + \ell_{c2}^2) + I_1 + I_2 \tag{7.120}$$

$$\Theta_2 = m_2\ell_1\ell_{c2} \tag{7.121}$$

$$\Theta_3 = m_2\ell_{c2}^2 + I_2 \tag{7.122}$$

then we can write the inertia matrix elements as

$$\begin{aligned} d_{11} &= \Theta_1 + 2\Theta_2 \cos(q_2) && (7.123)\\ d_{12} &= d_{21} = \Theta_3 + \Theta_2 \cos(q_2) && (7.124)\\ d_{22} &= \Theta_3 && (7.125) \end{aligned}$$

No additional parameters are required in the Christoffel symbols as these are functions of the elements of the inertia matrix. However, the gravitational torques generally require additional parameters. Setting

$$\begin{aligned} \Theta_4 &= m_1\ell_{c1} + m_2\ell_1 && (7.126)\\ \Theta_5 &= m_2\ell_2 && (7.127) \end{aligned}$$

we can write the gravitational terms g_1 and g_2 as

$$\begin{aligned} g_1 &= \Theta_4 g\cos(q_1) + \Theta_5 g\cos(q_1+q_2) && (7.128)\\ g_2 &= \Theta_5 g\cos(q_1+q_2) && (7.129) \end{aligned}$$

Substituting these into the equations of motion it is straightforward to write the dynamics in the form (7.119) where

$$Y(q,\dot{q},\ddot{q}) = \qquad (7.130)$$

$$\begin{bmatrix} \ddot{q}_1 & \cos(q_2)(2\ddot{q}_1+\ddot{q}_2)-\sin(q_2)(\dot{q}_1^2+2\dot{q}_1\dot{q}_2) & \ddot{q}_2 & g\cos(q_1) & g\cos(q_1+q_2) \\ 0 & \cos(q_2)\ddot{q}_1+\sin(q_2)\dot{q}_1^2 & \ddot{q}_1+\ddot{q}_2 & 0 & g\cos(q_1+q_2) \end{bmatrix}$$

and the parameter vector Θ is given by

$$\Theta = \begin{bmatrix} \Theta_1 \\ \Theta_2 \\ \Theta_3 \\ \Theta_4 \\ \Theta_5 \end{bmatrix} = \begin{bmatrix} m_1\ell_{c1}^2 + m_2(\ell_1^2 + \ell_{c2}^2) + I_1 + I_2 \\ m_2\ell_1\ell_{c2} \\ m_2\ell_{c2}^2 + I_2 \\ m_1\ell_{c1} + m_2\ell_1 \\ m_2\ell_2 \end{bmatrix} \qquad (7.131)$$

Thus, we have parameterized the dynamics using a five dimensional parameter space. Note that in the absence of gravity only three parameters are needed.

◇

7.6 NEWTON-EULER FORMULATION

In this section we present a method for analyzing the dynamics of robot manipulators known as the **Newton-Euler formulation**. This method leads to exactly the same final answers as the Lagrangian formulation presented in earlier sections, but the route taken is quite different. In particular, in the Lagrangian formulation we treat the manipulator as a whole and perform the analysis using a Lagrangian function (the difference between the kinetic energy and the potential energy). In contrast, in the Newton-Euler formulation we treat each link of the robot in turn, and write down the equations describing its linear motion and its angular motion. Of course, since each link is coupled to other links, these equations that describe each link contain coupling forces and torques that appear also in the equations that describe neighboring links. By doing a so-called forward-backward recursion, we are able to determine all of these coupling terms and eventually to arrive at a description of the manipulator as a whole. Thus, we see that the philosophy of the Newton-Euler formulation is quite different from that of the Lagrangian formulation.

At this stage the reader can justly ask whether there is a need for another formulation, and the answer is not clear. Historically, both formulations were evolved in parallel, and each was perceived as having certain advantages. For instance, it was believed at one time that the Newton-Euler formulation is better suited to recursive computation than the Lagrangian formulation. However, the current situation is that both of the formulations are equivalent in almost all respects. Thus, at present, the main reason for having another method of analysis at our disposal is that it might provide different insights.

In any mechanical system one can identify a set of generalized coordinates (which we introduced in Section 7.1 and labeled q) and corresponding generalized forces (also introduced in Section 7.1 and labeled τ). Analyzing the dynamics of a system means finding the relationship between q and τ. At this stage we must distinguish between two aspects. First, we might be interested in obtaining **closed-form equations** that describe the time evolution of the generalized coordinates, such as Equation (7.87). Second, we might be interested in knowing what generalized forces need to be applied in order to realize a *particular* time evolution of the generalized coordinates. The distinction is that in the latter case we only want to know what time dependent function $\tau(t)$ produces a particular trajectory $q(t)$ and may not care to know the general functional relationship between the two. It is perhaps fair to say that in the former type of analysis, the Lagrangian formulation is superior while in the latter case the Newton-Euler formulation is superior. Looking ahead to topics beyond the scope of the book, if one wishes to study more advanced mechanical phenomena such as elastic deformations of the links, then the Lagrangian formulation is clearly superior.

In this section we present the general equations that describe the Newton-Euler formulation. In the next section we illustrate the method by applying it to the planar elbow manipulator studied in Section 7.4 and show that the resulting equations are the same as Equation (7.87).

The facts of Newtonian mechanics that are pertinent to the present discussion can be stated as follows:

- Every action has an equal and opposite reaction. Thus, if body 1 applies a force f and torque τ to body 2, then body 2 applies a force $-f$ and torque $-\tau$ to body 1.

- The rate of change of the linear momentum equals the total force applied to the body.

- The rate of change of the angular momentum equals the total torque applied to the body.

Applying the second fact to the linear motion of a body yields the relationship

$$\frac{d(mv)}{dt} = f \tag{7.132}$$

where m is the mass of the body, v is the velocity of the center of mass with respect to an inertial frame, and f is the sum of external forces applied to the body. Since in robotic applications the mass is constant as a function of time, Equation (7.132) can be simplified to the familiar relationship

$$ma = f \tag{7.133}$$

where $a = \dot{v}$ is the acceleration of the center of mass.

Applying the third fact to the angular motion of a body gives

$$\frac{d(I_0\omega_0)}{dt} = \tau_0 \tag{7.134}$$

where I_0 is the moment of inertia of the body about an inertial frame whose origin is at the center of mass, ω_0 is the angular velocity of the body, and τ_0 is the sum of torques applied to the body. Now, there is an essential difference between linear motion and angular motion. Whereas the mass of a body is constant in most applications, its moment of inertia with respect an inertial frame may or may not be constant. To see this, suppose we attach a frame rigidly to the body, and let I denote the inertia matrix of the body with respect to this frame. Then I remains the same irrespective of whatever motion the body executes. However, the matrix I_0 is given by

$$I_0 = RIR^T \tag{7.135}$$

where R is the rotation matrix that transforms coordinates from the body attached frame to the inertial frame. Thus, there is no reason to expect that I_0 is constant as a function of time.

One possible way of overcoming this difficulty is to write the angular motion equation in terms of a frame rigidly attached to the body. This leads to

$$I\dot{\omega} + \omega \times (I\omega) = \tau \tag{7.136}$$

where I is the (constant) inertia matrix of the body with respect to the body attached frame, ω is the angular velocity, but expressed in the body attached frame, and τ is the total torque on the body, again expressed in the body attached frame. Let us now give a derivation of Equation (7.136) to demonstrate clearly where the term $\omega \times (I\omega)$ comes from; note that this term is called the **gyroscopic term**.

Let R denote the orientation of the frame rigidly attached to the body with respect to the inertial frame; note that it could be a function of time. Then Equation (7.135) gives the relation between I and I_0. Now, by the definition of the angular velocity, we know that

$$\dot{R}R^T = S(\omega_0) \tag{7.137}$$

In other words, the angular velocity of the body, *expressed in an inertial frame*, is given by Equation (7.137). Of course, the same vector, expressed in the body attached frame, is given by

$$\omega_0 = R\omega, \; \omega = R^T\omega_0 \tag{7.138}$$

Hence, the angular momentum, expressed in the inertial frame, is

$$h = I_0\omega_0 = RIR^TR\omega = RI\omega \tag{7.139}$$

Differentiating and noting that I is constant gives an expression for the rate of change of the angular momentum, expressed as a vector in the inertial frame

$$\dot{h} = \dot{R}I\omega + RI\dot{\omega} \tag{7.140}$$

Now, since

$$\dot{R} = S(\omega_0)R \tag{7.141}$$

we have, with respect to the inertial frame, that

$$\dot{h} \;=\; S(\omega_0)RI\omega + RI\dot{\omega} \tag{7.142}$$

With respect to the frame rigidly attached to the body, the rate of change of the angular momentum is

$$\begin{aligned} R^T\dot{h} &= R^TS(\omega_0)RI\omega + I\dot{\omega} \\ &= S(R^T\omega_0)I\omega + I\dot{\omega} \\ &= S(\omega)I\omega + I\dot{\omega} = \omega \times (I\omega) + I\dot{\omega} \end{aligned} \tag{7.143}$$

This establishes Equation (7.136). Of course we can, if we wish, write the same equation in terms of vectors expressed in an inertial frame. But we will see shortly that there is an advantage to writing the force and moment equations with respect to a frame attached to link i, namely that a great many vectors reduce to constant vectors, thus leading to significant simplifications in the equations.

Now, we derive the Newton-Euler formulation of the equations of motion of an n-link manipulator. For this purpose, we first choose frames $0, \ldots, n$, where frame 0 is an inertial frame, and frame i is rigidly attached to link i for $i \geq 1$. We also introduce several vectors, *which are all expressed in frame* i. The first set of vectors pertains to the velocities and accelerations of various parts of the manipulator.

$a_{c,i}$ = the acceleration of the center of mass of link i

$a_{e,i}$ = the acceleration of the end of link i (that is, the origin of frame $i+1$)

ω_i = the angular velocity of frame i w.r.t. frame 0

α_i = the angular acceleration of frame i w.r.t. frame 0

The next several vectors pertain to forces and torques.

g_i = the acceleration due to gravity (expressed in frame i)

f_i = the force exerted by link $i-1$ on link i

τ_i = the torque exerted by link $i-1$ on link i

R_{i+1}^i = the rotation matrix from frame $i+1$ to frame i

The final set of vectors pertain to physical features of the manipulator. *Note that each of the following vectors is constant as a function of* q. In other words, each of the vectors listed here is independent of the configuration of the manipulator.

$$
\begin{aligned}
m_i &= \text{the mass of link } i \\
I_i &= \text{the inertia matrix of link } i \text{ about a frame parallel} \\
&\quad \text{to frame } i \text{ whose origin is at the center of mass of link } i \\
r_{i,ci} &= \text{the vector from } o_i \text{ to the center of mass of link } i \\
r_{i+1,ci} &= \text{the vector from } o_1 \text{ to the center of mass of link } i \\
r_{i,i+1} &= \text{the vector from } o_i \text{ to } o_{i+1}
\end{aligned}
$$

Now, consider the free body diagram shown in Figure 7.12.

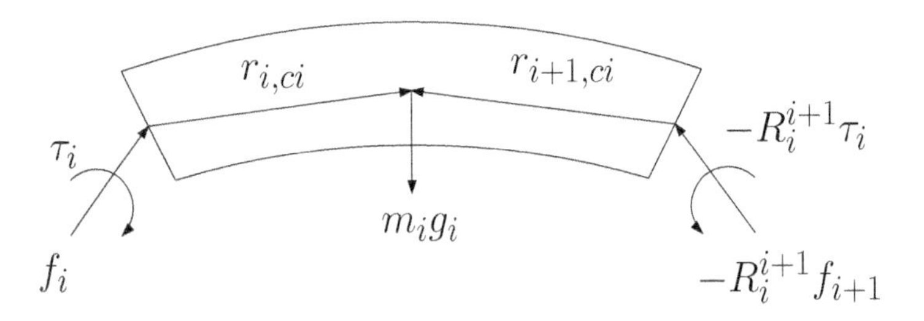

Figure 7.12: Forces and moments on link i.

This shows link i together with all forces and torques acting on it. Let us analyze each of the forces and torques shown in the figure. First, f_i is the force applied by link $i-1$ to link i. Next, by the law of action and reaction, link $i+1$ applies a force of $-f_{i+1}$ to link i, but this vector is expressed in frame $i+1$ according to our convention. In order to express the same vector in frame i, it is necessary to multiply it by the rotation matrix R_{i+1}^i. Similar explanations apply to the torques τ_i and $-R_{i+1}^i\tau_{i+1}$. The force m_ig_i is the gravitational force. Since all vectors in Figure 7.12 are expressed in frame i, the gravity vector g_i is in general a function of i.

Writing down the force balance equation for link i gives

$$
f_i - R_{i+1}^i f_{i+1} + m_i g_i = m_i a_{c,i} \tag{7.144}
$$

Next, we compute the moment balance equation for link i. For this purpose, it is important to note two things. First, the moment exerted by a force f about a point is given by $f \times r$, where r is the radial vector *from* the point where the force is applied to the point about which we are computing the moment. Second, in the moment equation below, the vector m_ig_i does not appear, since it is applied directly at the center of mass. Thus, we have

$$\tau_i - R^i_{i+1}\tau_{i+1} + f_i \times r_{i,ci} - (R^i_{i+1}f_{i+1}) \times r_{i+1,ci} \tag{7.145}$$
$$= \alpha_i + \omega_i \times (I_i\omega_i)$$

Now, we present the heart of the Newton-Euler formulation, which consists of finding the vectors $f_1, \ldots, f_n$ and $\tau_1, \ldots, \tau_n$ corresponding to a given set of vectors $q, \dot{q}, \ddot{q}$. In other words, we find the forces and torques in the manipulator that correspond to a given set of generalized coordinates and their first two derivatives. This information can be used to perform either type of analysis, as described above. That is, we can either use the equations below to find the f and τ corresponding to a particular trajectory $q(\cdot)$, or else to obtain closed-form dynamical equations. The general idea is as follows. Given $q, \dot{q}, \ddot{q}$, suppose we are somehow able to determine all of the velocities and accelerations of various parts of the manipulator, that is, all of the quantities $a_{c,i}$, ω_i, and α_i. Then we can solve Equation (7.144) and Equation (7.145) recursively to find all the forces and torques, as follows. First, set $f_{n+1} = 0$ and $\tau_{n+1} = 0$. This expresses the fact that there is no link $n+1$. Then we can solve Equation (7.144) to obtain

$$f_i = R^i_{i+1}f_{i+1} + m_i a_{c,i} - m_i g_i \tag{7.146}$$

By successively substituting $i = n, n-1, \ldots, 1$ we find all forces. Similarly, we can solve Equation (7.145) to obtain

$$\tau_i = \tag{7.147}$$
$$R^i_{i+1}\tau_{i+1} - f_i \times r_{i,ci} + (R^i_{i+1}f_{i+1}) \times r_{i+1,ci} + \alpha_i + \omega_i \times (I_i\omega_i)$$

By successively substituting $i = n, n-1, \ldots, 1$ we find all torques. Note that the above iteration runs in the direction of decreasing i.

Thus, the solution is complete once we find an easily computed relation between $q, \dot{q}, \ddot{q}$ and $a_{c,i}$, ω_i, and α_i. This can be obtained by a recursive procedure in the direction of *increasing* i. This procedure is given below for the case of revolute joint s; the corresponding relationships for prismatic joints are actually easier to derive.

In order to distinguish between quantities expressed with respect to frame i and the base frame, we use a superscript (0) to denote the latter. Thus, for example, ω_i denotes the angular velocity of frame i expressed in frame i, while $\omega_i^{(0)}$ denotes the same quantity expressed in an inertial frame.

Now, we have that

$$\omega_i^{(0)} = \omega_{i-1}^{(0)} + z_{i-1}\dot{q}_i \tag{7.148}$$

This merely expresses the fact that the angular velocity of frame i equals that of frame $i-1$ plus the added rotation from joint i. To get a relation between ω_i and ω_{i-1}, we need only express the above equation in frame i rather than the base frame, taking care to account for the fact that ω_i and ω_{i-1} are expressed in different frames. This leads to

$$\omega_i = (R_i^{i-1})^T \omega_{i-1} + b_i \dot{q}_i \tag{7.149}$$

where

$$b_i = (R_i^0)^T z_{i-1} \tag{7.150}$$

is the axis of rotation of joint i expressed in frame i.

Next let us compute the angular acceleration α_i. It is important to note here that

$$\alpha_i = (R_i^0)^T \dot{\omega}_i^{(0)} \tag{7.151}$$

In other words, α_i is the derivative of the angular velocity of frame i, but expressed in frame i. It is not true that $\alpha_i = \dot{\omega}_i$! We will encounter a similar situation with the velocity and acceleration of the center of mass. Now, we see directly from Equation (7.148) that

$$\dot{\omega}_i^{(0)} = \dot{\omega}_{i-1}^{(0)} + z_{i-1}\ddot{q}_i + \omega_i^{(0)} \times z_{i-1}\dot{q}_i \tag{7.152}$$

Expressing the above equation in frame i gives

$$\alpha_i = (R_{i-1}^i)^T \alpha_{i-1} + b_i \ddot{q}_i + \omega_i \times b_i \dot{q}_i \tag{7.153}$$

Now, we come to the linear velocity and acceleration terms. Note that, in contrast to the angular velocity, the linear velocity does not appear anywhere in the dynamic equations; however, an expression for the linear velocity is needed before we can derive an expression for the linear acceleration. From Section 4.5, we get that the velocity of the center of mass of link i is given by

$$v_{c,i}^{(0)} = v_{e,i-1}^{(0)} + \omega_i^{(0)} \times r_{i,ci}^{(0)} \tag{7.154}$$

To obtain an expression for the acceleration, we note that the vector $r_{i,ci}^{(0)}$ is constant in frame i. Thus

$$a_{c,i}^{(0)} = a_{e,i-1}^{(0)} \times r_{i,ci}^{(0)} + \omega_i^{(0)} \times (\omega_i^{(0)} \times r_{i,ci}^{(0)}) \tag{7.155}$$

Now

$$a_{c,i} = (R_i^0)^T a_{c,i}^{(0)} \tag{7.156}$$

Let us carry out the multiplication and use the familiar property

$$R(a \times b) = (Ra) \times (Rb) \tag{7.157}$$

We also have to account for the fact that $a_{e,i-1}$ is expressed in frame $i-1$ and transform it to frame i. This gives

$$a_{c,i} = (R_i^{i-1})^T a_{e,i-1} + \dot{\omega}_i \times r_{i,ci} + \omega_i \times (\omega_i \times r_{i,ci}) \tag{7.158}$$

Now, to find the acceleration of the end of link i, we can use Equation (7.158) with $r_{i,i+1}$ replacing $r_{i,ci}$. Thus

$$a_{e,i} = (R_i^{i-1})^T a_{e,i-1} + \dot{\omega}_i \times r_{i,i+1} + \omega_i \times (\omega_i \times r_{i,i+1}) \tag{7.159}$$

Now, the recursive formulation is complete. We can now state the Newton-Euler formulation as follows.

1. Start with the initial conditions

$$\omega_0 = 0, \alpha_0 = 0, a_{c,0} = 0, a_{e,0} = 0 \tag{7.160}$$

   and solve Equations (7.149), (7.153), and (7.159) and (7.158) (in that order) to compute ω_i, α_i, and $a_{c,i}$ for i increasing from 1 to n.

2. Start with the terminal conditions

$$f_{n+1} = 0, \qquad \tau_{n+1} = 0 \tag{7.161}$$

   and use Equation (7.146) and Equation (7.147) to compute f_i and τ_i for i decreasing from n to 1.

7.6.1 Planar Elbow Manipulator Revisited

In this section we apply the recursive Newton-Euler formulation derived in Section 7.6 to analyze the dynamics of the planar elbow manipulator of Figure 7.9, and show that the Newton-Euler method leads to the same equations as the Lagrangian method, namely Equation (7.87).

We begin with the forward recursion to express the various velocities and accelerations in terms of q_1, q_2 and their derivatives. Note that, in this simple case, it is quite easy to see that

$$\omega_1 = \dot{q}_1 k, \ \alpha_1 = \ddot{q}_1 k, \ \omega_2 = (q_1 + q_2)k, \ \alpha_2 = (\ddot{q}_1 + \ddot{q}_2)k \tag{7.162}$$

so that there is no need to use Equations (7.149) and (7.153). Also, the vectors that are independent of the configuration are as follows:

$$r_{1,c1} = \ell_{c1}i, \; r_{2,c1} = (\ell_1 - \ell_{c1})i, \; r_{1,2} = \ell_1 i \tag{7.163}$$

$$r_{2,c2} = \ell_{c2}i, \; r_{3,c2} = (\ell_2 - \ell_{c2})i, \; r_{2,3} = \ell_2 i \tag{7.164}$$

Forward Recursion Link 1

Using Equation (7.158) with $i = 1$ and noting that $a_{e,0} = 0$ gives

$$\begin{aligned} a_{c,1} &= \ddot{q}_1 k \times \ell_{c1} i + \dot{q}_1 k \times (\dot{q}_1 k \times \ell_{c1} i) \\ &= \ell_{c1}\ddot{q}_1 j - \ell_{c1}\dot{q}_1^2 i = \begin{bmatrix} -\ell_{c1}\dot{q}_1^2 \\ \ell_c \ddot{q}_1 \\ 0 \end{bmatrix} \end{aligned} \tag{7.165}$$

Notice how simple this computation is when we do it with respect to frame 1. Compare with the same computation in frame 0. Finally, we have

$$g_1 = -(R_0^1)^T g j = g \begin{bmatrix} \sin q_1 \\ -\cos q_1 \\ 0 \end{bmatrix} \tag{7.166}$$

where g is the acceleration due to gravity. At this stage we can economize a bit by not displaying the third components of these accelerations, since they are obviously always zero. Similarly, the third components of all forces will be zero while the first two components of all torques will be zero. To complete the computations for link 1, we compute the acceleration of the end of link 1. Clearly, this is obtained from Equation (7.165) by replacing ℓ_{c1} by ℓ_1. Thus

$$a_{e,1} = \begin{bmatrix} -\ell_1 \dot{q}_1^2 \\ \ell_1 \ddot{q}_1 \end{bmatrix} \tag{7.167}$$

Forward Recursion: Link 2

Once again we use Equation (7.158) and substitute for ω_2 from Equation (7.162); this yields

$$\begin{aligned} \alpha_{c,2} &= (R_1^2)^T a_{e,1} + (\ddot{q}_1 + \ddot{q}_2) k \times \ell_{c2} i \\ &+ (\dot{q}_1 + \dot{q}_2) k \times [(\dot{q}_1 + \dot{q}_2) k \times \ell_{c2} i] \end{aligned} \tag{7.168}$$

The only quantity in the above equation that is configuration dependent is the first one. This can be computed as

$$
\begin{aligned}
(R_1^2)^T a_{e,1} &= \begin{bmatrix} \cos q_2 & \sin q_2 \\ -\sin q_2 & \cos q_2 \end{bmatrix} \begin{bmatrix} -\ell_1 \dot{q}_1^2 \\ \ell_1 \ddot{q}_1 \end{bmatrix} \\
&= \begin{bmatrix} -\ell_1 \dot{q}_1^2 \cos q_2 + \ell_1 \ddot{q}_1 \sin \dot{q}_2 \\ \ell_1 \dot{q}_1^2 \sin q_2 + \ell_1 \ddot{q}_1 \cos q_2 \end{bmatrix} \qquad (7.169)
\end{aligned}
$$

Substituting into Equation (7.168) gives

$$
a_{c,2} = \begin{bmatrix} -\ell_1 \dot{q}_1^2 \cos q_2 + \ell_1 \ddot{q}_1 \sin q_2 - \ell_{c2}(\dot{q}_1 + \dot{q}_2)^2 \\ \ell_1 \dot{q}_1^2 \sin q_2 + \ell_1 \ddot{q}_1 \cos q_2 - \ell_{c2}(\ddot{q}_1 + \ddot{q}_2) \end{bmatrix} \qquad (7.170)
$$

The gravitational vector is

$$
g_2 = g \begin{bmatrix} \sin(q_1 + q_2) \\ -\cos(q_1 + q_2) \end{bmatrix} \qquad (7.171)
$$

Since there are only two links, there is no need to compute $a_{e,2}$. Hence, the forward recursions are complete at this point.

Backward Recursion: Link 2

Now, we carry out the backward recursion to compute the forces and joint torques. Note that, in this instance, the joint torques are the externally applied quantities, and our ultimate objective is to derive dynamical equations involving the joint torques. First we apply Equation (7.146) with $i = 2$ and note that $f_3 = 0$. This results in

$$
\begin{aligned}
f_2 &= m_2 a_{c,2} - m_2 g_2 & (7.172) \\
\tau_2 &= I_2 \alpha_2 + \omega_2 \times (I_2 \omega_2) - f_2 \times \ell_{c2} i & (7.173)
\end{aligned}
$$

Now, we can substitute for ω_2, α_2 from Equation (7.162), and for $a_{c,2}$ from Equation (7.170). We also note that the gyroscopic term equals zero, since both ω_2 and $I_2\omega_2$ are aligned with k. Now, the cross product $f_2 \times \ell_{c2} i$ is clearly aligned with k and its magnitude is just the second component of f_2. The final result is

$$
\begin{aligned}
\tau_2 &= I_2(\ddot{q}_1 + \ddot{q}_2)k + [m_2 \ell_1 \ell_{c2} \sin q_2 \dot{q}_1^2 + m_2 \ell_1 \ell_{c2} \cos q_2 \ddot{q}_1 \\
&\quad + m_2 \ell_{c2}^2 (\ddot{q}_1 + \ddot{q}_2) + m)2\ell_{c2} g \cos(q_1 + q_2)]k \qquad (7.174)
\end{aligned}
$$

We can see that the third component of the vector τ_2 in the above equation is the same as the second equation in (7.87).

Backward Recursion: Link 1

To complete the derivation, we apply Equation (7.146) and Equation (7.147) with $i = 1$. First, the force equation is

$$f_1 = m_1 a_{c,1} + R_1^2 f_2 - m_1 g_1 \tag{7.175}$$

and the torque equation is

$$\begin{aligned} \tau_1 &= R_1^2 \tau_2 - f_1 \times \ell_{c,1} i - (R_1^2 f_2) \times (\ell_1 - \ell_{c1}) i \\ &\quad + I_1 \alpha_1 + \omega_1 \times (I_1 \omega_1) \end{aligned} \tag{7.176}$$

Now, we can simplify things a bit. First, $R_1^2 \tau_2 = \tau_2$, since the rotation matrix does not affect the third components of vectors. Second, the gyroscopic term is again equal to zero. Finally, when we substitute for f_1 from Equation (7.175) into Equation (7.176), a little algebra gives

$$\begin{aligned} \tau_1 &= \tau_2 - m_1 a_{c,1} \times \ell_{c1} i + m_1 g_1 \times \ell_{c1} i \\ &\quad - (R_1^2 f_2) \times \ell_1 i + I_1 i + I_1 \alpha_1 \end{aligned} \tag{7.177}$$

Once again, all these products are quite straightforward, and the only difficult calculation is that of $R_1^2 f_2$. The final result is

$$\begin{aligned} \tau_1 &= \tau_2 + \left[m_1 \ell_{c1}^2 + m_1 \ell_{c1} g \cos q_1 + m_2 \ell_1 g \cos q_1 + I_1 \ddot{q}_1 \right. \\ &\quad \left. + m_2 \ell_1^2 \ddot{q}_1 - m_1 \ell_1 \ell_{c2} (\dot{q}_1 + \dot{q}_2)^2 \sin q_2 + m_2 \ell_1 \ell_{c2} (\ddot{q}_1 + \ddot{q}_2) \cos q_2 \right] k \end{aligned} \tag{7.178}$$

If we now substitute for τ_1 from Equation (7.174) and collect terms, we will get the first equation in Equation (7.87). The details are routine and are left to the reader.

7.7 SUMMARY

In this chapter we treated the dynamics of n-link rigid robots in detail. We derived the Euler-Lagrange equations from D'Alembert's principle and the principle of virtual work. These equations take the form

$$\frac{d}{dt} \frac{\partial \mathcal{L}}{\partial \dot{q}_k} - \frac{\partial \mathcal{L}}{\partial q_k} = \tau_k \; ; \; k = 1, \ldots n$$

where n is the number of degrees of freedom and $\mathcal{L} = \mathcal{K} - \mathcal{P}$ is the Lagrangian function; the difference of the Kinetic and Potential energies, which are written in terms of a set of generalized coordinates $(q_1, \ldots, q_n)$. The terms τ_k are generalized forces acting on the system.

We derived computable formulas for the kinetic and potential energies; the kinetic energy $\mathcal{K}$ is given as

$$\begin{aligned} K &= \frac{1}{2}\dot{q}^T \left[\sum_{i=1}^{n}\{m_i J_{v_i}(q)^T J_{v_i}(q) + J_{\omega_i}(q)^T R_i(q) I_i R_i(q)^T J_{\omega_i}(q)\}\right] \dot{q} \\ &= \frac{1}{2}\dot{q}^T D(q)\dot{q} \end{aligned}$$

where

$$D(q) = \left[\sum_{i=1}^{n}\{m_i J_{v_i}(q)^T J_{v_i}(q) + J_{\omega_i}(q)^T R_i(q) I_i R_i(q)^T J_{\omega_i}(q)\}\right]$$

is the $n \times n$ **inertia matrix** of the manipulator. The matrices I_i in the above formula are the link inertia tensors. The inertia tensor is computed in a body attached frame as

$$I = \begin{bmatrix} I_{xx} & I_{xy} & I_{xz} \\ I_{yx} & I_{yy} & I_{yz} \\ I_{zx} & I_{zy} & I_{zz} \end{bmatrix}$$

where

$$\begin{aligned} I_{xx} &= \int\int\int (y^2 + z^2)\rho(x, y, z)dx\ dy\ dz \\ I_{yy} &= \int\int\int (x^2 + z^2)\rho(x, y, z)dx\ dy\ dz \\ I_{zz} &= \int\int\int (x^2 + y^2)\rho(x, y, z)dx\ dy\ dz \end{aligned}$$

and

$$\begin{aligned} I_{xy} = I_{yx} &= -\int\int\int xy\rho(x, y, z)dx\ dy\ dz \\ I_{xz} = I_{zx} &= -\int\int\int xz\rho(x, y, z)dx\ dy\ dz \\ I_{yz} = I_{zy} &= -\int\int\int yz\rho(x, y, z)dx\ dy\ dz \end{aligned}$$

are the principle moments of inertia and cross-products of inertia, respectively, where the integration is taken over the region of space occupied by the body.

The formula for the potential energy of the ith link is

$$P_i = m_i g^T r_{ci}$$

where g is gravity vector expressed in the inertial frame and the vector r_{ci} gives the coordinates of the center of mass of link i in the inertial. The total potential energy of the n-link robot is therefore

$$P = \sum_{i=1}^{n} P_i = \sum_{i=1}^{n} m_i g^T r_{ci}$$

We then derived a special form of the Euler-Lagrange equations using the above expressions for the kinetic and potential energies as

$$\sum_{j=1}^{n} d_{kj}(q)\ddot{q}_j + \sum_{i=1}^{n}\sum_{j=1}^{n} c_{ijk}(q)\dot{q}_i\dot{q}_j + g_k(q) = \tau_k, \qquad k = 1, \ldots, n$$

where the terms

$$g_k = \frac{\partial P}{\partial q_k}$$

are gravitational generalized forces

$$c_{ijk} := \frac{1}{2}\left\{\frac{\partial d_{kj}}{\partial q_i} + \frac{\partial d_{ki}}{\partial q_j} - \frac{\partial d_{ij}}{\partial q_k}\right\}$$

and the terms c_{ijk} are Christoffel symbols of the first kind.

In vector-matrix form the Euler-Lagrange equations become

$$D(q)\ddot{q} + C(q,\dot{q})\dot{q} + g(q) = \tau$$

where the $(k,j)^{th}$ element of the matrix $C(q,\dot{q})$ is defined as

$$\begin{aligned} c_{kj} &= \sum_{i=1}^{n} c_{ijk}(q)\dot{q}_i \\ &= \sum_{i=1}^{n} \frac{1}{2}\left\{\frac{\partial d_{kj}}{\partial q_j} + \frac{\partial d_{ki}}{\partial q_j} - \frac{\partial d_{ij}}{\partial q_k}\right\}\dot{q}_i \end{aligned}$$

and the gravity vector $g(q)$ is given by

$$g(q) = [g_1(q), \ldots, g_n(q)]^T$$

Next, we derived some important properties of the Euler-Lagrange equations, namely, the properties of **skew symmetry**, **passivity**, and **linearity in the parameters**. The skew symmetry property states that the matrix

$N(q, \dot{q}) = \dot{D}(q) - 2C(q, \dot{q})$. The passivity property states that there exists a constant $\beta > 0$ such that

$$\int_0^T \dot{q}^T(\zeta)\tau(\zeta)d\zeta \geq -\beta, \quad \forall\, T > 0$$

The linearity-in-the-parameters-property states that there exists an $n \times \ell$ function, $Y(q, \dot{q}, \ddot{q})$, called the regressor, and an ℓ-dimensional vector Θ, called the parameter vector such that the Euler-Lagrange equations can be written

$$D(q) + C(q, \dot{q})\dot{q} + g(q) = Y(q, \dot{q}, \ddot{q})\Theta = \tau$$

We also derived bounds on the inertia matrix for an n-link manipulator as

$$\lambda_1(q)I_{n \times n} \leq D(q) \leq \lambda_n(q)I_{n \times n}$$

In case the robot contains only revolute joints, the functions λ_1 and λ_n can be chosen as positive constants.

Finally, we discussed the Newton-Euler formulation of robot dynamics. The Newton-Euler formulation is a recursive scheme that is equivalent to the Euler-Lagrange method but offers some advantages from the standpoint of online computation.

PROBLEMS

7-1 Verify Equation (7.18) by direct calculation, neglecting quadratic terms in δr_1 and δr_2.

7-2 Consider a rigid body undergoing a pure rotation with no external forces acting on it. The kinetic energy is then given as

$$K = \frac{1}{2}(I_{xx}\omega_x^2 + I_{yy}\omega_y^2 + I_{zz}\omega_z^2)$$

with respect to a coordinate frame located at the center of mass and whose coordinate axes are principal axes. Take as generalized coordinates the Euler angles ϕ, θ, ψ and show that the Euler-Lagrange equations of motion of the rotating body are

$$\begin{aligned} I_{xx}\dot{\omega}_x + (I_{zz} - I_{yy})\omega_y\omega_z &= 0 \\ I_{yy}\dot{\omega}_y + (I_{xx} - I_{zz})\omega_z\omega_x &= 0 \\ I_{zz}\dot{\omega}_z + (I_{yy} - I_{xx})\omega_x\omega_y &= 0 \end{aligned}$$

7-3 Find the moments of inertia and cross products of inertia of a uniform rectangular solid of sides a, b, c with respect to a coordinate system with origin at the one corner and axes along the edges of the solid.

7-4 Given the inertia matrix $D(q)$ defined by Equation (7.83) show that $\det D(q) \neq 0$ for all q.

7-5 Show that the inertia matrix $D(q)$ for an n-link robot is always positive definite.

7-6 Verify the expression (7.59) that was used to derive the Christoffel symbols.

7-7 Consider a 3-link cartesian manipulator,

(a) Compute the inertia tensor J_i for each link $i = 1, 2, 3$ assuming that the links are uniform rectangular solids of length 1, width $\frac{1}{4}$, and height $\frac{1}{4}$, and mass 1.

(b) Compute the 3×3 inertia matrix $D(q)$ for this manipulator.

(c) Show that the Christoffel symbols c_{ijk} are all zero for this robot. Interpret the meaning of this for the dynamic equations of motion.

(d) Derive the equations of motion in matrix form:

$$D(q)\ddot{q} + C(q, \dot{q})\dot{q} + g(q) \quad = \quad u$$

7-8 Derive the Euler-Lagrange equations for the planar RP robot in Figure 3.25.

7-9 Derive the Euler-Lagrange equations for the planar RPR robot in Figure 3.33.

7-10 Derive the Euler-Lagrange equations of motion for the three-link RRR robot of Figure 3.32. Explore the use of symbolic software, such as Maple or Mathematica, for this problem. See, for example, the *Robotica* package [96].

7-11 For each of the robots above, define a parameter vector, Θ, compute the regressor, $Y(q, \dot{q}, \ddot{q})$ and express the equations of motion as

$$Y(q, \dot{q}, \ddot{q})\Theta = \tau \tag{7.179}$$

7-12 Recall for a particle with kinetic energy $K = \frac{1}{2}m\dot{x}^2$, the **momentum** is defined as

$$p = m\dot{x} = \frac{dK}{d\dot{x}}$$

Therefore, for a mechanical system with generalized coordinates $q_1, \ldots, q_n$, we define the **generalized momentum** p_k as

$$p_k = \frac{\partial L}{\partial \dot{q}_k}$$

where L is the Lagrangian of the system. With $K = \frac{1}{2}\dot{q}^T D(q)\dot{q}$ and $L = K - V$ prove that

$$\sum_{k=1}^{n} \dot{q}_k p_k = 2K$$

7-13 There is another formulation of the equations of motion of a mechanical system that is useful, the so-called **Hamiltonian** formulation. Define the Hamiltonian function H by

$$H = \sum_{k-1}^{n} \dot{q}_k p_k - L$$

(a) Show that $H = K + V$.

(b) Using the Euler-Lagrange equations, derive Hamilton's equations

$$\dot{q}_k = \frac{\partial H}{\partial p_k}$$
$$\dot{p}_k = -\frac{\partial H}{\partial q_k} + \tau_k$$

where τ_k is the input generalized force.

(c) For two-link manipulator of Figure 7.8 compute Hamiltonian equations in matrix form. Note that Hamilton's equations are a system of first order differential equations as opposed to a second order system given by Lagrange's equations.

7-14 Given the Hamiltonian H for a rigid robot, show that

$$\frac{dH}{dt} = \dot{q}^T \tau$$

where τ is the external force applied at the joints. What are the units of $\frac{dH}{dt}$?

NOTES AND REFERENCES

A general reference for dynamics is [46]. More advanced treatments of dynamics can be found in [2] and [88]. The Lagrangian and recursive Newton-Euler formulations of the dynamic equations are given in [52]. These two approaches are shown to be equivalent in [116]. A detailed discussion of holonomic and nonholonomic constraints is found in [71]. The same reference also treats both the Lagrangian and Hamiltonian formulations of dynamics in detail. The properties of skew symmetry and passivity are discussed in [117], [72], and [99]. Parametrization of robot dynamics in terms of a minimal set of inertia parameters is treated in [43]. Identification of manipulator inertia parameters is discussed in [42].

Chapter 8

MULTIVARIABLE CONTROL

In Chapter 6 we discussed techniques to derive control laws for each joint of a manipulator based on a single-input/single-output model. Coupling effects among the joints were regarded as disturbances to the individual systems. In reality, the dynamic equations of a robot manipulator form a complex, nonlinear, and multivariable system. In this chapter, therefore, we treat the robot control problem in the context of nonlinear, multivariable control. This approach allows us to provide more rigorous analysis of the performance of control systems, and also allows us to design robust and adaptive nonlinear control laws that guarantee stability and tracking of arbitrary trajectories.

We first reformulate the manipulator dynamic equations in a form more suitable for the discussion to follow. Recall the robot equations of motion given by Equation (7.62) and the actuator dynamics given by Equation (6.15)

$$\sum_{j=1}^{n} d_{kj}(q)\ddot{q}_j + \sum_{i=1}^{n}\sum_{j=1}^{n} c_{ijk}(q)\dot{q}_i\dot{q}_j + g_k = \tau_k \tag{8.1}$$

$$J_{m_k}\ddot{\theta}_{m_k} + B_k\dot{\theta}_{m_k} = K_{m_k}/R_k V_k - \tau_k/r_k \tag{8.2}$$

for $k = 1, \ldots, n$ where $B_k = B_{m_k} + K_{b_k}K_{m_k}/R_k$. Multiplying Equation (8.2) by the gear ratio r_k and using the fact that the motor angles θ_{m_k} and the link angles q_k are related by

$$\theta_{m_k} = r_k q_k \tag{8.3}$$

we may write Equation (8.2) as

$$r_k^2 J_m \ddot{q}_k + r_k^2 B_k \dot{q}_k = r_k K_{m_k}/R V_k - \tau_k \tag{8.4}$$

Substituting Equation (8.4) into Equation (8.1) yields

$$r_k^2 J_{m_k} \ddot{q}_k + \sum_{j=1}^{n} d_{kj}\ddot{q}_j + \sum_{i,j=1}^{n} c_{ijk}\dot{q}_i\dot{q}_j + r_k^2 B_k \dot{q}_k + g_k = r_k \frac{K_m}{R} V_k \tag{8.5}$$

for $k = 1, \ldots, n$. In matrix form Equation (8.5) can be written as

$$M(q)\ddot{q} + C(q,\dot{q})\dot{q} + B\dot{q} + g(q) \quad = \quad u \tag{8.6}$$

where $M(q) = D(q) + J$ with J a diagonal matrix with diagonal elements $r_k^2 J_{m_k}$. The gravity vector $g(q)$ and the matrix $C(q,\dot{q})$ defining the Coriolis and centrifugal generalized forces are defined as before in Equations (7.64) and (7.65). The input vector u has components

$$u_k \quad = \quad r_k \frac{K_{m_k}}{R_k} V_k \ , \ \text{for } k = 1, \ldots, n$$

Note that u_k has units of torque.

Henceforth, we will take the friction coefficient matrix $B = 0$ for simplicity and use Equation (8.6) as the plant model for our subsequent development. We leave it as an exercise for the reader (Problem 8-1) to show that the properties of passivity, skew-symmetry, bounds on the inertia matrix, and linearity in the parameters continue to hold for the system (8.6).

8.1 PD CONTROL REVISITED

It is a rather remarkable fact that the simple PD control scheme for set-point control of rigid robots that we discussed in Chapter 6 can be rigorously shown to work in the general case of Equation (8.6).[1] An independent joint PD control scheme can be written in vector form as

$$u \quad = \quad -K_P\tilde{q} - K_D\dot{q} \tag{8.7}$$

where $\tilde{q} = q - q^d$ is the error between the desired (constant) joint displacement vector q^d and the actual joint displacement vector q, and K_P, K_D are diagonal matrices of (positive) proportional and derivative gains, respectively. We first show that, in the absence of gravity, that is, if $g(q)$ is zero in Equation (8.6), the PD control law given in Equation (8.7) achieves asymptotic tracking of the desired joint positions. This, in effect, reproduces the result derived previously but is more rigorous, in the sense that the nonlinear coupling terms are not approximated by a constant disturbance.

[1] The reader should review the discussion on Lyapunov stability in Appendix C.

To show that the control law given in Equation (8.7) achieves asymptotic tracking consider the Lyapunov function candidate

$$V = 1/2\dot{q}^T M(q)\dot{q} + 1/2\tilde{q}^T K_P \tilde{q} \tag{8.8}$$

The first term in Equation (8.8) is the kinetic energy of the robot and the second term accounts for the proportional feedback $K_P\tilde{q}$. Note that V represents the total energy that would result if the joint actuators were replaced by springs with stiffness constants represented by K_P and with equilibrium positions at q^d. Thus, V is a positive function except at the "goal" configuration $q = q^d$, $\dot{q} = 0$, at which point V is zero. The idea is to show that along any motion of the robot, the function V is decreasing to zero. This will imply that the robot is moving toward the desired goal configuration.

To show this we note that, since q^d is constant, the time derivative of V is given by

$$\dot{V} = \dot{q}^T M(q)\ddot{q} + 1/2\dot{q}^T \dot{M}(q)\dot{q} + \dot{q}^T K_P \tilde{q} \tag{8.9}$$

Solving for $M(q)\ddot{q}$ in Equation (8.6) with $g(q) = 0$ and substituting the resulting expression into Equation (8.9) yields

$$\begin{aligned} \dot{V} &= \dot{q}^T(u - C(q,\dot{q})\dot{q}) + 1/2\dot{q}^T \dot{M}(q)\dot{q} + \dot{q}^T K_P \tilde{q} \\ &= \dot{q}^T(u + K_P\tilde{q}) + 1/2\dot{q}^T(\dot{M}(q) - 2C(q,\dot{q}))\dot{q} \\ &= \dot{q}^T(u + K_P\tilde{q}) \end{aligned} \tag{8.10}$$

where in the last equality we have used the fact that $\dot{M} - 2C$ is skew symmetric. Substituting the PD control law (8.7) for u into the above yields

$$\dot{V} = -\dot{q}^T K_D \dot{q} \leq 0 \tag{8.11}$$

The above analysis shows that V is decreasing as long as $\dot{q}$ is not zero. This by itself is not enough to prove the desired result since it is conceivable that the manipulator could reach a position where $\dot{q} = 0$ but $q \neq q^d$. To show that this cannot happen we can use LaSalle's theorem (Appendix C). Suppose $\dot{V} \equiv 0$[2]. Then Equation (8.11) implies that $\dot{q} \equiv 0$ and hence $\ddot{q} \equiv 0$. From the equations of motion with PD control

$$M(q)\ddot{q} + C(q,\dot{q})\dot{q} = -K_P\tilde{q} - K_D\dot{q}$$

[2]The notation $\dot{V} \equiv 0$ means that the expression is *identically* equal to zero, not simply zero at one instant

we must then have

$$0 = -K_P\tilde{q}$$

which implies that $\tilde{q} = 0$. LaSalle's theorem then implies that the equilibrium is globally asymptotically stable.

In case there are gravitational terms present in Equation (8.6), then Equation (8.10) must be modified as

$$\dot{V} = \dot{q}^T(u - g(q) + K_P\tilde{q}) \tag{8.12}$$

The presence of the gravitational term in Equation (8.12) means that PD control alone cannot guarantee asymptotic tracking. In practice there will be a steady state error or offset. Assuming that the closed-loop system is stable, the robot configuration q that is achieved will satisfy

$$K_P(q - q^d) = g(q) \tag{8.13}$$

The physical interpretation of Equation (8.13) is that the configuration q must be such that the motor generates a steady state "holding torque" $K_P(q - q^d)$ sufficient to balance the gravitational torque $g(q)$. Thus, we see that the steady state error can be reduced by increasing the position gain K_P.

In order to remove this steady state error we can modify the PD control law as

$$u = -K_P\tilde{q} - K_D\dot{q} + g(q) \tag{8.14}$$

The modified control law given by Equation (8.14), in effect, cancels the gravitational terms and we achieve the same Equation (8.11) as before. The control law given by Equation (8.14) requires the computation at each instant of the gravitational terms $g(q)$ from the Lagrangian equations. In the case that these terms are unknown the control law (8.14) cannot be computed. We will say more about this and related issues later in the context of robust and adaptive control.

8.1.1 The Effect of Joint Flexibility

In Chapter 6 we considered the effect of joint flexibility and showed for a lumped model of a single-link robot that a PD control could be designed for set-point tracking. In this section we will discuss the analogous result in the general case of an n-link manipulator.

We first derive a model similar to Equation (8.6) to represent the dynamics of an n-link robot with joint flexibility. For simplicity, assume that the joints are revolute and are actuated by permanent magnet DC motors. We model the flexibility of the ith joint as a linear torsional spring with stiffness constant k_i, for $i = 1, \ldots, n$.

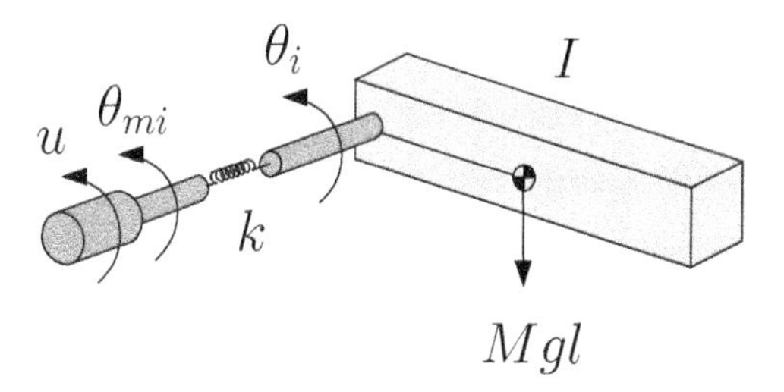

Figure 8.1: A single link of a flexible joint manipulator. The joint elasticity is represented by a torsional spring between the link angle θ_i and the motor shaft angle θ_{mi}.

Referring to Figure 8.1, we let $q_1 = [\theta_1, \ldots, \theta_n]^T$ be the vector of DH-joint variables and $q_2 = [\frac{1}{r_{m_1}}\theta_{m_1}, \ldots, \frac{1}{r_{m_n}}\theta_{m_n}]^T$ be the vector of motor shaft angles (reflected to the link side of the gears). Then $q_1 - q_2$ is the vector of elastic joint deflections. Neglecting the effect of link motion on the kinetic energy of the rotors, the kinetic and potential energies of the manipulator are given by

$$\mathcal{K} = \frac{1}{2}\dot{q}_1^T D(q_1)\dot{q}_1 + \frac{1}{2}\dot{q}_2^T J\dot{q}_2 \tag{8.15}$$

$$\mathcal{P} = P(q_1) + \frac{1}{2}(q_1 - q_2)^T K(q_1 - q_2) \tag{8.16}$$

where J and K are diagonal matrices of inertia and stiffness constants, respectively.

$$J = \begin{bmatrix} J_1 & 0 & \ldots & 0 \\ 0 & J_2 & \ldots & 0 \\ 0 & 0 & \ddots & 0 \\ 0 & 0 & \ldots & J_n \end{bmatrix}, \quad K = \begin{bmatrix} k_1 & 0 & \ldots & 0 \\ 0 & k_2 & \ldots & 0 \\ 0 & 0 & \ddots & 0 \\ 0 & 0 & \ldots & k_n \end{bmatrix} \tag{8.17}$$

Note that we have simply augmented the kinetic energy of the rigid joint robot with the kinetic energy of the actuators and we have likewise augmented the potential energy due to gravity of the rigid joint model with the elastic potential energy of the linear springs at the joints. It is now straightforward to compute the Euler-Lagrange equations for this system as (Problem 8-2)

$$\begin{aligned} D(q_1)\ddot{q}_1 + C(q_1, \dot{q}_1)\dot{q}_1 + g(q_1) + K(q_1 - q_2) &= 0 \\ J\ddot{q}_2 + K(q_2 - q_1) &= u \end{aligned} \tag{8.18}$$

For the problem of set point tracking with PD control, consider a control law of the form

$$u = -K_P\tilde{q}_2 - K_D\dot{q}_2 \tag{8.19}$$

where $\tilde{q}_2 = q_2 - q^d$ and q^d is a vector of constant set points. As in the case of the rigid joint model, suppose now that the gravity vector $g(q_1) = 0$. To show asymptotic tracking for the closed-loop system consider the Lyapunov function candidate

$$V = \frac{1}{2}\dot{q}_1^T D(q_1)\dot{q}_1 + \frac{1}{2}\dot{q}_2^T J\dot{q}_2 + \frac{1}{2}(q_1 - q_2)^T K(q_1 - q_2) + \frac{1}{2}\tilde{q}^T K_P\tilde{q} \tag{8.20}$$

We leave it as an exercise (Problem 8-3) to show that an application of LaSalle's theorem proves global asymptotic stability of the system.

In the absence of gravity it is easy to show (Problem 8-4) that the motor and link angles are equal in the steady state. Thus, one may choose the set point q^d as a vector of desired DH variables.

If gravity is present, then it is not apparent how one can implement gravity compensation in a manner similar to the rigid joint case. We will address this question later in the context of feedback linearization in Chapter 10.

8.2 INVERSE DYNAMICS

We now consider the application of more complex nonlinear control techniques for trajectory tracking of rigid manipulators. The first algorithm that we consider is known as the method of **inverse dynamics**. The method of inverse dynamics is, as we shall see in Chapter 10, a special case of the method of **feedback linearization**.

After presenting the basic idea of inverse dynamics, both in joint space and in task space, we will discuss the practical situation of uncertainty in the parameters defining the manipulator dynamics. The problem of parametric uncertainty naturally leads to a discussion of robust and adaptive control, which we will discuss in the remaining sections.

8.2.1 Joint Space Inverse Dynamics

Consider again the dynamic equations of an n-link rigid robot in matrix form

$$M(q)\ddot{q} + C(q, \dot{q})\dot{q} + g(q) \quad = \quad u \tag{8.21}$$

The idea of inverse dynamics is to seek a nonlinear feedback control law

$$u \quad = \quad f(q, \dot{q}, t) \tag{8.22}$$

which, when substituted into Equation (8.21), results in a linear closed-loop system. For general nonlinear systems, such a control law may be quite difficult or impossible to find. In the case of the manipulator dynamics given by Equations (8.21), however, the problem is actually easy. By inspecting Equation (8.21) we see that if we choose the control u according to the equation

$$u \quad = \quad M(q)a_q + C(q, \dot{q})\dot{q} + g(q) \tag{8.23}$$

then, since the inertia matrix M is invertible, the combined system given by Equations (8.21)–(8.23) reduces to

$$\ddot{q} \quad = \quad a_q \tag{8.24}$$

The term a_q represents a new input that is yet to be chosen. Equation (8.24) is known as the **double integrator system** as it represents n uncoupled double integrators. Equation (8.23) is called the **inverse dynamics control** and achieves a rather remarkable result, namely that the system given by Equation (8.24) is linear and decoupled. This means that each input a_{q_k} can be designed to control a SISO linear system. Moreover, assuming that a_{q_k} is a function only of q_k and $\dot{q}_k$, then the closed-loop system will be decoupled.

Since a_q can now be designed to control a linear second order system, an obvious choice is to set

$$a_q \quad = \quad \ddot{q}^d(t) - K_0\tilde{q} - K_1\dot{\tilde{q}} \tag{8.25}$$

where $\tilde{q} = q - q^d$, $\dot{\tilde{q}} = \dot{q} - \dot{q}^d$, K_0, K_1 are diagonal matrices with diagonal elements consisting of position and velocity gains, respectively and the reference trajectory

$$t \quad \rightarrow \quad (q^d(t), \dot{q}^d(t), \ddot{q}^d(t)) \tag{8.26}$$

defines the desired time history of joint positions, velocities, and accelerations. Note that Equation (8.25) is nothing more than a PD control with feedforward acceleration as defined in Chapter 6.

Substituting Equation (8.25) into Equation (8.24), results in

$$\ddot{\tilde{q}}(t) + K_1\dot{\tilde{q}}(t) + K_0\tilde{q}(t) = 0 \tag{8.27}$$

A simple choice for the gain matrices K_0 and K_1 is

$$K_0 = \begin{bmatrix} \omega_1^2 & 0 & \dots & 0 \\ 0 & \omega_2^2 & \dots & 0 \\ \vdots & & \ddots & \vdots \\ 0 & 0 & \dots & \omega_n^2 \end{bmatrix}, \quad K_1 = \begin{bmatrix} 2\omega_1 & 0 & \dots & 0 \\ 0 & 2\omega_2 & \dots & 0 \\ \vdots & & \ddots & \vdots \\ 0 & 0 & \dots & 2\omega_n \end{bmatrix} \tag{8.28}$$

which results in a decoupled closed-loop system with each joint response equal to the response of a critically damped linear second order system with natural frequency ω_i. As before, the natural frequency ω_i determines the speed of response of the joint, or equivalently, the rate of decay of the tracking error.

The inverse dynamics approach is extremely important as a basis for control and it is worthwhile to examine it from alternative viewpoints. We can give a second interpretation of the control law (8.23) as follows. Consider again the manipulator dynamics (8.21). Since $M(q)$ is invertible for $q \in \mathbb{R}^n$ we may solve for the acceleration $\ddot{q}$ of the manipulator as

$$\ddot{q} = M^{-1}\{u - C(q,\dot{q})\dot{q} - g(q)\} \tag{8.29}$$

Suppose we were able to specify the acceleration as the input to the system. That is, suppose we had actuators capable of producing directly a commanded acceleration (rather than indirectly by producing a force or torque). Then the dynamics of the manipulator, which is after all a position control device, would be given as

$$\ddot{q}(t) = a_q(t) \tag{8.30}$$

where $a_q(t)$ is the input acceleration vector. This is again the familiar double integrator system. Note that Equation (8.30) is not an approximation in any sense; rather it represents the actual open-loop dynamics of the system provided that the acceleration is chosen as the input. The control problem for the system (8.30) is now easy and the acceleration input a_q can be chosen as before according to Equation (8.25).

In reality, however, such "acceleration actuators" are not available to us and we must be content with the ability to produce a generalized force (torque) u_i at each joint i. Comparing Equations (8.29) and (8.30) we see that the torque input u and the acceleration input a_q of the manipulator are related by

$$M^{-1}\{u - C(q,\dot{q})\dot{q} - g(q)\} = a_q \tag{8.31}$$

Solving Equation (8.31) for the input torque $u(t)$ yields

$$u = M(q)a_q + C(q,\dot{q})\dot{q} + g(q) \tag{8.32}$$

which is the same as the previously derived expression (8.23). Thus, the inverse dynamics can be viewed as an input transformation that transforms the problem from one of choosing torque input commands to one of choosing acceleration input commands.

Note that the implementation of this control scheme requires the real-time computation of the inertia matrix and the vectors of Coriolis, centrifugal, and gravitational generalized forces. An important issue therefore in the control system implementation is the design of the computer architecture for the above computations.

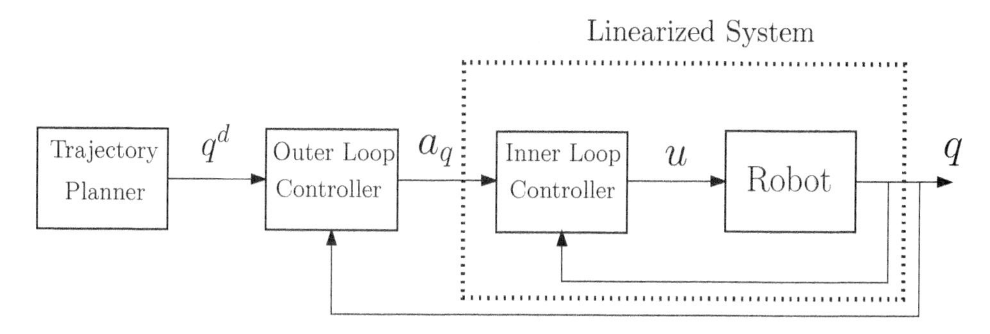

Figure 8.2: Inner-loop/outer-loop control architecture. The inner-loop control computes the vector u of input torques as a function of the measured joint positions and velocities and the given outer-loop control in order to compensate the nonlinearities in the plant model. The outer-loop control designed to track a given reference trajectory can then be based on a linear and decoupled plant model.

Figure 8.2 illustrates the so-called **inner-loop/outer-loop** control architecture. By this we mean that the computation of the nonlinear control law (8.23) is performed in an inner-loop with the vectors $q, \dot{q}$, and a_q as its inputs and u as output. The outer loop in the system is then the computation of the additional input term a_q. Note that the outer-loop control a_q is more in line with the notion of a feedback control in the usual sense of being error driven. The design of the outer-loop feedback control is in theory greatly simplified since it is designed for the plant represented by the dotted lines in Figure 8.2, which is now a linear system.

8.2.2 Task Space Inverse Dynamics

As an illustration of the importance of the inner-loop/outer-loop control architecture, we will show that tracking in task space can be achieved by modifying the outer-loop control a_q in Equation (8.24) while leaving the inner-loop control unchanged. Let $X \in R^6$ represent the end-effector pose using any minimal representation of $SO(3)$. Since X is a function of the joint variables $q\mathcal{Q}$ we have

$$\begin{aligned} \dot{X} &= J(q)\dot{q} && (8.33) \\ \ddot{X} &= J(q)\ddot{q} + \dot{J}(q)\dot{q} && (8.34) \end{aligned}$$

where $J = J_a$ is the analytical Jacobian of Equation (4.84). Given the double integrator system (8.24) in joint space we see that if a_q is chosen as

$$a_q = J^{-1}\left\{a_X - \dot{J}\dot{q}\right\} \tag{8.35}$$

the result is a double integrator system in task space coordinates

$$\ddot{X} = a_X \tag{8.36}$$

Given a task space trajectory $X^d(t)$ satisfying the same smoothness and boundedness assumptions as the joint space trajectory $q^d(t)$, we may choose a_X as

$$a_X = \ddot{X}^d - K_0(X - X^d) - K_1(\dot{X} - \dot{X}^d) \tag{8.37}$$

so that the task space tracking error $\tilde{X} = X - X^d$ satisfies

$$\ddot{\tilde{X}} + K_1\dot{\tilde{X}} + K_0\tilde{X} = 0 \tag{8.38}$$

Therefore, a modification of the outer-loop control achieves a linear and decoupled system directly in the task space coordinates without the need to compute a joint space trajectory and without the need to modify the nonlinear inner-loop control.

Note that we have used a minimal representation for the orientation of the end effector in order to specify a trajectory $X \in \mathbb{R}^6$. In general, if the end effector coordinates are given in $SE(3)$, then the Jacobian J in the above formulation will be the geometric Jacobian of Equation (4.42). In this case

$$\dot{X} = \begin{bmatrix} v \\ \omega \end{bmatrix} = \begin{bmatrix} \dot{x} \\ \omega \end{bmatrix} = J(q)\dot{q} \tag{8.39}$$

and the outer-loop control

$$a_q = J^{-1}(q) \left\{ \begin{bmatrix} a_x \\ a_\omega \end{bmatrix} - \dot{J}(q)\dot{q} \right\} \tag{8.40}$$

applied to Equation (8.24) results in the system

$$\ddot{x} = a_x \in \mathbb{R}^3 \tag{8.41}$$
$$\dot{\omega} = a_\omega \in \mathbb{R}^3 \tag{8.42}$$
$$\dot{R} = S(\omega)R, \quad R \in SO(3), \ S \in so(3) \tag{8.43}$$

Although, in this latter case, the dynamics have not been linearized to a double integrator, the outer-loop terms a_v and a_ω may still be used to define control laws to track end–effector trajectories in $SE(3)$.

In both cases we see that nonsingularity of the Jacobian is necessary to implement the outer-loop control. If the robot has more or fewer than six joints, then the Jacobians are not square. In this case, other schemes have been developed using, for example, the Jacobian pseudoinverse. See [25] for details.

8.3 ROBUST AND ADAPTIVE MOTION CONTROL

A drawback to the implementation of the inverse dynamics control methodology described in the previous section is that the parameters of the system must be known exactly. If the parameters are uncertain, for example when the manipulator picks up an unknown load, then the ideal performance of the inverse dynamics controller is no longer guaranteed. This section is concerned with the **robust** and **adaptive** motion control problem. The goal of both robust and adaptive control is to maintain performance in terms of stability, tracking error, or other specifications despite parametric uncertainty, external disturbances, unmodeled dynamics, or other uncertainties present in the system. In distinguishing between robust control and adaptive control, we follow the commonly accepted notion that a robust controller is a fixed controller designed to satisfy performance specifications over a given range of uncertainties, whereas an adaptive controller incorporates some sort of online parameter estimation. This distinction is important. For example, in a repetitive motion task, the tracking errors produced by a fixed robust controller would tend to be repetitive as well, whereas tracking errors produced by an adaptive controller might be expected to decrease over time as the plant and/or control parameters are updated based on runtime information. At the same time, adaptive controllers that perform well in the face of parametric uncertainty may not perform well in the face of other types of uncertainty such as external disturbances or unmodeled dynamics. Therefore, an understanding of the trade-offs involved is important in deciding whether to employ robust or adaptive control design methods in a given situation.

8.3.1 Robust Inverse Dynamics

The inverse dynamics approach relies on exact cancellation of nonlinearities in the robot equations of motion. The practical implementation of inverse dynamics control requires consideration of various sources of uncertainties such as modeling errors, unknown loads, and computation errors. Let us return to the Euler-Lagrange equations of motion

$$M(q)\ddot{q} + C(q,\dot{q})\dot{q} + g(q) = u \tag{8.44}$$

and write the inverse dynamics control input u as

$$u = \hat{M}(q)a_q + \hat{C}(q,\dot{q})\dot{q} + \hat{g}(q) \tag{8.45}$$

where the notation $\hat{(\cdot)}$ represents the computed or nominal value of $(\cdot)$ and indicates that the theoretically exact inverse dynamics control cannot be achieved in practice due to the uncertainties in the system. The error or mismatch $\tilde{(\cdot)} = (\cdot) - \hat{(\cdot)}$ is a measure of one's knowledge of the system parameters.

If we substitute Equation (8.45) into Equation (8.44) we obtain, after some algebra (Problem 8-8),

$$\ddot{q} = a_q + \eta(q,\dot{q},a_q) \tag{8.46}$$

where

$$\eta \;=\; M^{-1}(\tilde{M}a_q + \tilde{C}\dot{q} + \tilde{g}) \tag{8.47}$$

is called the **uncertainty**. We define E as

$$E := M^{-1}\tilde{M} = M^{-1}\hat{M} - I \tag{8.48}$$

which allows us to express the uncertainty η as

$$\eta = Ea_q + M^{-1}(\tilde{C}\dot{q} + \tilde{g}) \tag{8.49}$$

The system (8.46) is still nonlinear and coupled due to the uncertainty $\eta(q, \dot{q}, a_q)$ and, therefore, we have no guarantee that the outer-loop control given by Equation (8.25) will satisfy desired tracking performance specifications. In the next section we show how to modify the outer-loop control (8.25) to guarantee global convergence of the tracking error for the system (8.46).

Outer Loop Design via Lyapunov's Second Method

There are several approaches to treat the robust inverse dynamics problem outlined above. In this section we will discuss the so-called theory of **guaranteed stability of uncertain systems**, which is based on Lyapunov's second method. In this approach we set the outer-loop control a_q as

$$a_q = \ddot{q}^d(t) - K_0\tilde{q} - K_1\dot{\tilde{q}} + \delta a \tag{8.50}$$

where δa is an additional term to be designed. In terms of the tracking error

$$e = \begin{bmatrix} \tilde{q} \\ \dot{\tilde{q}} \end{bmatrix} = \begin{bmatrix} q - q^d \\ \dot{q} - \dot{q}^d \end{bmatrix} \tag{8.51}$$

we may write Equations (8.46) and (8.50) as

$$\dot{e} = Ae + B\{\delta a + \eta\} \tag{8.52}$$

where

$$A = \begin{bmatrix} 0 & I \\ -K_0 & -K_1 \end{bmatrix}, \quad B = \begin{bmatrix} 0 \\ I \end{bmatrix} \tag{8.53}$$

Thus, the double integrator is first stabilized by the linear feedback term $-K_0\tilde{q} - K_1\dot{\tilde{q}}$, and the additional control term δa should be designed to overcome the potentially destabilizing effect of the uncertainty η. The basic idea is to assume that we are able to compute a bound $\rho(e, t) \geq 0$ on the uncertainty η as

$$||\eta|| \leq \rho(e, t) \tag{8.54}$$

and design the additional input term δa to guarantee ultimate boundedness of the error trajectory $e(t)$ in Equation (8.52). Note that the bound ρ is in general a function of the tracking error e and time.

Returning to our expression for the uncertainty η and substituting for a_q from Equation (8.50) we have

$$\begin{aligned} \eta &= Ea_q + M^{-1}(\tilde{C}\dot{q} + \tilde{g}) \\ &= E\delta a + E(\ddot{q}^d - K_0\tilde{q} - K_1\dot{\tilde{q}}) + M^{-1}(\tilde{C}\dot{q} + \tilde{g}) \end{aligned} \tag{8.55}$$

Let us assume that we can find constants $\alpha < 1$, γ_1, and γ_2, together with possibly time-varying γ_3 such that

$$||\eta|| \leq \alpha||\delta a|| + \gamma_1||e|| + \gamma_2||e||^2 + \gamma_3 \tag{8.56}$$

We note that the condition $\alpha := ||E|| = ||M^{-1}\hat{M} - I|| < 1$ determines how close our estimate $\hat{M}$ must be to the true inertia matrix. Suppose that we have bounds on M^{-1} as

$$\underline{\mathrm{M}} \leq ||M^{-1}|| \leq \overline{M} \tag{8.57}$$

If we choose the estimated inertia matrix $\hat{M}$ as

$$\hat{M} = \frac{2}{\overline{M} + \underline{\mathrm{M}}} I \tag{8.58}$$

then it can be shown that

$$||M^{-1}\hat{M} - I|| \leq \frac{\overline{M} - \underline{\mathrm{M}}}{\overline{M} + \underline{\mathrm{M}}} < 1 \tag{8.59}$$

The point is that there is always a choice for $\hat{M}$ that satisfies the condition $||E|| < 1$.

Next, assume, for the moment, that $||\delta a|| \leq \rho(e, t)$ which must then be checked a posteriori. It follows that

$$||\eta|| \leq \alpha\rho(e, t) + \gamma_1||e|| + \gamma_2||e||^2 + \gamma_3 =: \rho(e, t) \tag{8.60}$$

which, since $\alpha < 1$, defines ρ as

$$\rho(e, t) = \frac{1}{1 - \alpha}(\gamma_1||e|| + \gamma_2||e||^2 + \gamma_3) \tag{8.61}$$

Since K_0 and K_1 are chosen so that the matrix A in Equation (8.52) is Hurwitz,[3] we may choose $Q > 0$ and let $P > 0$ be the unique symmetric positive definite matrix satisfying the Lyapunov equation

$$A^T P + PA = -Q \tag{8.62}$$

Defining the control δa according to

$$\delta a = \begin{cases} -\rho(e, t)\dfrac{B^T Pe}{||B^T Pe||} & ; \text{ if } \quad ||B^T Pe|| \neq 0 \\ \\ 0 & ; \text{ if } \quad ||B^T Pe|| = 0 \end{cases} \tag{8.63}$$

[3] A Hurwitz matrix is one that has all its eigenvalues in the open left half of the complex plane.

it follows that the Lyapunov function $V = e^TPe$ satisfies $\dot{V} < 0$ along solution trajectories of Equation (8.52). To show this result, we compute

$$\dot{V} = -e^TQe + 2e^TPB\{\delta a + \eta\} \tag{8.64}$$

For simplicity set $w = B^TPe$ and consider the second term $w^T\{\delta a + \eta\}$ in the above expression. If $w = 0$ this term vanishes and for $w \neq 0$ we have

$$\delta a = -\rho\frac{w}{||w||} \tag{8.65}$$

and hence, using the Cauchy-Schwartz inequality we have

$$\begin{aligned} w^T(-\rho\frac{w}{||w||} + \eta) &\leq -\rho||w|| + ||w||\,||\eta|| \\ &= ||w||(-\rho + ||\eta||) \leq 0 \end{aligned} \tag{8.66}$$

since $||\eta|| \leq \rho$. Therefore

$$\dot{V} \leq -e^TQe < 0 \tag{8.67}$$

and the result follows. Finally, note that $||\delta a|| \leq \rho$ as required.

Since the above control term δa is discontinuous on the subspace defined by $B^TPe = 0$, solution trajectories on this subspace are not well defined in the usual sense. One may define solutions in a generalized sense, the so-called **Filippov solutions** [37]. A detailed treatment of discontinuous control systems is beyond the scope of this text. In practice, the discontinuity in the control results in the phenomenon of **chattering**, where the control switches rapidly between the control values in (8.63).

One may implement a continuous approximation to the discontinuous control as

$$\delta a = \begin{cases} -\rho(e,t)\dfrac{B^TPe}{||B^TPc||} & ; \text{ if } \quad ||B^TPe|| > \epsilon \\ \\ -\dfrac{\rho(e,t)}{\epsilon}B^TPe & ; \text{ if } \quad ||B^TPe|| \leq \epsilon \end{cases} \tag{8.68}$$

In this case, since the control signal given by Equation (8.68) is continuous, a solution to the system (8.52) exists for any initial condition and we can prove the following result.

Theorem 3 *All trajectories of the system (8.52) are uniformly ultimately bounded (u.u.b.) using the continuous control law (8.68). (See Appendix C for the definition of uniform ultimate boundedness.)*

Proof: As before, choose $V(e) = e^T Pe$ and compute

$$\begin{aligned} \dot{V} &= -e^T Qe + 2w^T(\delta a + \eta) & (8.69) \\ &\leq -e^T Qe + 2w^T(\delta a + \rho\frac{w}{||w||}) & (8.70) \end{aligned}$$

with $||w|| = ||B^T Pe||$ as above. For $||w|| \geq \epsilon$ the argument proceeds as above and $\dot{V} < 0$. For $||w|| \leq \epsilon$ the second term above becomes

$$2w^T\left(-\frac{\rho}{\epsilon}w + \rho\frac{w}{||w||}\right) = -2\frac{\rho}{\epsilon}||w||^2 + 2\rho||w||$$

This expression attains a maximum value of $\epsilon\frac{\rho}{2}$ when $||w|| = \frac{\epsilon}{2}$. Thus, we have

$$\dot{V} \leq -e^T Qe + \epsilon\frac{\rho}{2} < 0 \tag{8.71}$$

provided

$$e^T Qe > \epsilon\frac{\rho}{2} \tag{8.72}$$

Using the relationship

$$\lambda_{min}(Q)||e||^2 \leq e^T Qe \leq \lambda_{max}(Q)||e||^2 \tag{8.73}$$

where $\lambda_{min}(Q)$, $\lambda_{max}(Q)$ denote the minimum and maximum eigenvalues, respectively, of the matrix Q, we have that $\dot{V} < 0$ if

$$\lambda_{min}(Q)||e||^2 > \epsilon\frac{\rho}{2} \tag{8.74}$$

or, equivalently

$$||e|| > \left(\frac{\epsilon\rho}{2\lambda_{min}(Q)}\right)^{\frac{1}{2}} =: \delta \tag{8.75}$$

Let S_δ denote the smallest level set of V containing $B(\delta)$, the ball of radius δ and let B_r denote the smallest ball containing S_δ. Then all solutions of the closed-loop system are u.u.b. with respect to B_r. The situation is shown in Figure 8.3. All trajectories will eventually enter the ball B_r; in fact, all trajectories will reach the boundary of S_δ since $\dot{V}$ is negative definite outside of S_δ.

Note that the radius of the ultimate boundedness set, and hence, the magnitude of the steady state tracking error, is proportional to the product of the uncertainty bound ρ and the constant ϵ. The constant ϵ is used to reduce or eliminate chattering and can be chosen only as large as necessary to eliminate chattering.

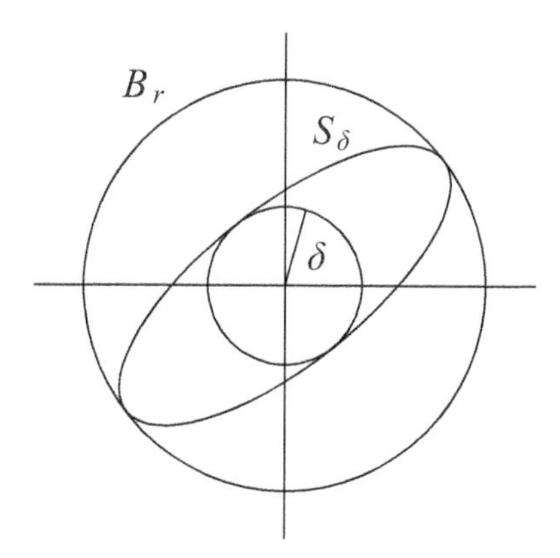

Figure 8.3: The uniform ultimate boundedness set. Since $\dot{V}$ is negative outside the ball B_δ, all trajectories will eventually enter the level set S_δ, the smallest level set of V containing B_δ. The system is thus u.u.b. with respect to B_r, the smallest ball containing S_δ.

8.3.2 Adaptive Inverse Dynamics

Once the linear parametrization property for manipulators became widely known in the mid-1980's, the first globally convergent adaptive control results began to appear. These first results were based on the inverse dynamics approach discussed above. Consider the plant given by Equation (8.44) and the control given by Equation (8.45) as above, but now suppose that the parameters appearing in Equation (8.45) are not fixed as in the robust control approach, but are time-varying estimates of the true parameters. Substituting Equation (8.45) into Equation (8.44) and setting

$$a_q = \ddot{q}^d - K_1(\dot{q} - \dot{q}^d) - K_0(q - q^d) \tag{8.76}$$

it can be shown (Problem 8-10), using the linear parametrization property, that

$$\ddot{\tilde{q}} + K_1\dot{\tilde{q}} + K_0\tilde{q} = \hat{M}^{-1}Y(q, \dot{q}, \ddot{q})\tilde{\theta} \tag{8.77}$$

where Y is the regressor function and $\tilde{\theta} = \hat{\theta} - \theta$, where $\hat{\theta}$ is the estimate of the parameter vector θ. In state space we write the system (8.77) as

$$\dot{e} = Ae + B\Phi\tilde{\theta} \tag{8.78}$$

where

$$A = \begin{bmatrix} 0 & I \\ -K_0 & -K_1 \end{bmatrix}, \quad B = \begin{bmatrix} 0 \\ I \end{bmatrix}, \quad \Phi = \hat{M}^{-1}Y(q, \dot{q}, \ddot{q}) \tag{8.79}$$

with K_0 and K_1 chosen as before as diagonal matrices of positive gains so that A is a Hurwitz matrix. Let P be the unique symmetric, positive definite matrix P satisfying the matrix Lyapunov equation

$$A^T P + PA = -Q \tag{8.80}$$

and choose the parameter update law as

$$\dot{\hat{\theta}} = -\Gamma^{-1}\Phi^T B^T Pe \tag{8.81}$$

where Γ is a constant, symmetric, positive definite matrix. Then, global convergence to zero of the tracking error with all internal signals remaining bounded can be shown using the Lyapunov function

$$V = e^T Pe + \tilde{\theta}^T \Gamma \tilde{\theta} \tag{8.82}$$

To see this we calculate $\dot{V}$ as (Problem 8-11)

$$\dot{V} = -e^T Qe + 2\tilde{\theta}^T \{\Phi^T B^T Pe + \Gamma\dot{\hat{\theta}}\} \tag{8.83}$$

the latter term following since θ is constant, that is, $\dot{\tilde{\theta}} = \dot{\hat{\theta}}$. Using the parameter update law (8.81) we have

$$\dot{V} = -e^T Qe \tag{8.84}$$

From this it follows that the position tracking errors converge to zero asymptotically and the parameter estimation errors remain bounded. We will not go through the details of the proof in this section. The argument is similar to that of the passivity-based, adaptive control approach of the next section so we will defer the details until later. In order to implement this adaptive inverse dynamics scheme, however, one notes that the acceleration $\ddot{q}$ is needed in the parameter update law and that $\hat{M}$ must be invertible. The need for the joint acceleration in the parameter update law presents a serious challenge to its implementation. Acceleration sensors are noisy and introduce additional cost whereas calculating the acceleration by numerical differentiation of position or velocity signals is not feasible in most cases. The invertibility of $\hat{M}$ can be enforced in the algorithm by resetting the parameter estimate whenever $\hat{\theta}$ would otherwise result in $\hat{M}$ becoming singular. The passivity-based approaches that we treat next remove both of these impediments.

8.4 PASSIVITY-BASED MOTION CONTROL

The inverse dynamics based approaches in the previous section rely on cancellation of nonlinearities in the system dynamics. In this section we discuss control techniques based on the passivity or skew symmetry property of the Euler-Lagrange equations. These methods do not rely on cancellation of nonlinearities and hence, do not lead to a linear closed-loop system even in the exact case of no uncertainty. However, as we shall see, the passivity-based methods have other advantages with respect to robust and adaptive control.

To motivate the robust and adaptive methods to follow consider again the Euler-Lagrange equations

$$M(q)\ddot{q} + C(q,\dot{q})\dot{q} + g(q) = u \tag{8.85}$$

and choose the control input according to

$$u = M(q)a + C(q,\dot{q})v + g(q) - Kr \tag{8.86}$$

where the quantities v, a, and r are given as

$$\begin{aligned} v &= \dot{q}^d - \Lambda\tilde{q} \\ a &= \dot{v} = \ddot{q}^d - \Lambda\dot{\tilde{q}} \\ r &= \dot{q} - v = \dot{\tilde{q}} + \Lambda\tilde{q} \end{aligned}$$

where K and Λ are diagonal matrices of constant, positive gains. Substituting the control law (8.86) into the plant model (8.85) leads to

$$M(q)\dot{r} + C(q,\dot{q})r + Kr = 0 \tag{8.87}$$

Note that, in contrast to the inverse dynamics control approach, the closed-loop system (8.87) is still a coupled nonlinear system. Stability and asymptotic convergence of the tracking error to zero therefore are not obvious and require additional analysis. Consider the Lyapunov function candidate

$$V = \frac{1}{2}r^T M(q) r + \tilde{q}^T \Lambda K \tilde{q} \tag{8.88}$$

Calculating $\dot{V}$ yields

$$\begin{aligned} \dot{V} &= r^T M\dot{r} + \tfrac{1}{2}r^T\dot{M}r + 2\tilde{q}^T\Lambda K\dot{\tilde{q}} \\ &= -r^T K r + 2\tilde{q}^T\Lambda K\dot{\tilde{q}} + \tfrac{1}{2}r^T(\dot{M} - 2C)r \end{aligned} \tag{8.89}$$

Using the skew symmetry property and the definition of r, Equation (8.89) reduces to

$$\begin{aligned} \dot{V} &= -\tilde{q}^T \Lambda^T K \Lambda \tilde{q} - \dot{\tilde{q}}^T K \dot{\tilde{q}} \\ &= -e^T Q e \end{aligned} \quad (8.90)$$

where

$$Q = \begin{bmatrix} \Lambda^T K \Lambda & 0 \\ 0 & K \end{bmatrix} \quad (8.91)$$

Therefore the equilibrium $e = 0$ in error space is globally asymptotically stable.

In fact, assuming constant norm bounds on the inertia matrix $M(q)$, as happens, for example, when all joints are revolute, it is fairly straightforward to prove global exponential stability of the tracking error. In the case of the inverse dynamics control considered previously, we concluded exactly the same result in a much simpler way. Therefore the advantages, if any, of the passivity-based control over the inverse dynamics control are unclear at this point. We will see in the next two sections that the real advantage of the passivity-based approach occurs for the robust and adaptive control problems. In the robust control approach considered next we will see that the assumption $||E|| = ||M^{-1}\hat{M} - I|| < 1$ can be eliminated and that the computation of the uncertainty bounds is greatly simplified. In the adaptive control approach we will see that the requirements on acceleration measurement and boundedness of the estimated inertia matrix $\hat{M}$ can be eliminated. Thus, the passivity-based approach has some very important advantages over the inverse dynamics approach for the robust and adaptive control problems.

8.4.1 Passivity-Based Robust Control

In this section we use the passivity-based approach above to derive an alternative robust control algorithm that exploits both the skew symmetry property and linearity in the parameters and leads to a much easier design in terms of computation of uncertainty bounds. We modify the control input (8.86) as

$$u = \hat{M}(q)a + \hat{C}(q, \dot{q})v + \hat{g}(q) - Kr \quad (8.92)$$

where K, Λ, v, a, and r are given as before. In terms of the linear parametrization of the robot dynamics, the control (8.92) becomes

$$u = Y(q, \dot{q}, a, v)\hat{\theta} - Kr \quad (8.93)$$

and the combination of Equation (8.92) with Equation (8.44) yields

$$M(q)\dot{r} + C(q,\dot{q})r + Kr = Y(\hat{\theta} - \theta) \tag{8.94}$$

We now choose the term $\hat{\theta}$ in Equation (8.93) as

$$\hat{\theta} = \theta_0 + \delta\theta \tag{8.95}$$

where θ_0 is a fixed nominal parameter vector and $\delta\theta$ is an additional control term. The system (8.94) then becomes

$$M(q)\dot{r} + C(q,\dot{q})r + Kr = Y(q,\dot{q},a,v)(\tilde{\theta} + \delta\theta) \tag{8.96}$$

where $\tilde{\theta} = \theta_0 - \theta$ is a constant vector and represents the parametric uncertainty in the system. If the uncertainty can be bounded by finding a nonnegative constant $\rho \geq 0$ such that

$$||\tilde{\theta}|| = ||\theta - \theta_0|| \leq \rho \tag{8.97}$$

then the additional term $\delta\theta$ can be designed according to

$$\delta\theta = \begin{cases} -\rho\dfrac{Y^T r}{||Y^T r||} & ; \text{ if } \quad ||Y^T r|| > \epsilon \\ \\ -\frac{\rho}{\epsilon} Y^T r & ; \text{ if } \quad ||Y^T r|| \leq \epsilon \end{cases} \tag{8.98}$$

Using the same Lyapunov function candidate from Equation (8.88) above, we may show uniform ultimate boundedness of the tracking error. Carrying out the details of the calculation of $\dot{V}$ yields

$$\dot{V} \quad = \quad -e^T Q e + r^T Y(\tilde{\theta} + \delta\theta) \tag{8.99}$$

Uniform ultimate boundedness of the tracking error follows with the control $\delta\theta$ from Equation (8.98) exactly as in the proof of Theorem 3. The details are left as an exercise (Problem 8-13).

Comparing this approach with the approach in Section 8.3.1 we see that finding a constant bound ρ for the constant vector $\tilde{\theta}$ is much simpler than finding a time-varying bound for η in Equation (8.47). The bound ρ in this case depends only on the inertia parameters of the manipulator, while $\rho(x,t)$ in Equation (8.54) depends on the manipulator state vector, the reference trajectory and, in addition, requires some assumptions on the estimated inertia matrix $\hat{M}(q)$.

8.4.2 Passivity-Based Adaptive Control

In the adaptive approach the vector $\hat{\theta}$ in Equation (8.93) is taken to be a time-varying estimate of the true parameter vector θ. Combining the control law, Equation (8.92), with Equation (8.44) yields

$$M(q)\dot{r} + C(q,\dot{q})r + Kr = Y\tilde{\theta} \tag{8.100}$$

The parameter estimate $\hat{\theta}$ may be computed using standard methods of adaptive control such as gradient or least squares. For example, using the gradient update law

$$\dot{\hat{\theta}} = -\Gamma^{-1}Y^T(q,\dot{q},a,v)r \tag{8.101}$$

together with the Lyapunov function

$$V = \frac{1}{2}r^T M(q)r + \tilde{q}^T \Lambda K \tilde{q} + \frac{1}{2}\tilde{\theta}^T \Gamma \tilde{\theta} \tag{8.102}$$

results in global convergence of the tracking errors to zero and boundedness of the parameter estimates.

To show this, we first note an important difference between the adaptive control approach and the robust control approach from the previous section. In the robust approach the states of the system are $\tilde{q}$ and $\dot{\tilde{q}}$. In the adaptive control approach, the fact that $\tilde{\theta}$ satisfies the differential equation (8.101)[4] means that the complete state vector now includes $\tilde{\theta}$ and the state equations are given by the coupled system, Equations (8.100) and (8.101). For this reason we included the positive definite term $\frac{1}{2}\tilde{\theta}^T\Gamma\tilde{\theta}$ in the Lyapunov function (8.102).[5]

If we now compute $\dot{V}$ along trajectories of the system (8.100), we obtain

$$\dot{V} = -\tilde{q}^T \Lambda^T K \Lambda \tilde{q} - \dot{\tilde{q}}^T K \dot{\tilde{q}} + \tilde{\theta}^T\{\Gamma\dot{\hat{\theta}} + Y^T r\} \tag{8.103}$$

Substituting the expression for $\dot{\hat{\theta}}$ from the gradient update law (8.101) into Equation (8.103) yields

$$\dot{V} = -\tilde{q}^T \Lambda^T K \Lambda \tilde{q} - \dot{\tilde{q}}^T K \dot{\tilde{q}} = -e^T Q e \leq 0 \tag{8.104}$$

where e and Q are defined as before, showing that the closed-loop system is stable in the sense of Lyapunov.

---

[4] Note that $\dot{\tilde{\theta}} = \dot{\hat{\theta}}$ since the parameter vector θ is constant.

[5] Similar remarks hold for the adaptive inverse dynamics approach in the previous section.

Note that we have claimed only that the Lyapunov function is negative semi-definite, not negative definite since $\dot{V}$ does not contain any terms that are negative definite in $\tilde{\theta}$. In fact, this situation is common in such adaptive control schemes and is a fundamental reason for several difficulties that arise in adaptive control such as lack of robustness to external disturbances and lack of parameter convergence. A detailed discussion of these and other problems in adaptive control is outside the scope of this text.

Returning to the problem at hand, although we conclude only stability in the sense of Lyapunov for the closed-loop system (8.100) and (8.101), further analysis will allow us to draw stronger conclusions. First, note that since $\dot{V}$ is nonincreasing from Equation (8.104), the value of $V(t)$ can be no greater than its value at $t = 0$. Since V consists of a sum of nonnegative terms, this means that each of the terms r, $\tilde{q}$, and $\tilde{\theta}$ are bounded as functions of time.

With regard to the tracking error, $\tilde{q}$, $\dot{\tilde{q}}$, we also note that $\dot{V}$ is quadratic in the error vector $e(t)$. Integrating both sides of Equation (8.104) gives

$$V(t) - V(0) = -\int_0^t e^T(\sigma)Qe(\sigma)d\sigma < \infty \tag{8.105}$$

As a consequence, the tracking error vector $e(t)$ is a so-called **square integrable function**. Such functions, under some mild additional restrictions, must tend to zero as $t \rightarrow \infty$. Specifically, we may appeal to the following known, as Barbalat's Lemma.

Lemma 8.1 *Suppose $f : \mathbb{R} \mapsto \mathbb{R}$ is a square integrable function and further suppose that its derivative $\dot{f}$ is bounded. Then $f(t) \rightarrow 0$ as $t \rightarrow \infty$.*

We note that, since both $r = \dot{\tilde{q}} + \Lambda\tilde{q}$ and $\tilde{q}$ have already been shown to be bounded, it follows that $\dot{\tilde{q}}$ is also bounded. Therefore, we have that $\tilde{q}$ is square integrable and its derivative is bounded. Hence, the tracking error $\tilde{q} \rightarrow 0$ as $t \rightarrow \infty$.

To show that the velocity tracking error also converges to zero, one must appeal to the equations of motion (8.100), from which one may argue that the acceleration $\ddot{q}$ is bounded. It follows that the velocity error $\dot{\tilde{q}}$ asymptotically converges to zero provided that the reference acceleration $\ddot{q}^d(t)$ is bounded.

8.5 SUMMARY

In this chapter we discussed the nonlinear control problem for robot manipulators. We developed models of both rigid and flexible joint robots. We then developed several control algorithms and discussed pros and cons of each as well as implementation aspects. Among the algorithms we discussed were PD control, inverse dynamics, and passivity-based control. Moreover we showed how to formulate robust and adaptive versions of the latter two approaches.

Plant Models

The rigid and flexible joint robot models are, respectively:

$$M(q)\ddot{q} + C(q,\dot{q})\dot{q} + g(q) = u$$

$$\begin{aligned} D(q_1)\ddot{q}_1 + C(q_1,\dot{q}_1)\dot{q}_1 + g(q_1) + K(q_1 - q_2) &= 0 \\ J\ddot{q}_2 + K(q_2 - q_1) &= u \end{aligned}$$

PD Control

A PD control law in joint space is of the form

$$u = -K_P\tilde{q} - K_D\dot{q}$$

Global asymptotic tracking for the rigid model can be shown using the Lyapunov function candidate below together with LaSalle's theorem in case the gravity vector $g(q) = 0$.

$$V = 1/2\dot{q}^T M(q)\dot{q} + 1/2\tilde{q}^T K_P\tilde{q}$$

PD Control with gravity compensation

With gravity present, the PD plus gravity compensation algorithm below also results in global asymptotic tracking for the rigid model.

$$u = -K_P\tilde{q} - K_D\dot{q} + g(q)$$

Joint Space Inverse Dynamics

The inverse dynamics control law consists of the following two expressions, the first being the **inner-loop** control and the second being the **outer-loop** control.

$$u = M(q)a_q + C(q,\dot{q})\dot{q} + g(q)$$

$$a_q = \ddot{q}^d(t) - K_0\tilde{q} - K_1\dot{\tilde{q}}$$

The inverse dynamics algorithm results in a closed-loop system that is linear and decoupled.

Task Space Inverse Dynamics

We showed that the modified outer-loop term below results in a linear, decoupled system in task space coordinates X, where X is a minimal representation of $SE(3)$ and J is the so-called analytical Jacobian.

$$a_q = J^{-1}\left\{a_X - \dot{J}\dot{q}\right\}$$

$$a_X = \ddot{X}^d - K_0(X - X^d) - K_1(\dot{X} - \dot{X}^d)$$

Robust Inverse Dynamics

We presented a Lyapunov-based approach for robust inverse dynamics control.

$$u = \hat{M}(q)a_q + \hat{C}(q,\dot{q})\dot{q} + \hat{g}(q)$$

where $\hat{M}$, $\hat{C}$, and $\hat{g}(q)$ are the nominal values of M, C, and g. From this we derived the state space model

$$\dot{e} = Ae + B\{\delta a + \eta\}$$

where η represents the uncertainty resulting from inexact cancellation of nonlinearities and

$$A = \begin{bmatrix} 0 & I \\ -K_0 & -K_1 \end{bmatrix} \; ; \; B = \begin{bmatrix} 0 \\ I \end{bmatrix}$$

The additional control input δa was chosen as

$$\delta a = \begin{cases} -\rho(e,t)\dfrac{B^TPe}{||B^TPe||} & ; \text{ if } \quad ||B^TPe|| > \epsilon \\[2ex] -\dfrac{\rho(e,t)}{\epsilon}B^TPe & ; \text{ if } \quad ||B^TPe|| \leq \epsilon \end{cases}$$

and shown to achieve uniform ultimate boundedness of all trajectories. This is a practical notion of asymptotic stability in the sense that the tracking errors can be made small.

Adaptive Inverse Dynamics

The adaptive version of the inverse dynamics control results in a system of the form

$$\dot{e} = Ae + B\Phi\tilde{\theta}$$

$$\dot{\hat{\theta}} = -\Gamma^{-1}\Phi^T B^T Pe$$

where θ represents the unknown parameters (masses, moments of inertia, etc.) The second equation above is used to estimate the parameters online. The Lyapunov function candidate below can be used to show asymptotic convergence of the tracking errors to zero and boundedness of the parameter estimation error.

$$V = e^T Pe + \frac{1}{2}\tilde{\theta}^T\Gamma\tilde{\theta}$$

Passivity-Based Robust Control

Following the treatment of inverse dynamics we introduced the notion of passivity-based control. This approach exploits the passivity property of the robot dynamics rather than attempting to cancel the nonlinearities as in the inverse dynamics approach. We presented an algorithm of the form

$$u = \hat{M}(q)a + \hat{C}(q,\dot{q})v + \hat{g}(q) - Kr$$

where the quantities v, a, and r are given as

$$\begin{aligned} v &= \dot{q}^d - \Lambda\tilde{q} \\ a &= \dot{v} = \ddot{q}^d - \Lambda\dot{\tilde{q}} \\ r &= \dot{q} - v = \dot{\tilde{q}} + \Lambda\tilde{q} \end{aligned}$$

and K is a diagonal matrix of positive gains. This results in a closed-loop system

$$M(q)\dot{r} + C(q,\dot{q})r + Kr = Y(\hat{\theta} - \theta)$$

In the robust passivity-based approach the term $\hat{\theta}$ is chosen as

$$\hat{\theta} = \theta_0 + \delta\theta$$

where θ_0 is a fixed nominal parameter vector and $\delta\theta$ is an additional control term. The additional term $\delta\theta$ can be designed according to

$$\delta\theta = \begin{cases} -\rho\dfrac{Y^T r}{||Y^T r||} & ; \text{ if } \quad ||Y^T r|| > \epsilon \\ \\ -\frac{\rho}{\epsilon}Y^T r & ; \text{ if } \quad ||Y^T r|| \leq \epsilon \end{cases}$$

where ρ is a bound on the parameter uncertainty. Uniform ultimate boundedness of the tracking errors follows using the Lyapunov function candidate

$$V = \frac{1}{2}r^T M(q)r + \tilde{q}^T \Lambda K \tilde{q}$$

Passivity-Based Adaptive Control

In the adaptive version of this approach we derived the system

$$M(q)\dot{r} + C(q, \dot{q})r + Kr = Y\tilde{\theta}$$

$$\dot{\tilde{\theta}} = -\Gamma^{-1}Y^T(q, \dot{q}, a, v)r$$

and used the Lyapunov function candidate

$$V = \frac{1}{2}r^T M(q)r + \tilde{q}^T \Lambda K \tilde{q} + \frac{1}{2}\tilde{\theta}^T \Gamma \tilde{\theta}$$

to show global convergence of the tracking errors to zero and boundedness of the parameter estimates.

PROBLEMS

8-1 Verify the properties of skew symmetry, passivity and linearity in the parameters for the system given by Equation (8.6). Compute bounds on the inertia matrix, $M(q)$, in terms of bounds on $D(q)$. Show that $M(q)$ is positive definite.

8-2 Form the Lagrangian for an n-link manipulator with joint flexibility using Equations (8.15) and (8.16). From this derive the equations of motion (8.18).

8-3 Complete the proof of stability of PD control for the flexible joint robot without gravity terms using the Lyapunov function candidate (8.20) and LaSalle's theorem.

8-4 Given the flexible joint model defined by Equation (8.18) with the PD control law (8.19), show that $q_1 = q_2$ in the steady state. What are the steady state values of q_1 and q_2 if the gravity term is present? How could one define the reference position q^d in the case that gravity is present?

8-5 Suppose that the PD control law given by Equation (8.19) is implemented using the link variables

$$u = K_p\tilde{q} - K_D\dot{q}_1$$

where $\tilde{q} = q_1 - q^d$. Show that the equilibrium $\tilde{q} = 0 = \dot{q}_1$ is unstable. Hint: Use Lyapunov's First Method, that is, show that the equilibrium is unstable for the linearized system.

8-6 Simulate an inverse dynamics control law for a two-link elbow manipulator whose equations of motion were derived in Chapter 7. Investigate what happens if there are constraints on the input torque.

8-7 For the system of Problem 8-6 what happens to the response of the system if the coriolis and centrifugal terms are dropped from the inverse dynamics control law in order to facilitate computation? What happens if incorrect values are used for the link masses? Investigate via computer simulation.

8-8 Carry out the details to derive the uncertain system (8.46) and (8.47).

8-9 Add an outer-loop correction term δa to the control law of Problem 8-7 to overcome the effects of uncertainty. Base your design on the second method of Lyapunov as in Section 8.3.1.

8-10 Derive the error equation (8.77) using the linearity in the parameters property of the robot dynamics.

8-11 Verify the expression for $\dot{V}$ in Equation (8.83).

8-12 Consider the coupled nonlinear system

$$\begin{aligned} \ddot{y}_1 + 3y_1y_2 + y_2^2 &= u_1 + y_2u_2 \\ \ddot{y}_2 + \cos y_1\dot{y}_2 + 3(y_1 - y_2) &= u_2 - 3(\cos y_1)^2y_2u_1 \end{aligned}$$

where u_1, u_2 are the inputs and y_1, y_2 are the outputs.

a) What is the dimension of the state space?

b) Choose state variables and write the system as a system of first order differential equations in state space.

c) Find an inverse dynamics control so that the closed-loop system is linear and decoupled, with each subsystem having natural frequency 10 radians and damping ratio 1/2.

8-13 Complete the proof of uniform ultimate boundedness for the passivity-based robust control law given by Equation (8.92) applied to the rigid robot model.

8-14 Prove the inequality (8.59).

NOTES AND REFERENCES

Many of the fundamental theoretical problems in motion control of robot manipulators were solved during an intense period of research from about the mid-1980's until the early-1990's during which time researchers first began to exploit the structural properties of manipulator dynamics such as feedback linearizability, skew symmetry and passivity, multiple time-scale behavior, and other properties. For a more advanced treatment of some of these topics, the reader is referred to [124] and [25].

The literature on robot control is vast and we have given only the basic results in several of the main areas of control. In the area of PD and PID control of manipulators, the earliest results are contained in [132]. These results were based on the Hamiltonian formulation of robot dynamics and effectively exploited the passivity property. The use of energy as a Lyapunov function is described in [65].

The problem of joint flexibility was first brought to the forefront of robotics research in [97], [130], and [131]. The model presented here to describe the dynamics of flexible joint robots is due to [122].

The inverse dynamics approach to control is also called the **method of computed torque** in the literature. The earliest results on computed torque appeared in [86] and [104]. A related approach known as the method of **resolved motion acceleration control** is due to [82]. In [69] these various control schemes are all compared to the method of inverse dynamics and shown to be essentially equivalent.

The robust inverse dynamics control approach here follows closely the general methodology in [21]. The earliest application of this method to the manipulator control problem was in [24] and [125]. This technique is closely related to the so-called **method of sliding modes** which has been applied to manipulator control in [118]. A very complete survey of robust control of robots up to about 1990 is found in [1]. Other results in robust control from an operator theoretic viewpoint are [127] and [49]. The passivity-based robust control result here is due to [123].

The adaptive inverse dynamics control result presented here is due to [22]. Other notable results in this area appeared in [90]. The first results in passivity-based adaptive control of manipulators was in [55] and [117]. The Lyapunov stability proof presented here is due to [121]. A unifying treatment of adaptive manipulator control from a passivity perspective was presented in [99]. Other works based on passivity are [14] and [10].

One of the problems with the adaptive control approaches considered here is the so-called **parameter drift** problem. The Lyapunov stability proofs presented show that the parameter estimates are bounded but there is no guarantee that the estimated parameters converge to their true values. It can be shown that the estimated parameters converge to the true parameters provided the reference trajectory satisfies the condition of **persistency of excitation**

$$\alpha I \leq \int_{t_0}^{t_0+T} Y^T(q^d, \dot{q}^d, \ddot{q}^d) Y(q^d, \dot{q}^d, \ddot{q}^d) dt \leq \beta I \tag{8.106}$$

for all t_0, where α, β, and T are positive constants.

Chapter 9

FORCE CONTROL

In previous chapters we considered the problem of tracking motion trajectories using a variety of elementary and advanced control methods. These position control schemes are adequate for tasks such as materials transfer, spray painting, or spot welding where the manipulator is not interacting significantly with objects in the workplace (hereafter referred to as the **environment**). However, tasks such as assembly, grinding, and deburring, which involve extensive contact with the environment, are often better handled by controlling the **forces**[1] of interaction between the manipulator and the environment rather than simply controlling the position of the end effector. For example, consider an application where the manipulator is required to wash a window, or to write with a felt tip marker. In both cases a pure position control scheme is unlikely to work. Slight deviations of the end effector from a planned trajectory would cause the manipulator either to lose contact with the surface or to press too strongly on the surface. For a highly rigid structure such as a robot, a slight position error could lead to extremely large forces of interaction with disastrous consequences (broken window, smashed pen, damaged end effector, etc.). The above applications are typical in that they involve both force control and trajectory control. In the window washing application, for example, one clearly needs to control the forces normal to the plane of the window and position in the plane of the window.

A force control strategy is one that modifies position trajectories based on the sensed forces. There are three main types of sensors for force feedback, **wrist force** sensors, **joint torque** sensors, and **tactile** or hand sensors. A wrist force sensor such as that shown in Figure 9.1 usually consists of an

---

[1]Hereafter we use *force* to mean force and/or torque and *position* to mean position and/or orientation.

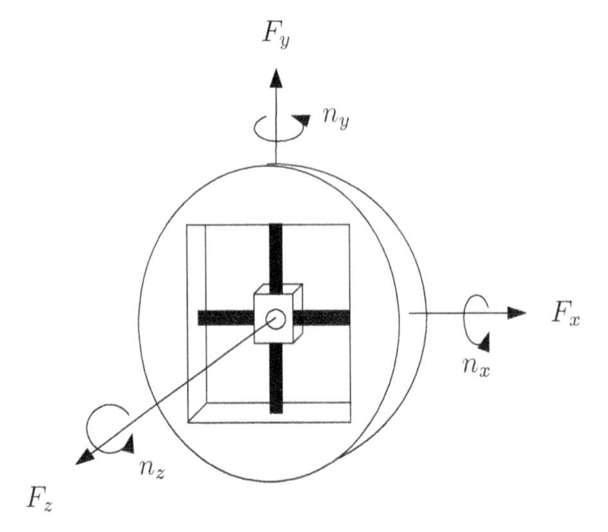

Figure 9.1: A wrist force sensor.

array of strain gauges and can delineate the three components of the vector force along the three axes of the sensor coordinate frame, and the three components of the torque about these axes. A joint torque sensor consists of strain gauges located on the actuator shaft. Tactile sensors are usually located on the fingers of the gripper and are useful for sensing gripping force and for shape detection. For the purposes of controlling the end-effector/environment interactions, the six-axis wrist sensor usually gives the best results and we shall henceforth assume that the manipulator is equipped with such a device.

9.1 COORDINATE FRAMES AND CONSTRAINTS

Force control tasks can be thought of in terms of constraints imposed by the robot/environment interaction. A manipulator moving through free space within its workspace is unconstrained in motion and can exert no forces since there is no source of reaction force from the environment. A wrist force sensor in such a case would record only the inertial forces due to any acceleration of the end effector. As soon as the manipulator comes in contact with the environment, say a rigid surface as shown in Figure 9.2, one or more degrees of freedom in motion may be lost since the manipulator cannot move through the environment surface. At the same time, the manipulator can exert forces against the environment.

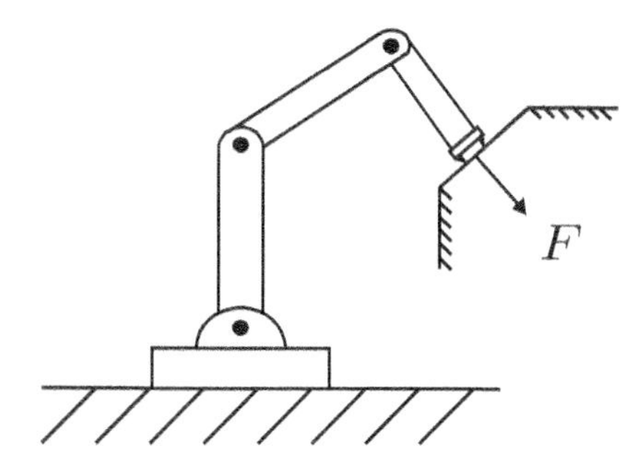

Figure 9.2: Robot end effector in contact with a rigid surface.

9.1.1 Reciprocal Bases

In order to describe the robot/environment interaction, let $\xi = [v^T, \omega^T]^T$ represent the instantaneous linear and angular velocity of the end effector and let $F = [f^T, n^T]^T$ represent the instantaneous force and moment acting on the end effector. The vectors ξ and F are each elements of six dimensional vector spaces, which we denote by $\mathcal{M}$ and $\mathcal{F}$, the motion and force spaces, respectively. The vectors ξ and F are called **Twists** and **Wrenches**, respectively, in more advanced texts [93] although we will continue to refer to them simply as velocity and force for simplicity.

Definition 9.1

1. *If* $\{e_1, \ldots, e_6\}$ *is a basis for the vector space* $\mathcal{M}$*, and* $\{f_1, \ldots, f_6\}$ *is a basis for* $\mathcal{F}$*, we say that these basis vectors are* **reciprocal** *provided*

$$\begin{aligned} e_i^T f_j &= 0 \quad \text{if } i \neq j \\ e_i^T f_j &= 1 \quad \text{if } i = j \end{aligned} \tag{9.1}$$

2. *A twist* $\xi \in \mathcal{M}$ *and a wrench* $F \in \mathcal{F}$ *are called* **reciprocal** *if*

$$\xi^T F = v^T f + \omega^T n = 0 \tag{9.2}$$

The advantage of using reciprocal basis vectors is that the product $\xi^T F$ is then **invariant** with respect to a linear change of basis from one reciprocal coordinate system to another. Thus, the **reciprocity condition** given by Equation (9.2) is invariant with respect to choice of reciprocal bases of $\mathcal{M}$ and $\mathcal{F}$. We shall see in specific cases below that the reciprocity relation given by Equation (9.2) may be used to design reference inputs to execute motion and force control tasks.

Since $\mathcal{M}$ and $\mathcal{F}$ are each six dimensional vector spaces, it is tempting to identify each with $\mathbb{R}^6$. However, it turns out that inner product expressions such as $\xi_1^T\xi_2$ or $F_1^TF_2$ for vectors ξ_i, F_i belonging to $\mathcal{M}$ and $\mathcal{F}$, respectively, are not necessarily well defined. For example, the expression

$$\xi_1^T\xi_2 = v_1^Tv_2 + \omega_1^T\omega_2 \tag{9.3}$$

is not invariant with respect to either choice of units or basis vectors in $\mathcal{M}$. It is possible to define inner product like operations, that is, symmetric, bilinear forms on $\mathcal{M}$ and $\mathcal{F}$ that have the necessary invariance properties. These are the so-called **Klein form**, $KL(\xi_1, \xi_2)$, and **Killing form**, $KI(\xi_1, \xi_2)$, defined according to

$$KL(\xi_1, \xi_2) = v_1^T\omega_2 + \omega_1^Tv_2 \tag{9.4}$$

$$KI(\xi_1, \xi_2) = \omega_1^T\omega_2 \tag{9.5}$$

However, a detailed discussion of these concepts is beyond the scope of this text. As the reader may suspect, the need for a careful treatment of these concepts is related to the geometry of $SO(3)$ as we have seen before in other contexts.

Example 9.1

Suppose that

$$\begin{aligned}\xi_1 &= [1, 1, 1, 2, 2, 2]^T \\ \xi_2 &= [2, 2, 2, -1, -1, -1]^T\end{aligned}$$

where the linear velocity is in meters/sec and angular velocity is in radians/sec. Then clearly, $\xi_1^T\xi_2 = 0$ and so one could infer that ξ_1 and ξ_2 are orthogonal vectors in $\mathcal{M}$. However, suppose now that the linear velocity is represented in units of centimeters/sec. Then

$$\begin{aligned}\xi_1 &= [1\times10^2, 1\times10^2, 1\times10^2, 2, 2, 2]^T \\ \xi_2 &= [2\times10^2, 2\times10^2, 2\times10^2, -1, -1, -1]^T\end{aligned}$$

and clearly $\xi_1^T\xi_2 \neq 0$. Thus, the usual notion of orthogonality is not meaningful in $\mathcal{M}$. It is easy to show that the equality $KL(\xi_1, \xi_2) = 0$ (respectively, $KI(\xi_1, \xi_2) = 0$) is independent of the units or the basis chosen to represent ξ_1 and ξ_2. For example, the condition $KI(\xi_1, \xi_2) = 0$ means that the axes of rotation defining ω_1 and ω_2 are orthogonal.

◇

9.1.2 Natural and Artificial Constraints

In this section we discuss so-called **natural constraints**, which are defined using the reciprocity condition given by Equation (9.2). We then discuss the notion of **Artificial Constraints**, which are used to define reference inputs for motion and force control tasks.

We begin by defining a so-called **compliance frame** $o_c x_c y_c z_c$ (also called a **constraint frame**) in which the task to be performed is easily described. For example in the window washing application we can define a frame at the tool with the z_c-axis along the surface normal direction. The task specification is then expressed in terms of maintaining a constant force in the z_c direction while following a prescribed trajectory in the x_c-y_c plane. Such a position constraint in the z_c direction, arising from the presence of a rigid surface, is a natural constraint. The force that the robot exerts against the rigid surface in the z_c direction, on the other hand, is not constrained by the environment. A desired force in the z_c direction would then be considered as an artificial constraint that must be maintained by the control system.

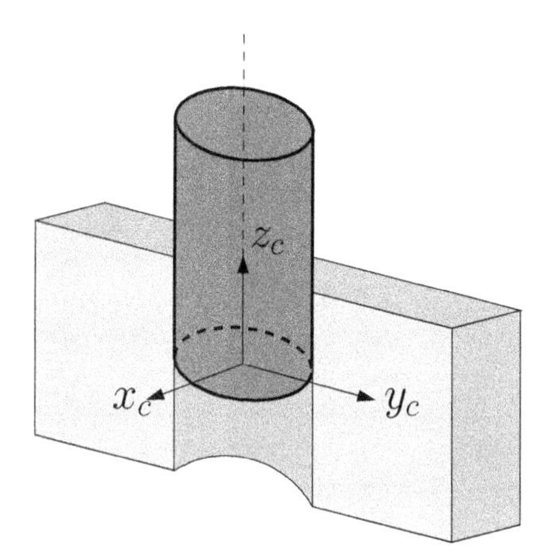

Natural Constraints	Artificial Constraints
$v_x = 0$	$f_x = 0$
$v_y = 0$	$f_y = 0$
$f_z = 0$	$v_z = v_d$
$\omega_x = 0$	$n_x = 0$
$\omega_y = 0$	$n_y = 0$
$n_z = 0$	$\omega_z = 0$

Figure 9.3: Inserting a peg into a hole.

Figure 9.3 shows a typical task, that of inserting a peg into a hole. With respect to a compliance frame $o_c x_c y_c z_c$, as shown at the end of the peg, we may take the standard orthonormal basis in $\mathbb{R}^6$ for both $\mathcal{M}$ and $\mathcal{F}$, in which case

$$\xi^T F = v_x f_x + v_y f_y + v_z f_z + \omega_x n_x + \omega_y n_y + \omega_z n_z \tag{9.6}$$

If we assume that the walls of the hole and the peg are perfectly rigid and there is no friction, it is easy to see that

$$\begin{array}{lll} v_x = 0 & v_y = 0 & f_z = 0 \\ \omega_x = 0 & \omega_y = 0 & n_z = 0 \end{array} \tag{9.7}$$

and thus, the reciprocity condition $\xi^T F = 0$ is satisfied. These relationships given by Equation (9.7) are termed **natural constraints**. Examining Equation (9.6) we see that the variables

$$f_x \quad f_y \quad v_z \quad n_x \quad n_y \quad \omega_z \tag{9.8}$$

are unconstrained by the environment. In other words, given the natural constraints from Equation (9.7), the reciprocity condition $\xi^T F = 0$ holds for all values of the above variables in Equation (9.8). We may therefore arbitrarily assign reference values, called **artificial constraints**, for these variables that must then be enforced by the control system to carry out the task at hand. For example, in the peg-in-hole task, we may define artificial constraints as

$$\begin{array}{lll} f_x = 0 & f_y = 0 & v_z = v^d \\ n_x = 0 & n_y = 0 & \omega_z = 0 \end{array} \tag{9.9}$$

where v^d is the desired speed of insertion of the peg in the z-direction.

Figure 9.4 shows natural and artificial constraints for the task of turning a crank.

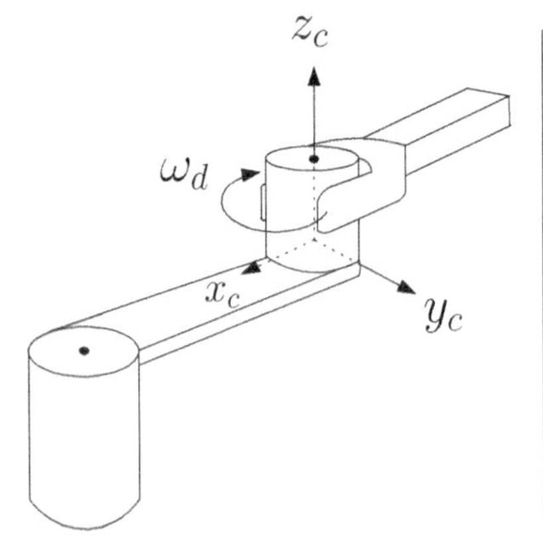

Natural Constraints	Artificial Constraints
$v_x = 0$	$f_x = 0$
$f_y = 0$	$v_y = 0$
$v_z = 0$	$f_z = 0$
$\omega_x = 0$	$n_x = 0$
$\omega_y = 0$	$n_y = 0$
$n_z = 0$	$\omega_z = \omega_d$

Figure 9.4: Turning a crank.

9.2 NETWORK MODELS AND IMPEDANCE

The reciprocity condition $\xi^T F = 0$ means that the forces of constraint do no work in directions compatible with motion constraints and holds under the ideal conditions of no friction and perfect rigidity of both the robot and the environment. In practice, compliance and friction in the robot/environment interface will alter the strict separation between motion constraints and force constraints.

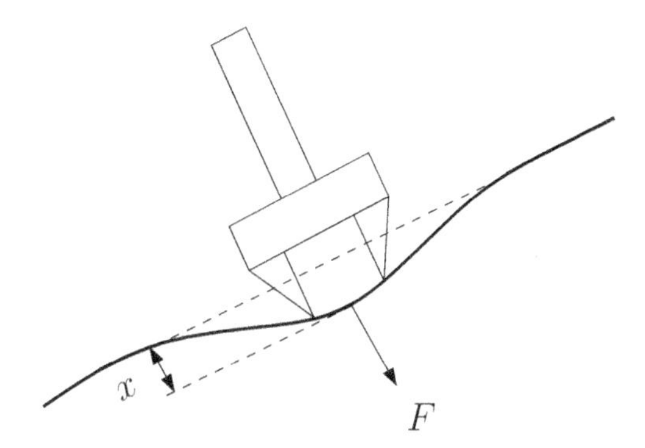

Figure 9.5: Compliant environment.

For example, consider the situation in Figure 9.5. Since the environment deforms in response to a force, there is clearly both motion and force normal to the surface. Thus, the product $\xi(t)F(t)$ along this direction will not be zero. Let k represent the stiffness of the surface so that $F = kx$. Then

$$\begin{aligned} \int_0^t \xi(u)F(u)du &= \int_0^t \dot{x}(u)kx(u)du = k\int_0^t \frac{d}{du}\frac{1}{2}kx^2(u)du \\ &= \frac{1}{2}k(x^2(t) - x^2(0)) \end{aligned}$$

is the change of the potential energy due to the material deformation. The environment stiffness k determines the amount of force needed to produce a given motion. The higher the value of k the more the environment "impedes" the motion of the end effector.

In this section we introduce the notion of **mechanical impedance**, which captures the relation between force and motion. We introduce so-called **network models**, which are particularly useful for modeling the interaction between the robot and the environment.

We model the robot and the environment as **one port networks** as shown in Figure 9.6. The dynamics of the robot and environment determine the relations among the **port variables**, V_r, F_r, and V_e, F_e, respectively.

The forces F_r, F_e are known as **effort** or **across** variables while the velocities V_r, V_e are known as **flow** or **through** variables. With this description, the

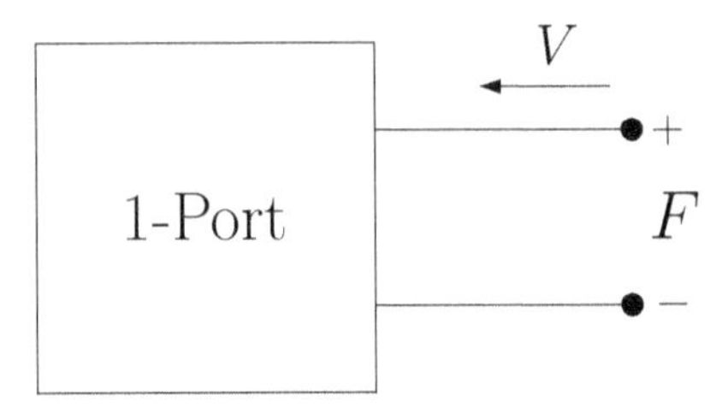

Figure 9.6: One-port network.

product of the port variables, $V^T F$, represents instantaneous **power** and the integral of this product

$$\int_0^t V^T(\sigma)F(\sigma)d\sigma$$

is the **energy** dissipated by the Network over the time interval $[0, t]$.

The robot and the environment are then coupled through their interaction ports, as shown in Figure 9.7, which describes the energy exchange between the robot and the environment.

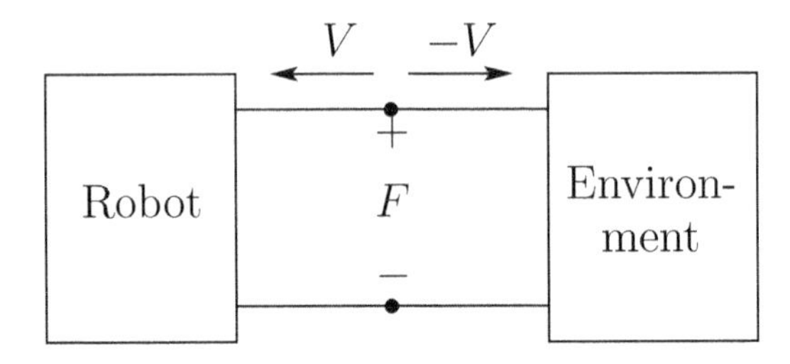

Figure 9.7: Robot/environment interaction.

9.2.1 Impedance Operators

The relationship between the effort and flow variables may be described in terms of an **impedance operator**. For linear, time invariant systems, we may utilize the s-domain or Laplace domain to define the impedance.

Definition 9.2 *Given the one-port network in Figure 9.6 the impedance, $Z(s)$ is defined as the ratio of the Laplace transform of the effort to the*

Laplace transform of the flow,

$$Z(s) = \frac{F(s)}{V(s)} \tag{9.10}$$

Example 9.2

Suppose a mass-spring-damper system is described by the differential equation

$$M\ddot{x} + B\dot{x} + Kx = F \tag{9.11}$$

Taking the Laplace transforms of both sides (assuming zero initial conditions) it follows that

$$Z(s) = F(s)/V(s) = Ms + B + K/s \tag{9.12}$$

◇

9.2.2 Classification of Impedance Operators

It seems intuitive that different types of environments would dictate different control strategies. For example, as we have seen, pure position control would be difficult in contact with a very stiff environment as in the window washing example. Similarly, interaction forces would be difficult to control if the environment is very soft. In this section we introduce terminology to classify robot and environment impedance operators that will prove useful in the analysis to follow.

Definition 9.3 *An impedance $Z(s)$ in the Laplace variable s is said to be*

1. **Inertial** *if and only if* $|Z(0)| = 0$
2. **Resistive** *if and only if* $|Z(0)| = B$ *for some constant* $0 < B < \infty$
3. **Capacitive** *if and only if* $|Z(0)| = \infty$

Example 9.3

Figure 9.8 shows examples of environment types. Figure 9.8(a) shows a mass on a frictionless surface. The impedance is $Z(s) = Ms$, which is inertial. Figure 9.8(b) shows a mass moving in a viscous medium with resistance B. Then $Z(s) = Ms + B$, which is resistive. Figure 9.8(c) shows a linear spring with stiffness K. Then $Z(s) = K/s$, which is capacitive.

◇

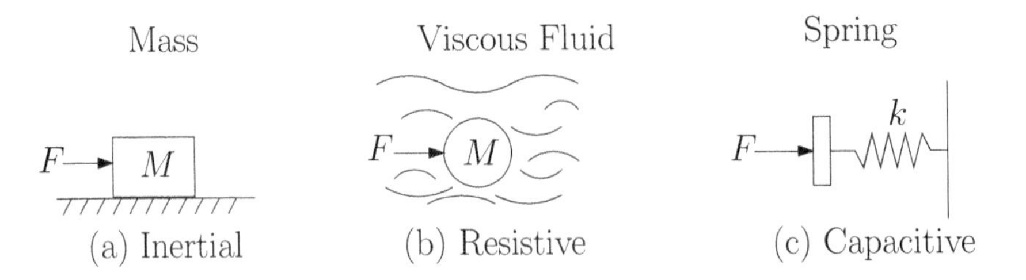

Figure 9.8: Examples of (a) inertial, (b) resistive, and (c) capacitive environments.

9.2.3 Thévenin and Norton Equivalents

In linear circuit theory it is common to use so-called **Thévenin** and **Norton** equivalent circuits for analysis and design. It is easy to show that any one-port network consisting of passive elements (resistors, capacitors, inductors) and current or voltage sources can be represented either as an impedance $Z(s)$ in series with an effort source (Thévenin Equivalent) or as an impedance $Z(s)$ in parallel with a flow source (Norton Equivalent). The independent sources F_s and V_s may be used to represent reference signal generators for force and velocity, respectively, or they may represent external disturbances.

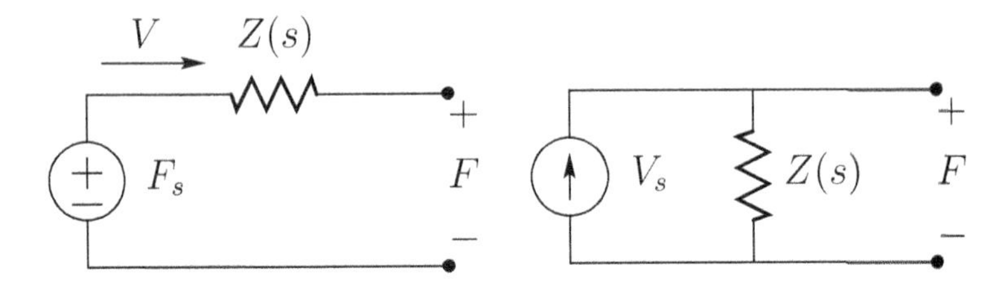

Figure 9.9: Thévenin and Norton equivalent networks.

9.3 TASK SPACE DYNAMICS AND CONTROL

Since a manipulator task, such as grasping an object, or inserting a peg into a hole, is typically specified relative to the end-effector frame, it is natural to derive the control algorithm directly in the task space rather than joint space.

9.3.1 Task Space Dynamics

When the manipulator is in contact with the environment, the dynamic equations of Chapter 7 must be modified to include the reaction torque

$J^T F_e$ corresponding to the end-effector force F_e, where J is the manipulator Jacobian. The modified equations of motion of the manipulator in joint space are, therefore,

$$M(q)\ddot{q} + C(q,\dot{q})\dot{q} + g(q) + J^T(q)F_e = u \tag{9.13}$$

Let us consider a modified inverse dynamics control law of the form

$$u = M(q)a_q + C(q,\dot{q})\dot{q} + g(q) + J^T(q)a_f \tag{9.14}$$

where a_q and a_f are outer-loop controls with units of acceleration and force, respectively. Using the relationship between joint space and task space variables derived in Chapter 8

$$\ddot{x} = J(q)\ddot{q} + \dot{J}(q)\dot{q} \tag{9.15}$$

$$a_x = J(q)a_q + \dot{J}(q)\dot{q} \tag{9.16}$$

we substitute Equations (9.14)–(9.16) into Equation (9.13) to obtain

$$\ddot{x} = a_x + W(q)(F_e - a_f) \tag{9.17}$$

where $W(q) = J(q)M^{-1}(q)J^T(q)$ is called the **mobility tensor**. There is often a conceptual advantage to separating the position and force control terms by assuming that a_x is a function only of position and velocity and a_f is a function only of force. However, for simplicity, we shall take $a_f = F_e$ to cancel the environment force and thus recover the task space double integrator system

$$\ddot{x} = a_x \tag{9.18}$$

and we will assume that any additional force feedback terms are included in the outer-loop term a_x. This entails no loss of generality as long as the Jacobian (hence $W(q)$) is invertible. This will become clear later in this chapter.

9.3.2 Impedance Control

In this section we discuss the notion of **impedance control**. We begin with an example that illustrates in a simple way the effect of force feedback

Example 9.4

Consider the one-dimensional system in Figure 9.10 consisting of a mass M on a frictionless surface subject to an environmental force F and control input u. The equation of motion of the system is

$$M\ddot{x} = u - F \tag{9.19}$$

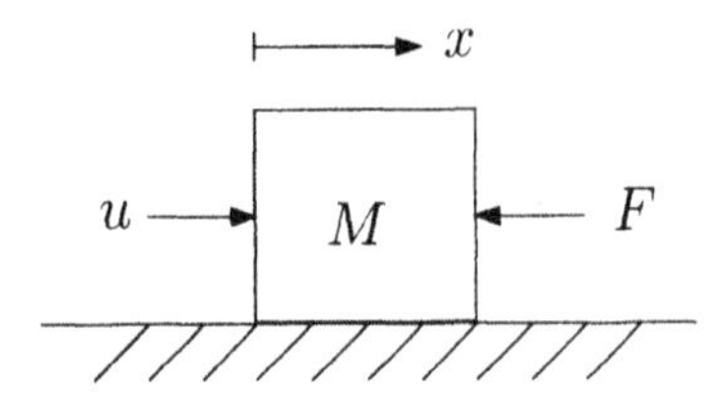

Figure 9.10: One dimensional system.

With $u = 0$, the object "appears to the environment" as a pure inertia with mass M. Suppose the control input u is chosen as a force feedback term $u = -mF$. Then the closed-loop system is

$$M\ddot{x} = -(1+m)F \implies \frac{M}{1+m}\ddot{x} = -F \tag{9.20}$$

Hence, the object now appears to the environment as an inertia with mass $\frac{M}{1+m}$. Thus, the force feedback has the effect of changing the **apparent inertia** *of the system.*

◇

The idea behind impedance control is to regulate the apparent inertia, damping, and stiffness, through force feedback as in the above example. For instance, in a grinding operation, it may be useful to reduce the apparent stiffness of the end effector normal to the part so that excessively large normal forces are avoided.

Next we show that impedance control may be realized within our standard inner-loop/outer-loop control architecture by a suitable choice of the outer-loop term a_x in Equation (9.18). Let $x^d(t)$ be a reference trajectory defined in task space coordinates and let M_d, B_d, K_d, be 6×6 matrices specifying desired inertia, damping, and stiffness, respectively. Let $\tilde{x}(t) = x(t) - x^d(t)$ be the tracking error in task space and set

$$a_x = \ddot{x}^d - M_d^{-1}(B_d\dot{\tilde{x}} + K_d\tilde{x} + F) \tag{9.21}$$

where F is the measured environmental force. Substituting Equation (9.21) into Equation (9.18) yields the closed-loop system

$$M_d\ddot{\tilde{x}} + B_d\dot{\tilde{x}} + K_d\tilde{x} = -F \tag{9.22}$$

which results in desired impedance properties of the end effector. Note that for $F = 0$ tracking of the reference trajectory, $x^d(t)$, is achieved, whereas for nonzero environmental force, tracking is not necessarily achieved. We will address this difficulty in the next section.

9.3.3 Hybrid Impedance Control

In this section we introduce the notion of **hybrid impedance control.** We again take as our starting point the linear, decoupled system given by Equation (9.18). The impedance control formulation in the previous section is independent of the environment dynamics. It is reasonable to expect that stronger results may be obtained by incorporating a model of the environment dynamics into the design. For example, we will illustrate below how one may control the manipulator impedance while simultaneously regulating either position or force, which is not possible with the pure impedance control law given by Equation (9.21).

We consider a one-dimensional system representing one component of the outer-loop system (9.18)

$$\ddot{x}_i = a_{x_i} \tag{9.23}$$

and we henceforth drop the subscript i for simplicity. We assume that the impedance Z_e of the environment in this direction is fixed and known, a priori. The impedance of the robot Z_r is of course determined by the control input. The Hybrid Impedance Control design proceeds as follows based on the classification of the environment impedance into inertial, resistive, or capacitive impedances:

1. If the environment impedance $Z_e(s)$ is capacitive, use a Norton network representation. Otherwise, use a Thévenin network representation.[2]

2. Choose a desired robot impedance $Z_r(s)$ and represent it as the **dual** to the environment impedance. Thévenin and Norton networks are considered dual to one another.

3. Couple the robot and environment one-ports and design the outer-loop control a_x to achieve the desired impedance of the robot while tracking a reference position or force.

We illustrate this procedure on two examples, a capacitive environment and an inertial environment, respectively.

Example 9.5 Capacitive Environment

In the case that the environment impedance is capacitive we have the robot/environment interconnection as shown in Figure 9.11 where the environment one-port is the Norton network and the robot one-port is the

[2] In fact, for a resistive environment, either representation may be used.

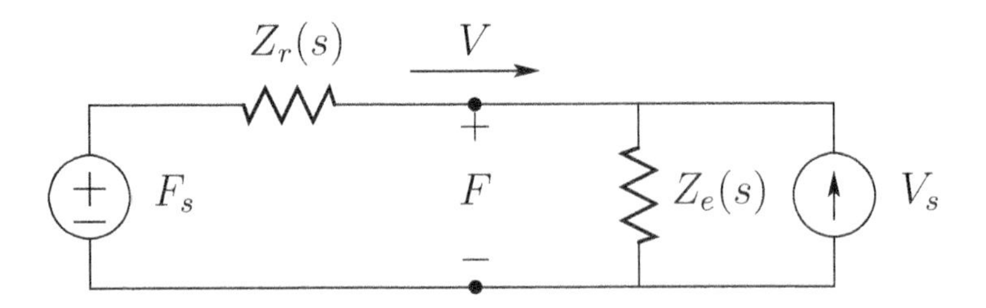

Figure 9.11: Capacitive environment case.

Thévenin network. Suppose that $V_s = 0$, that is, suppose there are no environmental disturbances, and that F_s represents a reference force. From the circuit diagram it is straightforward to show that

$$\frac{F}{F_s} = \frac{Z_e(s)}{Z_e(s) + Z_r(s)} \tag{9.24}$$

Then the steady state force error e_{ss} to a step reference force $F_s = \frac{F^d}{s}$ is given by the final value theorem as

$$e_{ss} = \frac{-Z_r(0)}{Z_r(0) + Z_e(0)} = 0 \tag{9.25}$$

since $Z_e(0) = \infty$ (capacitive environment) and $Z_r \neq 0$ (non-capacitive robot).

The implications of the above calculation are that we can track a constant force reference value, while simultaneously specifying a given impedance Z_r for the robot.

In order to realize this result we need to design outer-loop control term a_x in Equation (9.23) using only position, velocity, and force feedback. This imposes a practical limitation on the achievable robot impedance functions, Z_r.

Suppose Z_r^{-1} has relative degree one. This means that

$$Z_r(s) = M_c s + Z_{rem}(s) \tag{9.26}$$

where $Z_{rem}(s)$ is a proper rational function. We now choose the outer-loop term a_x as

$$a_x = -\frac{1}{M_c} Z_{rem} \dot{x} + \frac{1}{m_c}(F_s - F) \tag{9.27}$$

Substituting this into the double integrator system $\ddot{x} = a_x$ yields

$$Z_r(s)\dot{x} = F_s - F \tag{9.28}$$

Thus, we have shown that, for a capacitive environment, force feedback can be used to regulate contact force and specify a desired robot impedance.
◇

Example 9.6 Inertial Environment
In the case that the environment impedance is inertial we have the robot/ environment interconnection as shown in Figure 9.12, where the environ-

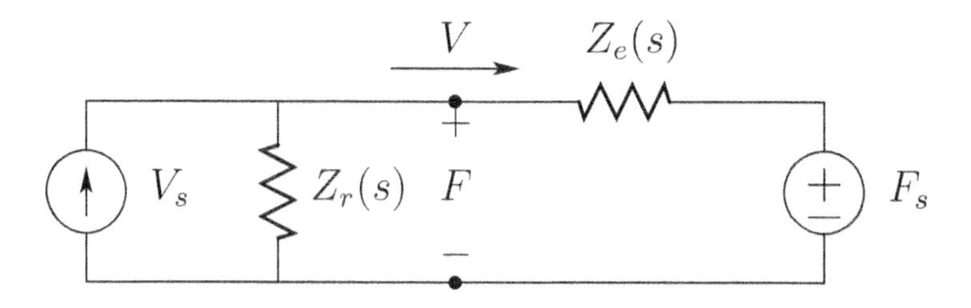

Figure 9.12: Inertial environment case.

ment one-port is a Thévenin network and the robot one-port is a Norton network. Suppose that $F_s = 0$ *and that* V_s *represents a reference velocity. From the circuit diagram it is straightforward to see that*

$$\frac{V}{V_s} = \frac{Z_r(s)}{Z_e(s) + Z_r(s)} \tag{9.29}$$

Then the steady state force error e_{ss} *to a step reference velocity command* $V_s = \frac{V^d}{s}$ *is given by the final value theorem as*

$$e_{ss} = \frac{-Z_e(0)}{Z_r(0) + Z_e(0)} = 0 \tag{9.30}$$

since $Z_e(0) = 0$ *(inertial environment) and* $Z_r \neq 0$ *(non-inertial robot).*
To achieve this non-inertia robot impedance we take, as before,

$$Z_r(s) = M_c s + Z_{rem}(s) \tag{9.31}$$

and set

$$a_x = \ddot{x}^d + \frac{1}{M_c} Z_{rem}(\dot{x}^d - \dot{x}) + \frac{1}{M_c} F \tag{9.32}$$

Then, substituting this into the double integrator equation $\ddot{x} = a_x$ *yields*

$$Z_r(s)(\dot{x}^d - x) = F \tag{9.33}$$

Thus, we have shown that, for an inertial environment, position control can be used to regulate a motion reference and specify a desired robot impedance.
◇

9.4 SUMMARY

This chapter covers some of the basic ideas in robot force control. A force control strategy is one that modifies position trajectories based on the sensed force.

Natural and Artificial Constraints

We first described so-called natural and artificial constraints using the notion of reciprocity. Given six dimensional velocity (or twist) and force (or wrench) vectors V and F, respectively, an ideal robot/environment contact task satisfies

$$V^T F = v_x f_x + v_y f_y + v_z f_z + \omega_x n_x + \omega_y n_y + \omega_z n_z = 0$$

In general, the chosen task imposes environmental constraints on six of the above variables. These are the natural constraints. The remaining variables can be arbitrarily assigned artificial constraints that are then maintained by the control system in order to complete the task.

Network Models and Impedance

We next introduced the notion of mechanical impedance to model the realistic case that the robot and environment are not perfectly rigid. The impedance is a measure of the ratio of force and velocity and is analogous to electrical impedance as a ratio of voltage and current. For this reason we introduced one-port network models of mechanical systems and modeled the robot/environment interaction as a connection of one-port networks.

Task Space Dynamics and Control

When the manipulator is in contact with the environment, the dynamic equations must be modified to include the reaction torque $J^T F_e$ corresponding to the end-effector force F_e. Thus, the equations of motion of the manipulator in joint space are given by

$$M(q)\ddot{q} + C(q, \dot{q})\dot{q} + g(q) + J^T(q)F_e = u$$

We therefore introduced a modified inverse dynamics control law of the form

$$u = M(q)a_q + C(q, \dot{q})\dot{q} + g(q) + J^T(q)a_f$$

where a_q and a_f are outer-loop controls with units of acceleration and force, respectively. The resulting system can be written as

$$\ddot{x} = a_x + W(q)(F_e - a_f)$$

where

$$a_x = J(q)a_q + \dot{J}(q)\dot{q}$$

is the outer-loop control in task space and $W(q) = J(q)M^{-1}(q)J^T(q)$ is the mobility tensor.

Impedance Control

Using this model we introduced the notions of impedance control and hybrid impedance control. The impedance control methodology is to design the outer-loop control terms a_x and a_f according to

$$\begin{aligned} a_x &= \ddot{x}^d - M_d^{-1}(B_d\dot{e} + K_d e + F_e) \\ a_f &= F_e \end{aligned}$$

to achieve the closed-loop system

$$M_d\ddot{e} + B_d\dot{e} + K_d e = -F_e \qquad (9.34)$$

which results in desired impedance properties of the end effector.

Hybrid Impedance Control

Using our network models we introduced a classification of robot/environment impedance operators $Z(s)$ as

1. Inertial if and only if $|Z(0)| = 0$

2. Resistive if and only if $|Z(0)| = B$ for some constant $0 < B < \infty$

3. Capacitive if and only if $|Z(0)| = \infty$

Using this impedance classification scheme we were able to derive so-called hybrid impedance control laws that allowed us both to regulate impedance and to regulate position and force.

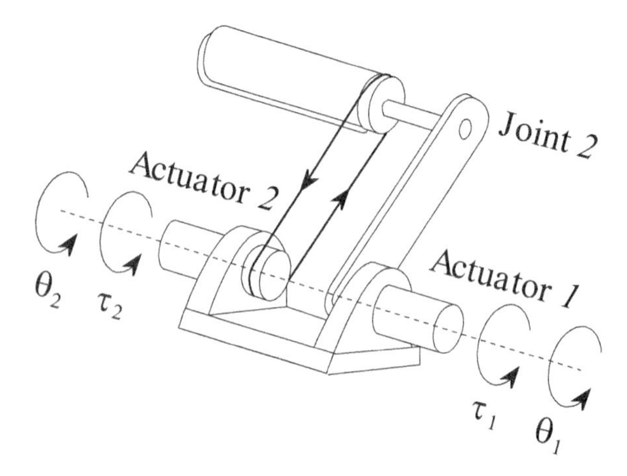

Figure 9.13: Two-link manipulator with remotely driven link.

PROBLEMS

9-1 Given the two-link planar manipulator of Figure 4.11, find the joint torques τ_1 and τ_2 corresponding to the end-effector force $[-1, -1]^T$.

9-2 Consider the two-link planar manipulator with remotely driven links shown in Figure 9.13. Find an expression for the motor torques needed to balance a force F at the end effector. Assume that the motor gear ratios are r_1, r_2, respectively.

9-3 What are the natural and artificial constraints for the task of inserting a square peg into a square hole? Sketch the compliance frame for this task.

9-4 Describe the natural and artificial constraints associated with the task of opening a box with a hinged lid. Sketch the compliance frame.

9-5 Discuss the task of opening a long two-handled drawer. How would you go about performing this task with two manipulators? Discuss the problem of coordinating the motion of the two arms. Define compliance frames for the two arms and describe the natural and artificial constraints.

9-6 Given the following tasks, classify the environments as either inertial, capacitive, or resistive according to Definition 9.3.

1. Turning a crank
2. Inserting a peg in a hole

3. Polishing the hood of a car
4. Cutting cloth
5. Shearing a sheep
6. Placing stamps on envelopes
7. Cutting meat

NOTES AND REFERENCES

Among the earliest results in robot force control was the work of Mason [89], who introduced the notion of natural and artificial constraints and Raibert and Craig [107], who introduced the notion of hybrid position/force control based on the decomposition of the force control problem into position controlled and force controlled directions relative to a compliance frame. Our use of the reciprocity condition is an outgrowth of this early work. The use of twists and wrenches to define global geometric notions in this context was introduced to the robotics community in [30].

The notion of impedance control is due to Hogan [51]. The hybrid impedance control concept and the classification of environments as inertial, resistive, or capacitive is taken from Anderson and Spong [3]. Alternate formulations of force control can be found in [25].

Chapter 10

GEOMETRIC NONLINEAR CONTROL

In this chapter we present some basic, but fundamental, ideas from **geometric nonlinear control theory**. We first give some background from differential geometry to set the notation and define basic quantities, such as **manifold**, **vector field**, **Lie bracket**, and so forth that we will need later. The main tool that we will use in this chapter is the **Frobenius theorem**, which we introduce in Section 10.1.2.

We then discuss the notion of **feedback linearization of nonlinear systems**. This approach generalizes the concept of inverse dynamics of rigid manipulators discussed in Chapter 8. The idea of feedback linearization is to construct a nonlinear control law as an **inner-loop control** which, in the ideal case, exactly linearizes the nonlinear system after a suitable state space change of coordinates. The designer can then design the **outer-loop control** in the new coordinates to satisfy the traditional control design specifications such as tracking and disturbance rejection.

In the case of rigid manipulators the inverse dynamics control of Chapter 8 and the feedback linearizing control are the same. However, as we shall see, the full power of the feedback linearization technique for manipulator control becomes apparent if one includes in the dynamic description of the manipulator the transmission dynamics, such as elasticity resulting from shaft windup and gear elasticity.

We also give an introduction to modeling and controllability of nonholonomic systems. We treat systems such as mobile robots and other systems subject to constraints arising from conservation of angular momentum or rolling contact. We discuss the controllability of a particular class of such systems, known as **driftless systems**. We present a result known as

Chow's theorem, which gives a sufficient condition for controllability of driftless systems.

10.1 BACKGROUND

In recent years an impressive volume of literature has emerged in the area of differential geometric methods for nonlinear systems, treating not only feedback linearization but also other problems such as disturbance decoupling, estimation, observers, and adaptive control. It is our intent here to give only that portion of the theory that finds an immediate application to robot control, and even then to give only the simplest versions of the results.

10.1.1 Manifolds, Vector Fields, and Distributions

The fundamental notion in differential geometry is that of a **differentiable manifold** (manifold for short) which is a topological space that is locally diffeomorphic[1] to Euclidean space $\mathbb{R}^m$. For our purposes here a manifold may be thought of as a subset of $\mathbb{R}^n$ defined by the zero set of a smooth vector valued function[2] $h : \mathbb{R}^n \rightarrow \mathbb{R}^p$, for $p < n$,

$$\begin{aligned} h_1(x_1, \ldots, x_n) &= 0 \\ &\vdots \\ h_p(x_1, \ldots, x_n) &= 0 \end{aligned}$$

We assume that the differentials $dh_1, \ldots, dh_p$ are linearly independent at each point, in which case the dimension of the manifold is $m = n - p$. Given an m-dimensional manifold M we may attach at each point $x \in M$ a **tangent space** $T_x M$, which is an m-dimensional vector space specifying the set of possible velocities (directional derivatives) at x.

Example 10.1

Consider the unit sphere S^2 in $\mathbb{R}^3$ defined by

$$h(x, y, z) = x^2 + y^2 + z^2 - 1 = 0$$

[1] A **diffeomorphism** is simply a differentiable function whose inverse exists and is also differentiable. We shall assume both the function and its inverse to be infinitely differentiable. Such functions are customarily referred to as C^∞ **diffeomorphisms**.

[2] Our definition amounts to the special case of an **embedded submanifold** of dimension $m = n - p$ in $\mathbb{R}^n$.

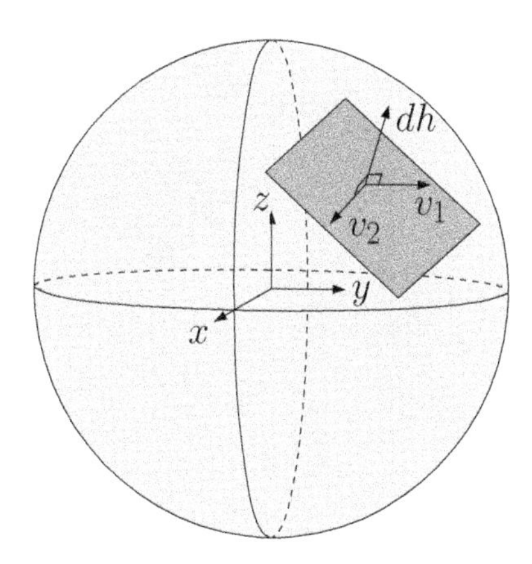

Figure 10.1: The sphere as a two-dimensional manifold in R^3.

S^2 is a two-dimensional submanifold of $\mathbb{R}^3$. At points in the upper hemisphere $z = \sqrt{1-x^2-y^2}$ and the tangent space is spanned by the vectors

$$\begin{aligned} v_1 &= [1, 0, -x/\sqrt{1-x^2-y^2}]^T \\ v_2 &= [0, 1, -y/\sqrt{1-x^2-y^2}]^T \end{aligned}$$

The differential of h is

$$dh = (2x, 2y, 2z) = (2x, 2y, 2\sqrt{1-x^2-y^2})$$

which is easily shown to be normal to the tangent plane at x, y, z.
◇

Definition 10.1 *A* **smooth vector field** *on a manifold M is an infinitely differentiable function $f : M \to T_xM$ represented as a column vector*

$$f(x) = \begin{bmatrix} f_1(x) \\ \vdots \\ f_m(x) \end{bmatrix}$$

Another useful notion is that of **cotangent space** and **covector field.** The cotangent space T_x^*M is the dual space of the tangent space. It is an m-dimensional vector space specifying the set of possible differentials of functions at x. Mathematically, T_x^*M is the space of all linear functionals on T_xM, that is, the space of functions from T_xM to $\mathbb{R}$.

Definition 10.2 *A* **smooth covector field** *is a function $w : M \to T_x^*M$ which is infinitely differentiable, represented as a row vector,*

$$w(x) = \begin{bmatrix} w_1(x), \ldots, w_m(x) \end{bmatrix}$$

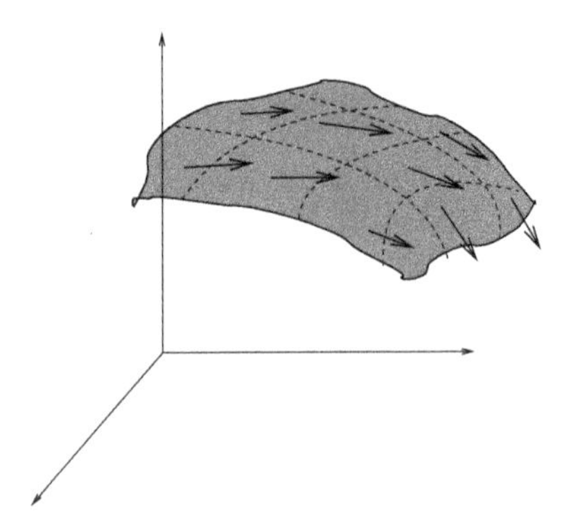

Figure 10.2: Pictoral representation of a vector field on a manifold.

Henceforth, whenever we use the term function, vector field, or covector field, it is assumed to be smooth. Since T_xM and T_x^*M are m-dimensional vector spaces, they are isomorphic and the only distinction we will make between vectors and covectors below is whether or not they are represented as column vectors or row vectors.

We may also have multiple vector fields defined simultaneously on a given manifold. Such a set of vector fields will span a subspace of the tangent space at each point. Likewise, we will consider multiple covector fields spanning a subspace of the cotangent space at each point. These notions give rise to so-called **distributions** and **codistributions**.

Definition 10.3

1. *Let $X_1(x), \ldots, X_k(x)$ be vector fields on M that are linearly independent at each point. A* **distribution** Δ *is the linear span (at each $x \in M$)*

$$\Delta = \text{span } \{X_1(x), \ldots, X_k(x)\} \tag{10.1}$$

2. *Likewise, let $w_1(x), \ldots, w_k(x)$ be covector fields on M, which are linearly independent at each point. By a* **Codistribution** Ω *we mean the linear span (at each $x \in M$)*

$$\Omega = \text{span}\{w_1(x), \ldots, w_k(x)\} \tag{10.2}$$

A distribution therefore assigns a vector space $\Delta(x)$ to each point $x \in M$. $\Delta(x)$ is a k-dimensional subspace of the m-dimensional tangent space T_xM. A codistribution likewise defines a k-dimensional subspace $\Omega(x)$ at each x of the m-dimensional cotangent space T_x^*M.

Vector fields are used to define differential equations and their associated flows. We restrict our attention here to nonlinear systems of the form

$$\begin{aligned} \dot{x} &= f(x) + g_1(x)u_1 + \cdots + g_m(x)u_m \\ &= f(x) + G(x)u \end{aligned} \tag{10.3}$$

where $f(x)$, $g_1(x), \ldots, g_m(x)$ are smooth vector fields on M, and where we define $G(x) = [g_1(x), \ldots, g_m(x)]$ and $u = [u_1, \ldots, u_m]^T$. For simplicity we will assume that $M = \mathbb{R}^n$.

Definition 10.4 *Let f and g be two vector fields on $\mathbb{R}^n$. The* **Lie bracket** *of f and g, denoted by $[f, g]$, is a vector field defined by*

$$[f, g] = \frac{\partial g}{\partial x} f - \frac{\partial f}{\partial x} g \tag{10.4}$$

where $\frac{\partial g}{\partial x}$ (respectively, $\frac{\partial f}{\partial x}$) denotes the $n \times n$ Jacobian matrix whose ij^{th} entry is $\frac{\partial g_i}{\partial x_j}$ (respectively, $\frac{\partial f_i}{\partial x_j}$).

Example 10.2

Suppose that vector fields $f(x)$ and $g(x)$ on $\mathbb{R}^3$ are given as

$$f(x) = \begin{bmatrix} x_2 \\ \sin x_1 \\ x_1 x_3^2 \end{bmatrix} \qquad g(x) = \begin{bmatrix} 0 \\ x_2^2 \\ 1 \end{bmatrix}$$

Then the vector field $[f, g]$ is computed according to Equation (10.4) as

$$\begin{aligned} [f, g] &= \begin{bmatrix} 0 & 0 & 0 \\ 0 & 2x_2 & 0 \\ 0 & 0 & 0 \end{bmatrix} \begin{bmatrix} x_2 \\ \sin x_1 \\ x_1 + x_3^2 \end{bmatrix} - \begin{bmatrix} 0 & 1 & 0 \\ \cos x_1 & 0 & 0 \\ 1 & 0 & 2x_3 \end{bmatrix} \begin{bmatrix} 0 \\ x_2^2 \\ 1 \end{bmatrix} \\ &= \begin{bmatrix} -x_2^2 \\ 2x_2 \sin x_1 \\ -2x_3 \end{bmatrix} \end{aligned}$$

◇

We also denote $[f, g]$ as $ad_f(g)$ and define $ad_f^k(g)$ inductively by

$$ad_f^k(g) = [f, ad_f^{k-1}(g)] \tag{10.5}$$

with $ad_f^0(g) = g$.

Definition 10.5 *Let $f : \mathbb{R}^n \rightarrow \mathbb{R}^n$ be a vector field on $\mathbb{R}^n$ and let $h : \mathbb{R}^n \rightarrow \mathbb{R}$ be a scalar function. The* **Lie derivative** *of h with respect to f, denoted $L_f h$, is defined as*

$$L_f h = \frac{\partial h}{\partial x} f(x) = \sum_{i=1}^{n} \frac{\partial h}{\partial x_i} f_i(x) \tag{10.6}$$

The Lie derivative is simply the directional derivative of h in the direction of f. We denote by $L_f^2 h$ the Lie derivative of $L_f h$ with respect to f, that is,

$$L_f^2 h = L_f(L_f h) \tag{10.7}$$

In general we define

$$L_f^k h = L_f(L_f^{k-1} h) \quad \text{for } k = 1, \ldots, n \tag{10.8}$$

with $L_f^0 h = h$.

The following technical lemma gives an important relationship between the Lie bracket and Lie derivative and is crucial to the subsequent development.

Lemma 10.1 *Let $h : \mathbb{R}^n \rightarrow \mathbb{R}$ be a scalar function and f and g be vector fields on $\mathbb{R}^n$. Then we have the following identity*

$$L_{[f,g]} h = L_f L_g h - L_g L_f h \tag{10.9}$$

Proof: Expand Equation (10.9) in terms of the coordinates $x_1, \ldots, x_n$ and equate both sides. The ith component $[f, g]_i$ of the vector field $[f, g]$ is given as

$$[f, g]_i = \sum_{j=1}^{n} \frac{\partial g_i}{\partial x_j} f_j - \sum_{j=1}^{n} \frac{\partial f_i}{\partial x_j} g_j$$

Therefore, the left-hand side of Equation (10.9) is

$$\begin{aligned}
L_{[f,g]} h &= \sum_{i=1}^{n} \frac{\partial h}{\partial x_i} [f, g]_i \\
&= \sum_{i=1}^{n} \frac{\partial h}{\partial x_i} \left(\sum_{j=1}^{n} \frac{\partial g_i}{\partial x_j} f_j - \sum_{j=1}^{n} \frac{\partial f_i}{\partial x_j} g_j \right) \\
&= \sum_{i=1}^{n} \sum_{j=1}^{n} \frac{\partial h}{\partial x_i} \left(\frac{\partial g_i}{\partial x_j} f_j - \frac{\partial f_i}{\partial x_j} g_j \right)
\end{aligned}$$

If the right-hand side of Equation (10.9) is expanded similarly it can be shown, with a little algebraic manipulation, that the two sides are equal. The details are left as an exercise (Problem 10-1).

10.1.2 The Frobenius Theorem

In this section we present a basic result in differential geometry known as the **Frobenius theorem**. The Frobenius theorem can be thought of as an existence theorem for solutions to certain systems of first order partial differential equations. Although a rigorous proof of this theorem is beyond the scope of this text, we can gain an intuitive understanding of it by considering the following system of partial differential equations

$$\frac{\partial z}{\partial x} = f(x, y, z) \tag{10.10}$$

$$\frac{\partial z}{\partial y} = g(x, y, z) \tag{10.11}$$

In this example there are two partial differential equations in a single dependent variable z. A solution to Equations (10.10) and (10.11) is a function $z = \phi(x, y)$ satisfying

$$\frac{\partial \phi}{\partial x} = f(x, y, \phi(x, y)) \tag{10.12}$$

$$\frac{\partial \phi}{\partial y} = g(x, y, \phi(x, y)) \tag{10.13}$$

We can think of the function $z = \phi(x, y)$ as defining a surface in $\mathbb{R}^3$ as in Figure 10.3. The function $\Phi : \mathbb{R}^2 \to \mathbb{R}^3$ defined by

$$\Phi(x, y) = (x, y, \phi(x, y)) \tag{10.14}$$

then characterizes both the surface and the solution of Equations (10.10) and (10.11). At each point (x, y) the tangent plane to the surface is spanned by two vectors found by taking partial derivatives of Φ in the x and y directions, respectively, that is, by

$$\begin{aligned} X_1 &= [1, 0, f(x, y, \phi(x, y))]^T \\ X_2 &= [0, 1, g(x, y, \phi(x, y))]^T \end{aligned} \tag{10.15}$$

The vector fields X_1 and X_2 are linearly independent and span a two-dimensional subspace at each point. Notice that X_1 and X_2 are completely specified by Equations (10.10) and (10.11). Geometrically, one can now

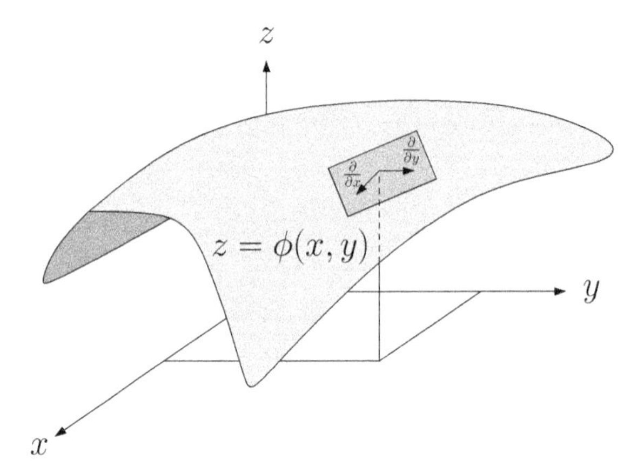

Figure 10.3: Integral manifold in $\mathbb{R}^3$.

think of the problem of solving this system of first order partial differential equations as the problem of finding a surface in $\mathbb{R}^3$ whose tangent space at each point is spanned by the vector fields X_1 and X_2. Such a surface, if it can be found, is called an **integral manifold** for Equations (10.10) and (10.11). If such an integral manifold exists then the set of vector fields, equivalently, the system of partial differential equations, is called **completely integrable**.

Let us reformulate this problem in yet another way. Suppose that $z = \phi(x, y)$ is a solution of Equations (10.10) and (10.11). Then it is a simple computation (Problem 10-2) to check that the function

$$h(x, y, z) = z - \phi(x, y) \tag{10.16}$$

satisfies the system of partial differential equations

$$\begin{aligned} L_{X_1} h &= 0 \\ L_{X_2} h &= 0 \end{aligned} \tag{10.17}$$

Conversely, suppose a scalar function h can be found satisfying (10.17), and suppose that we can solve the equation

$$h(x, y, z) = 0 \tag{10.18}$$

for z, as $z = \phi(x, y)$.[3] Then it can be shown (Problem 10-3) that ϕ satisfies Equations (10.10) and (10.11). Hence, complete integrability of the

[3]The **implicit function theorem** states that Equation (10.18) can be solved for z as long as $\frac{\partial h}{\partial z} \neq 0$.

set of vector fields $\{X_1, X_2\}$ is equivalent to the existence of h satisfying Equation (10.17). With the preceding discussion as background we state the following.

Definition 10.6 *A distribution $\Delta = span\{X_1, \ldots, X_m\}$ on $\mathbb{R}^n$ is said to be* **completely integrable** *if and only if there are $n - m$ linearly independent functions $h_1, \ldots, h_{n-m}$ satisfying the system of partial differential equations*

$$L_{X_i} h_j = 0 \quad \textit{for } 1 \leq i \leq m \; , \; 1 \leq j \leq n - m \tag{10.19}$$

Another important concept is the notion of **involutivity** as defined next.

Definition 10.7 *A distribution $\Delta = span\{X_1, \ldots, X_m\}$ is said to be* **involutive** *if and only if there are scalar functions $\alpha_{ijk} : \mathbb{R}^n \rightarrow \mathbb{R}$ such that*

$$[X_i, X_j] \;=\; \sum_{k=1}^{m} \alpha_{ijk} X_k \text{for all } i, j, k \tag{10.20}$$

Involutivity simply means that if one forms the Lie bracket of any pair of vector fields in Δ then the resulting vector field can be expressed as a linear combination of the original vector fields $X_1, \ldots, X_m$. An involutive distribution is thus closed under the operation of taking Lie brackets. Note that the coefficients in this linear combination are allowed to be smooth functions on $\mathbb{R}^n$.

In the simple case of Equations (10.10) and (10.11) one can show that involutivity of the set $\{X_1, X_2\}$ defined by Equation (10.19) is equivalent to interchangeability of the order of partial derivatives of h, that is, $\dfrac{\partial^2 h}{\partial x \partial y} = \dfrac{\partial^2 h}{\partial y \partial x}$. The Frobenius Theorem, stated next, gives the conditions for the existence of a solution to the system of partial differential Equations (10.19).

Theorem 4 Frobenius *A distribution Δ is completely integrable if and only if it is involutive.*

The importance of the Frobenius theorem is that it allows one to determine whether or not a given distribution is integrable without having to actually solve the partial differential equations. The involutivity condition can, in principle, be computed from the given vector fields alone.

10.2 FEEDBACK LINEARIZATION

To introduce the idea of feedback linearization consider the following simple system,

$$\dot{x}_1 = a\sin(x_2) \tag{10.21}$$
$$\dot{x}_2 = -x_1^2 + u \tag{10.22}$$

Note that we cannot simply choose u in the above system to cancel the nonlinear term $a\sin(x_2)$. However, if we first change variables by setting

$$y_1 = x_1 \tag{10.23}$$
$$y_2 = a\sin(x_2) = \dot{x}_1 \tag{10.24}$$

then, by the chain rule, y_1 and y_2 satisfy

$$\begin{aligned} \dot{y}_1 &= y_2 \\ \dot{y}_2 &= a\cos(x_2)(-x_1^2 + u) \end{aligned} \tag{10.25}$$

We see that the nonlinearities can now be cancelled by the control input

$$u = \frac{1}{a\cos(x_2)}v + x_1^2 \tag{10.26}$$

which results in the linear system in the (y_1, y_2) coordinates

$$\begin{aligned} \dot{y}_1 &= y_2 \\ \dot{y}_2 &= v \end{aligned} \tag{10.27}$$

The term v has the interpretation of an outer-loop control and can be designed to place the poles of the second order linear system given by Equation (10.27) in the coordinates (y_1, y_2). For example, the outer-loop control

$$v = -k_1 y_1 - k_2 y_2 \tag{10.28}$$

applied to Equation (10.27) results in the closed-loop system

$$\begin{aligned} \dot{y}_1 &= y_2 \\ \dot{y}_2 &= -k_1 y_1 - k_2 y_2 \end{aligned} \tag{10.29}$$

which has characteristic polynomial

$$p(s) = s^2 + k_2 s + k_1 \tag{10.30}$$

and hence, the closed-loop poles of the system with respect to the coordinates (y_1, y_2) are completely specified by the choice of k_1 and k_2. Figure 10.4

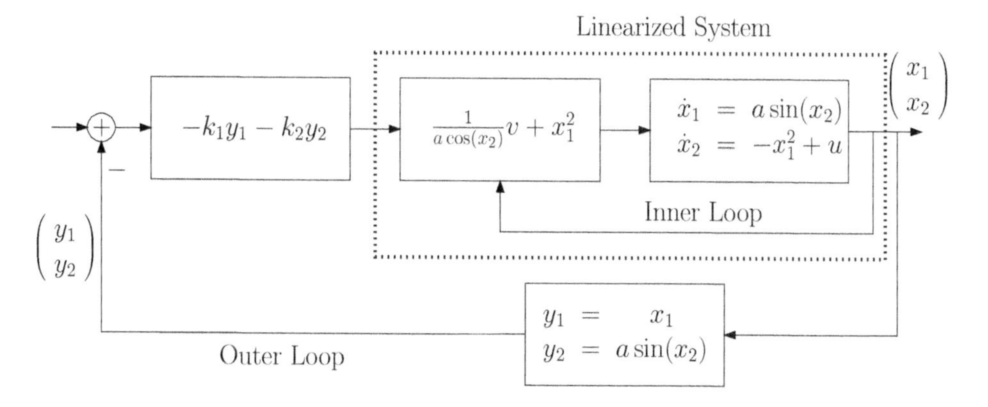

Figure 10.4: Inner-loop/outer-loop control architecture for feedback linearization.

illustrates the inner-loop/outer-loop implementation of the above control strategy. The response in the y variables is easy to determine. The corresponding response of the system in the original coordinates (x_1, x_2) can be found by inverting the transformation given by Equations (10.23) and (10.24). The result is

$$\begin{aligned} x_1 &= y_1 \\ x_2 &= \sin^{-1}(y_2/a) \qquad -a < y_2 < +a \end{aligned} \tag{10.31}$$

This example illustrates several important features of feedback linearization. The first thing to note is the local nature of the result. We see from Equations (10.23) and (10.24) that the transformation and the control make sense only in the region $-\infty < x_1 < \infty$, $-\frac{\pi}{2} < x_2 < \frac{\pi}{2}$. Second, in order to control the linear system given by Equation (10.27), the coordinates (y_1, y_2) must be available for feedback. This can be accomplished by measuring them directly if they are physically meaningful variables, or by computing them from the measured (x_1, x_2) coordinates using the transformation given by Equations (10.23) and (10.24). In the latter case the parameter a must be known precisely.

In Section 10.3 we give necessary and sufficient conditions under which a general single-input nonlinear system can be transformed into a linear system using a nonlinear change of variables and nonlinear feedback as in the above example.

10.3 SINGLE-INPUT SYSTEMS

The idea of feedback linearization is easiest to understand in the context of single-input systems. In this section we give necessary and sufficient conditions for single-input nonlinear system to be locally feedback linearizable. As an illustration, we apply this result to the control of a single-link manipulator with joint elasticity.

Definition 10.8 *A single-input nonlinear system*

$$\dot{x} = f(x) + g(x)u \tag{10.32}$$

where $f(x)$ and $g(x)$ are vector fields on $\mathbb{R}^n$, $f(0) = 0$, and $u \in \mathbb{R}$, is said to be **feedback linearizable** *if there exists a diffeomorphism $T : U \to \mathbb{R}^n$, defined on an open region U in $\mathbb{R}^n$ containing the origin, and nonlinear feedback*

$$u = \alpha(x) + \beta(x)v \tag{10.33}$$

with $\beta(x) \neq 0$ on U such that the transformed state

$$y = T(x) \tag{10.34}$$

satisfies the linear system of equations

$$\dot{y} = Ay + bv \tag{10.35}$$

where

$$A = \begin{bmatrix} 0 & 1 & 0 & & & 0 \\ 0 & 0 & 1 & & & \cdot \\ \cdot & \cdot & \cdot & \cdot & & \cdot \\ \cdot & \cdot & \cdot & & \cdot & \cdot \\ \cdot & \cdot & \cdot & & & 1 \\ 0 & 0 & \cdot & \cdot & 0 & 0 \end{bmatrix}, \quad b = \begin{bmatrix} 0 \\ 0 \\ \cdot \\ \cdot \\ \cdot \\ 1 \end{bmatrix} \tag{10.36}$$

The nonlinear transformation given by (10.34) and the nonlinear control law (10.33), when applied to the nonlinear system (10.32), result in a linear controllable system (10.35). The diffeomorphism $T(x)$ can be thought of as a nonlinear change of coordinates in the state space. The idea of feedback linearization is that, if one first changes to the coordinate system $y = T(x)$, then there exists a nonlinear control law to cancel the nonlinearities in the system. The feedback linearization is said to be **global** if the region U is all of $\mathbb{R}^n$.

We next derive necessary and sufficient conditions on the vector fields f and g in Equation (10.32) for the existence of such a transformation. Let us set

$$y = T(x) \tag{10.37}$$

and see what conditions the transformation $T(x)$ must satisfy. Differentiating both sides of Equation (10.37) with respect to time yields

$$\dot{y} = \frac{\partial T}{\partial x}\dot{x} \tag{10.38}$$

where $\frac{\partial T}{\partial x}$ is the Jacobian matrix of the transformation $T(x)$. Using Equations (10.32) and (10.35), Equation (10.38) can be written as

$$\frac{\partial T}{\partial x}(f(x) + g(x)u) = Ay + bv \tag{10.39}$$

In component form with

$$T = \begin{bmatrix} T_1 \\ \cdot \\ \cdot \\ \cdot \\ T_n \end{bmatrix}, \quad A = \begin{bmatrix} 0 & 1 & 0 & & & 0 \\ 0 & 0 & 1 & & & \cdot \\ \cdot & \cdot & \cdot & \cdot & & \cdot \\ \cdot & \cdot & \cdot & & \cdot & \cdot \\ \cdot & \cdot & \cdot & & & 1 \\ 0 & 0 & \cdot & \cdot & 0 & 0 \end{bmatrix}, \quad b = \begin{bmatrix} 0 \\ 0 \\ \cdot \\ \cdot \\ \cdot \\ 1 \end{bmatrix} \tag{10.40}$$

we see that the first equation in Equation (10.39) is

$$L_f T_1 + L_g T_1 u = T_2 \tag{10.41}$$

Similarly, the other components of T satisfy

$$\begin{aligned} L_f T_2 + L_g T_2 u &= T_3 \\ &\vdots \\ L_f T_n + L_g T_n u &= v \end{aligned} \tag{10.42}$$

Since we assume that $T_1, \ldots, T_n$ are independent of u while v is not independent of u we conclude from (10.42) that

$$L_g T_1 = L_g T_2 = \cdots = L_g T_{n-1} = 0 \tag{10.43}$$

$$L_g T_n \neq 0 \tag{10.44}$$

This leads to the system of partial differential equations

$$L_f T_i = T_{i+1} \; ; \quad i = 1, \ldots, n-1 \tag{10.45}$$

together with

$$L_f T_n + L_g T_n u = v \tag{10.46}$$

Using Lemma 10.1 together with Equations (10.43) and (10.44) we can derive a system of partial differential equations in terms of T_1 alone as follows. Using $h = T_1$ in Lemma 10.1 we have

$$L_{[f,g]} T_1 = L_f L_g T_1 - L_g L_f T_1 = 0 - L_g T_2 = 0 \tag{10.47}$$

Thus, we have shown

$$L_{[f,g]} T_1 = 0 \tag{10.48}$$

By proceeding inductively it can be shown (Problem 10-4) that

$$L_{ad_f^k g} T_1 \;=\; 0 \quad k = 0, 1, \ldots n-2 \tag{10.49}$$

$$L_{ad_f^{n-1} g} T_1 \;\neq\; 0 \tag{10.50}$$

If we can find T_1 satisfying the system of partial differential equations (10.49), then $T_2, \ldots, T_n$ are found inductively from Equation (10.45) and the control input u is found from (10.46) as

$$u = \frac{1}{L_g T_n}(v - L_f T_n) \tag{10.51}$$

We have thus reduced the problem to solving the system given by Equation (10.49) for T_1. When does such a solution exist?

First note that the vector fields g, $ad_f(g), \ldots, ad_f^{n-1}(g)$ must be linearly independent. If not, that is, if for some index i

$$ad_f^i(g) \;=\; \sum_{k=0}^{i-1} \alpha_k ad_f^k(g) \tag{10.52}$$

then $ad_f^{n-1}(g)$ would be a linear combination of g, $ad_f(g), \ldots, ad_f^{n-2}(g)$ and Equation (10.50) could not hold. Now, by the Frobenius theorem, Equation (10.49) has a solution if and only if the distribution $\Delta = \text{span}\{g, ad_f(g), \ldots, ad_f^{n-2}(g)\}$ is involutive. Putting this together we have shown the following

Theorem 5 *The nonlinear system*

$$\dot{x} \;=\; f(x) + g(x)u \tag{10.53}$$

is feedback linearizable if and only if there exists an open region U containing the origin in $\mathbb{R}^n$ in which the following conditions hold:

1. *The vector fields* $\{g, ad_f(g), \ldots, ad_f^{n-1}(g)\}$ *are linearly independent in* U.

2. *The distribution* $\Delta = \text{span}\{g, ad_f(g), \ldots, ad_f^{n-2}(g)\}$ *is involutive in* U.

Example 10.3 Flexible Joint Robot

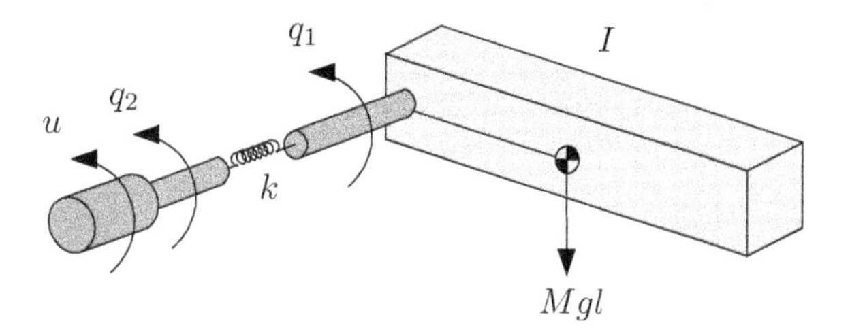

Figure 10.5: Single-link, flexible joint robot.

Consider the single link, flexible joint manipulator shown in Figure 10.5. Ignoring damping for simplicity, the equations of motion are

$$\begin{aligned} I\ddot{q}_1 + Mgl\sin(q_1) + k(q_1 - q_2) &= 0 \\ J\ddot{q}_2 + k(q_2 - q_1) &= u \end{aligned} \tag{10.54}$$

Note that since the nonlinearity enters into the first equation the control u *cannot simply be chosen to cancel it as in the case of the rigid manipulator equations. In state space we set*

$$\begin{array}{ll} x_1 = q_1 & x_2 = \dot{q}_1 \\ x_3 = q_2 & x_4 = \dot{q}_2 \end{array} \tag{10.55}$$

and write the system (10.54) as

$$\begin{aligned} \dot{x}_1 &= x_2 \\ \dot{x}_2 &= -\frac{MgL}{I}\sin(x_1) - \frac{k}{I}(x_1 - x_3) \\ \dot{x}_3 &= x_4 \\ \dot{x}_4 &= \frac{k}{J}(x_1 - x_3) + \frac{1}{J}u \end{aligned} \tag{10.56}$$

The system is thus of the form (10.32) with

$$f(x) = \begin{bmatrix} x_2 \\ -\frac{MgL}{I}\sin(x_1) - \frac{k}{I}(x_1 - x_3) \\ x_4 \\ \frac{k}{J}(x_1 - x_3) \end{bmatrix}, \qquad g(x) = \begin{bmatrix} 0 \\ 0 \\ 0 \\ \frac{1}{J} \end{bmatrix} \tag{10.57}$$

Therefore, with $n = 4$, the necessary and sufficient conditions for feedback linearization of this system are that

$$rank\left[g, ad_f(g), ad_f^2(g), ad_f^3(g)\right] = 4 \tag{10.58}$$

and that the distribution

$$\Delta = \text{span}\{g, ad_f(g), ad_f^2(g)\} \tag{10.59}$$

is involutive. Performing the indicated calculations it is easy to check that (Problem 10-8)

$$[g, ad_f(g), ad_f^2(g), ad_f^3(g)] = \begin{bmatrix} 0 & 0 & 0 & -\frac{k}{IJ} \\ 0 & 0 & \frac{k}{IJ} & 0 \\ 0 & -\frac{1}{J} & 0 & \frac{k}{J^2} \\ \frac{1}{J} & 0 & -\frac{k}{J^2} & 0 \end{bmatrix} \tag{10.60}$$

which has rank 4 for k, I, $J \neq 0$. Also, since the vector fields $\{g, ad_f(g), ad_f^2(g)\}$ are constant, the distribution Δ is involutive. To see this it suffices to note that the Lie bracket of two constant vector fields is zero. Hence, the Lie bracket of any two members of the set of vector fields in Equation (10.59) is zero which is trivially a linear combination of the vector fields themselves. It follows that the system given by Equation (10.54) is feedback linearizable. The new coordinates

$$y_i = T_i \qquad i = 1, \ldots, 4 \tag{10.61}$$

are found from the conditions given by Equation (10.49), with $n = 4$, that is

$$\begin{aligned} L_g T_1 &= 0 \\ L_{[f,g]} T_1 &= 0 \\ L_{ad_f^2 g} T_1 &= 0 \\ L_{ad_f^3 g} T_1 &\neq 0 \end{aligned} \tag{10.62}$$

Carrying out the above calculations leads to the system of equations (Problem 10-9)

$$\frac{\partial T_1}{\partial x_2} = 0 \; ; \; \frac{\partial T_1}{\partial x_3} = 0 \; ; \; \frac{\partial T_1}{\partial x_4} = 0 \tag{10.63}$$

and

$$\frac{\partial T_1}{\partial x_1} \neq 0 \tag{10.64}$$

From this we see that the function T_1 should be a function of x_1 alone. Therefore, we take the simplest solution

$$y_1 = T_1 = x_1 \tag{10.65}$$

and compute from Equation (10.45) (Problem 10-10)

$$\begin{aligned} y_2 &= T_2 = L_f T_1 = x_2 \\ y_3 &= T_3 = L_f T_2 = -\frac{MgL}{I}\sin(x_1) - \frac{k}{I}(x_1 - x_3) \\ y_4 &= T_4 = L_f T_3 = -\frac{MgL}{I}\cos(x_1)x_2 - \frac{k}{I}(x_2 - x_4) \end{aligned} \tag{10.66}$$

The feedback linearizing control input u is found from the condition

$$u = \frac{1}{L_g T_4}(v - L_f T_4) \tag{10.67}$$

as (Problem 10-11)

$$u = \frac{IJ}{k}(v - a(x)) = \beta(x)v + \alpha(x) \tag{10.68}$$

where

$$\begin{aligned} a(x) = & \frac{MgL}{I}\sin(x_1)\left(x_2^2 + \frac{MgL}{I}\cos(x_1) + \frac{k}{I}\right) \\ & + \frac{k}{I}(x_1 - x_3)\left(\frac{k}{I} + \frac{k}{J} + \frac{MgL}{I}\cos(x_1)\right) \end{aligned} \tag{10.69}$$

Therefore in the coordinates $y_1, \ldots, y_4$ with the control law given by Equation (10.68) the system becomes

$$\begin{aligned} \dot{y}_1 &= y_2 \\ \dot{y}_2 &= y_3 \\ \dot{y}_3 &= y_4 \\ \dot{y}_4 &= v \end{aligned} \tag{10.70}$$

or, in matrix form,

$$\dot{y} = Ay + bv \qquad (10.71)$$

where

$$A = \begin{bmatrix} 0 & 1 & 0 & 0 \\ 0 & 0 & 1 & 0 \\ 0 & 0 & 0 & 1 \\ 0 & 0 & 0 & 0 \end{bmatrix}, \qquad b = \begin{bmatrix} 0 \\ 0 \\ 0 \\ 1 \end{bmatrix} \qquad (10.72)$$

It is interesting to note that the above feedback linearization is actually global. In order to see this we need only compute the inverse of the change of variables given by Equations (10.65)and (10.66). By inspection we see that

$$\begin{aligned} x_1 &= y_1 \\ x_2 &= y_2 \\ x_3 &= y_1 + \frac{I}{k}\left(y_3 + \frac{MgL}{I}\sin(y_1)\right) \\ x_4 &= y_2 + \frac{I}{k}\left(y_4 + \frac{MgL}{I}\cos(y_1)y_2\right) \end{aligned} \qquad (10.73)$$

The inverse transformation is well defined and differentiable everywhere and, hence, the feedback linearization for the system given by Equation (10.54) holds globally. The transformed variables $y_1, \ldots, y_4$ are themselves physically meaningful. We see that

$$\begin{aligned} y_1 = x_1 &= \text{link position} \\ y_2 = x_2 &= \text{link velocity} \\ y_3 = \dot{y}_2 &= \text{link acceleration} \\ y_4 = \dot{y}_3 &= \text{link jerk} \end{aligned} \qquad (10.74)$$

Since the motion trajectory of the link is typically specified in terms of these quantities they are natural variables to use for feedback.

◇

Example 10.4

One way to execute a step change in the link position, while keeping the manipulator motion smooth, is to require a constant jerk during the motion. This can be accomplished by a cubic polynomial trajectory using the methods of Chapter 5.5. Therefore, let us specify a trajectory

$$q_1^d(t) = y_1^d = a_0 + a_1 t + a_2 t^2 + a_3 t^3 \qquad (10.75)$$

so that

$$\begin{aligned} y_2^d &= \dot{y}_1^d = a_1 + 2a_2 t + 3a_3 t^2 \\ y_3^d &= \dot{y}_2^d = 2a_2 + 6a_3 t \\ y_4^d &= \dot{y}_3^d = 6a_3 \end{aligned}$$

Then a linear control law that tracks this trajectory, which is essentially equivalent to the feedforward/feedback scheme of Chapter 8, is given by

$$v = -k_0(y_1 - y_1^d) - k_1(y_2 - y_2^d) - k_2(y_3 - y_3^d) - k_3(y_4 - y_4^d) \quad (10.76)$$

Applying this control law to the fourth order linear system given by Equation (10.68) we see that the tracking error $e(t) = y_1 - y_1^d$ *satisfies the fourth order linear equation*

$$\frac{d^4 e}{dt^4} + k_3 \frac{d^3 e}{dt^3} + k_2 \frac{d^2 e}{dt^2} + k_1 \frac{de}{dt} + k_0 e = 0 \quad (10.77)$$

and, hence, the error dynamics are completely determined by the choice of gains $k_0, \ldots, k_3$.

◇

Notice that the feedback control law given by Equation (10.76) is stated in terms of the variables $y_1, \ldots, y_4$. Thus, it is important to consider how these variables are to be determined so that they may be used for feedback in case they cannot be measured directly. Although the first two variables, representing the link position and velocity, are easy to measure, the remaining variables, representing link acceleration and jerk, are difficult to measure with any degree of accuracy using present technology. One could measure the original variables $x_1, \ldots, x_4$ which represent the motor and link positions and velocities, and compute $y_1, \ldots, y_4$ using the transformation Equations (10.65) and (10.66). In this case the parameters appearing in the transformation equations would have to be known precisely.

10.4 FEEDBACK LINEARIZATION FOR N-LINK ROBOTS

In the general case of an n-link manipulator the dynamic equations represent a multi-input nonlinear system. The conditions for feedback linearization of multi-input systems are more difficult to state, but the conceptual idea is the same as the single-input case. That is, one seeks a coordinate system in which the nonlinearities can be exactly cancelled by one or more of the inputs. In the multi-input system we can also decouple the system, that is, linearize the system in such a way that the resulting linear system is composed of subsystems, each of which is affected by only a single one of the outer-loop control inputs. Since we are concerned only with the application of these ideas to manipulator control we will not need the most general results in multi-input feedback linearization. Instead, we will use the physical insight gained by our detailed derivation of this result in the single-link case to derive a feedback linearizing control both for n-link rigid manipulators and for n-link manipulators with elastic joints.

Example 10.5

We will first verify what we have stated previously, namely that for an n-link rigid manipulator the feedback linearizing control is identical to the inverse dynamics control of Chapter 8. To see this, consider the rigid robot equations of motion given by Equation (8.6), which we write in state space as

$$\begin{aligned} \dot{x}_1 &= x_2 \\ \dot{x}_2 &= -M(x_1)^{-1}(C(x_1, x_2)x_2 + g(x_1)) + M(x_1)^{-1}u \end{aligned} \tag{10.78}$$

with $x_1 = q$, $x_2 = \dot{q}$. In this case a feedback linearizing control is found by simply inspecting Equation (10.78) as

$$u = M(x_1)v + C(x_1, x_2)x_2 + g(x_1) \tag{10.79}$$

Substituting Equation (10.79) into Equation (10.78) yields

$$\begin{aligned} \dot{x}_1 &= x_2 \\ \dot{x}_2 &= v \end{aligned} \tag{10.80}$$

Equation (10.80) represents a set of n second order systems of the form

$$\begin{aligned} \dot{x}_{1i} &= x_{2i} \\ \dot{x}_{2i} &= v_i\ , \quad i = 1, \ldots, n \end{aligned} \tag{10.81}$$

Comparing Equation (10.79) with Equation (8.23) we see indeed that the feedback linearizing control for a rigid manipulator is precisely the inverse dynamics control of Chapter 8.

◇

Example 10.6

Including the joint flexibility in the dynamic description of an n-link robot results in a Lagrangian system with $2n$ degrees of freedom. Recall the Euler-Lagrange equations of motion for the flexible joint robot from Chapter 8

$$\begin{aligned} D(q_1)\ddot{q}_1 + C(q_1, \dot{q}_1)\dot{q}_1 + g(q_1) + K(q_1 - q_2) &= 0 \\ J\ddot{q}_2 - K(q_1 - q_2) &= u \end{aligned} \tag{10.82}$$

In state space, which is now $\mathbb{R}^{4n}$, we define state variables in block form

$$\begin{array}{ll} \dot{x}_1 = q_1 & x_2 = \dot{q}_1 \\ \dot{x}_3 = q_2 & x_4 = \dot{q}_2 \end{array} \tag{10.83}$$

Then from Equation (10.82) we have:

$$\begin{aligned} \dot{x}_1 &= x_2 \\ \dot{x}_2 &= -D(x_1)^{-1}\{h(x_1, x_2) + K(x_1 - x_3)\} \\ \dot{x}_3 &= x_4 \\ \dot{x}_4 &= J^{-1}K(x_1 - x_3) + J^{-1}u \end{aligned} \tag{10.84}$$

where we define $h(x_1, x_2) = C(x_1, x_2)x_2 + g(x_1)$ for simplicity. This system is of the form

$$\dot{x} = f(x) + G(x)u \tag{10.85}$$

In the single-link case we saw that the system could be linearized by nonlinear feedback if we took as state variables the link position, velocity, acceleration, and jerk. Following the single-input example, we can attempt to do the same thing in the multi-link case and derive a feedback linearizing transformation blockwise as follows. Set

$$\begin{aligned} &y_1 = T_1(x_1) = x_1 \\ &y_2 = T_2(x) = \dot{y}_1 = x_2 \\ &y_3 = T_3(x) = \dot{y}_2 = \dot{x}_2 = -D^{-1}\{h(x_1, x_2) + K(x_1 - x_3)\} \\ &y_4 = T_4(x) = \dot{y}_3 = -\frac{d}{dt}[D^{-1}]\{h(x_1, x_2) + K(x_1 - x_3)\} \\ &-D^{-1}\left\{\frac{\partial h}{\partial x_1}x_2 + \frac{\partial h}{\partial x_2}[-D^{-1}(h(x_1, x_2) + K(x_1 - x_3))] + K(x_2 - x_4)\right\} \\ &= a_4(x_1, x_2, x_3) + D(x_1)^{-1}Kx_4 \end{aligned} \tag{10.86}$$

where for simplicity we define the function a_4 to be everything in the definition of y_4 except the last term, which is $D^{-1}Kx_4$. Note that x_4 appears only in this last term so that a_4 depends only on x_1, x_2, x_3.

As in the single-link case, the above mapping is a global diffeomorphism. Its inverse can be found by inspection to be

$$\begin{aligned} x_1 &= y_1 \\ x_2 &= y_2 \\ x_3 &= y_1 + K^{-1}(D(y_1)y_3 + h(y_1, y_2)) \\ x_4 &= K^{-1}D(y_1)(y_4 - a_4(y_1, y_2, y_3)) \end{aligned} \tag{10.87}$$

The linearizing control law can now be found from the condition

$$\dot{y}_4 = v \tag{10.88}$$

Computing $\dot{y}_4$ *from Equation (10.86) yields*

$$\begin{aligned} v &= \frac{\partial a_4}{\partial x_1}x_2 - \frac{\partial a_4}{\partial x_2}D^{-1}(h + K(x_1 - x_3)) + \frac{\partial a_4}{\partial x_3}x_4 \\ &+ \frac{d}{dt}[D^{-1}]Kx_4 + D^{-1}K(J^{-1}K(x_1 - x_3) + J^{-1}u) \\ &= a(x) + b(x)u \end{aligned} \tag{10.89}$$

where $a(x)$ *denotes all the terms in Equation (10.89) but the last term, which involves the input* u*, and* $b(x) = D^{-1}(x)KJ^{-1}$.

Solving the above expression for u *yields*

$$u = b(x)^{-1}(v - a(x)) = \alpha(x) + \beta(x)v \tag{10.90}$$

where $\beta(x) = JK^{-1}D(x)$ *and* $\alpha(x) = -b(x)^{-1}a(x)$.

With the nonlinear change of coordinates given by Equation (10.86) and nonlinear feedback given by Equation (10.90) the transformed system now has the linear block form

$$\begin{aligned} \dot{y} &= \begin{bmatrix} 0 & I & 0 & 0 \\ 0 & 0 & I & 0 \\ 0 & 0 & 0 & I \\ 0 & 0 & 0 & 0 \end{bmatrix} y + \begin{bmatrix} 0 \\ 0 \\ 0 \\ I \end{bmatrix} v \\ &= Ay + Bv \end{aligned} \tag{10.91}$$

where $I = n \times n$ *identity matrix,* $0 = n \times n$ *zero matrix,* $y^T = (y_1^T, y_2^T, y_3^T, y_4^T) \in \mathbb{R}^{4n}$*, and* $v \in \mathbb{R}^n$*. The system (10.91) represents a set of* n *decoupled quadruple integrators. The outer-loop design can now proceed as before, because not only is the system linearized, but it consists of* n *subsystems each identical to the fourth order system (10.70).*

$\diamond$

10.5 NONHOLONOMIC SYSTEMS

In this section we return to a discussion of systems subject to constraints. A constraint on a mechanical system restricts its motion by limiting the set of paths that the system can follow. We briefly discussed so-called **holonomic constraints** in Chapter 7 when we derived the Euler-Lagrange equations of motion. Our treatment of force control in Chapter 9 dealt with unilateral constraints defined by the environmental contact. In this section we expand upon the notion of systems subject to constraints and discuss **nonholonomic systems**.

Let $\mathcal{Q}$ denote the configuration space of a given system and let $q = [q_1, \ldots, q_n]^T \in \mathcal{Q}$ denote the vector of generalized coordinates defining the system configuration. We recall the following definition.

Definition 10.9 *A set of $k < n$ constraints*

$$h_i(q_1, \ldots, q_n) = 0 \ , \qquad i = 1, \ldots, k \tag{10.92}$$

is called **holonomic**, *where each h_i is a smooth mapping from $Q \mapsto \mathbb{R}$.*

We assume that the constraints are independent so that the differentials

$$\begin{aligned} dh_1 &= \left[\frac{\partial h_1}{\partial q_1}, \ldots, \frac{\partial h_1}{\partial q_n} \right] \\ &\vdots \\ dh_k &= \left[\frac{\partial h_k}{\partial q_1}, \ldots, \frac{\partial h_k}{\partial q_n} \right] \end{aligned}$$

are linearly independent covectors. Note that in order to satisfy these constraints the motion of the system must lie on the hypersurface defined by

$$h_i(q(t)) = 0, \ i = 1, \ldots, k \qquad \text{for all } t > 0 \tag{10.93}$$

As a consequence, by differentiating the functions in Equation (10.93), we have

$$< dh_i, \dot{q} >= 0 \qquad i = 1, \ldots, k \tag{10.94}$$

which says that the differentials dh_i are orthogonal to the velocity $\dot{q}$.

It frequently happens that constraints are expressed, not as constraints on the configuration as in Equation (10.92), but as constraints on the velocity, namely,

$$< w_i, \dot{q} >= 0 \ , \ i = 1, \ldots, k \tag{10.95}$$

where $w_i(q)$ are covectors. Constraints of the form given by Equation (10.95) are known as **Pfaffian** constraints. The crucial question in such cases is, therefore, when can the covectors $w_1, \ldots, w_k$ be expressed as differentials of smooth functions, $h_1, \ldots, h_k$? We express this as

Definition 10.10 *Constraints of the form*

$$< w_i, \dot{q} >= 0 \qquad i = 1, \ldots, k \tag{10.96}$$

are holonomic if there exists smooth functions $h_1, \ldots, h_k$ *such that*

$$w_i(q) = dh_i(q) \qquad i = 1, \ldots, k \tag{10.97}$$

and **nonholonomic** *otherwise, that is, if no such functions* $h_1, \ldots, h_k$ *exist.*

We can begin to see a connection with our earlier discussion of integrability and the Frobenius Theorem if we think of Equation (10.97) as a set of partial differential equations in the (unknown) functions h_i. Indeed, the term **integrable constraint** is frequently used interchangeably with holonomic constraint for this reason.

10.5.1 Involutivity and Holonomy

Now, given a set of Pfaffian constraints (10.95), let Ω be the codistribution defined by the covectors $w_1, \ldots, w_k$ and let $\{g_1, \ldots, g_m\}$ for $m = n - k$ be a basis for the distribution Δ that **annihilates** Ω, that is, such that

$$< w_i, g_j >= 0 \qquad \text{for each } i, j \tag{10.98}$$

We use the notation $\Delta = \Omega^\perp$.[4] Notice from Equation (10.97) that

$$0 =< w_i, g_j >=< dh_i, g_j > \qquad \text{for each } i, j \tag{10.99}$$

Using our previous notation for Lie derivative, the above system of equations may be written as

$$L_{g_j} h_i = 0 \qquad i = 1, \ldots, k; \; , \; j = 1, \ldots, m \tag{10.100}$$

The following theorem thus follows immediately from the Frobenius Theorem.

Theorem 6 *Let* Ω *be the codistribution defined by covectors* $w_1, \ldots, w_k$. *Then the constraints* $< w_i, \dot{q} >= 0$, $i = 1, \ldots, k$, *are holonomic if and only if the distribution* $\Delta = \Omega^\perp$ *is involutive.*

[4]This is pronounced “omega perp”.

10.5.2 Driftless Control Systems

It is important to note that the velocity vector $\dot{q}$ of the system is orthogonal to the covectors w_i according to Equation (10.96) and, hence, lies in the distribution $\Delta = \Omega^\perp$. This means that the velocity vector $\dot{q}$ can be expressed as a linear combination of the basis vectors $g_1, \ldots, g_m$. In other words, we may write

$$\dot{q} = g_1(q)u_1 + \cdots + g_m(q)u_m \tag{10.101}$$

for suitable coefficients $u_1, \ldots, u_m$. In many systems of interest, the coefficients u_i in Equation (10.101) have the interpretation of control inputs. In such cases, Equation (10.101) defines a useful model for control design. Equation (10.101) is called **driftless** because $\dot{q} = 0$ when the control inputs $u_1, \ldots, u_m$ are zero. In the next section we give some examples of driftless systems arising from nonholonomic constraints, followed by a discussion of controllability of driftless systems and Chow's Theorem in Section 10.6.

10.5.3 Examples of Nonholonomic Systems

Nonholonomic constraints arise in two primary ways:

1. In so-called **rolling without slipping** constraints. For example, the translational and rotational velocities of a rolling wheel are not independent if the wheel rolls without slipping. Examples include
   - a unicycle
   - an automobile, tractor/trailer, or wheeled mobile robot
   - manipulation of rigid objects
2. In systems where angular momentum is conserved. Examples include
   - space robots
   - satellites
   - gymnastic robots

Example 10.7 The Unicycle

The unicycle is equivalent to a wheel rolling on a plane and is thus the simplest example of a nonholonomic system. Refering to Figure 10.6 we see that the configuration of the unicycle is defined by the variables x, y, θ, and ϕ, where x and y denote the Cartesian position of the ground contact point,

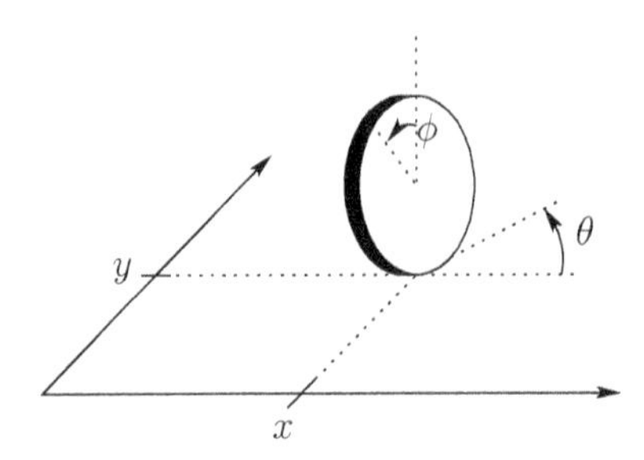

Figure 10.6: The unicycle.

θ denotes the heading angle, and ϕ denotes the angle of the wheel measured from the vertical. The rolling without slipping condition means that

$$\begin{aligned} \dot{x} - r\dot{\phi}\cos\theta &= 0 \\ \dot{y} - r\dot{\phi}\sin\theta &= 0 \end{aligned} \tag{10.102}$$

where r is the radius of the wheel. These constraints can be written in the form (10.96) with $q = [x, y, \theta, \phi]^T$ and

$$\begin{aligned} w_1 &= \begin{bmatrix} 1 & 0 & 0 & -r\cos\theta \end{bmatrix} \\ w_2 &= \begin{bmatrix} 0 & 1 & 0 & -r\sin\theta \end{bmatrix} \end{aligned} \tag{10.103}$$

Since the dimension of the configuration space is $n = 4$ and there are two constraint equations, we need to find two function g_1, g_2 orthogonal to w_1, w_2. It is easy to see that

$$g_1 = \begin{bmatrix} 0 \\ 0 \\ 1 \\ 0 \end{bmatrix} \; ; \; g_2 = \begin{bmatrix} r\cos\theta \\ r\sin\theta \\ 0 \\ 1 \end{bmatrix} \tag{10.104}$$

are both orthogonal to w_1 and w_2. Thus, we can write

$$\dot{q} = g_1(q)u_1 + g_2(q)u_2 \tag{10.105}$$

where u_1 is the turning rate and u_2 is the rate of rolling.

◇

We can now check to see if rolling without slipping constraint on the unicycle is holonomic or nonholonomic using Theorem 6. It is easy to show (Problem 10-18) that the Lie bracket $[g_1, g_2]$ is given by

$$[g_1, g_2] = \begin{bmatrix} -r\sin\theta \\ r\cos\theta \\ 0 \\ 0 \end{bmatrix} \tag{10.106}$$

which is not in the distribution $\Delta = \text{span}\{g_1, g_2\}$. Therefore, the constraints on the unicycle are nonholonomic. We shall see the consequences of this fact in the next section when we discuss controllability of driftless systems.

Example 10.8 The Kinematic Car

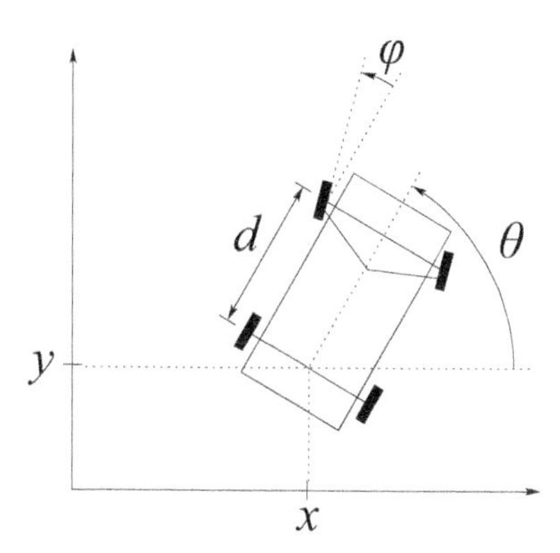

Figure 10.7: The kinematic car.

Figure 10.7 shows a simple representation of a car, or mobile robot with steerable front wheels. The configuration of the car can be described by $q = [x, y, \theta, \phi]^T$*, where* x *and* y *are the point at the center of the rear axle,* θ *is the heading angle, and* ϕ *is the steering angle as shown in the figure. The rolling without slipping constraints are found by setting the sideways velocity of the front and rear wheels to zero. This leads to*

$$\begin{aligned} \sin\theta\ \dot{x} - \cos\theta\ \dot{y} &= 0 \\ \sin(\theta+\phi)\ \dot{x} - \cos(\theta+\phi)\ \dot{y} - d\cos\phi\ \dot{\theta} &= 0 \end{aligned} \tag{10.107}$$

which can be written as

$$\begin{aligned} \begin{bmatrix} \sin\theta & \cos\theta & 0 & 0 \end{bmatrix} \dot{q} &= \ < w_1, \dot{q} > = 0 \\ \begin{bmatrix} \sin(\theta+\phi) & -\cos(\theta+\phi) & -d\cos\phi & 0 \end{bmatrix} \dot{q} &= \ < w_2, \dot{q} > = 0 \end{aligned} \tag{10.108}$$

It is thus straightforward to find vectors

$$g_1 = \begin{bmatrix} 0 \\ 0 \\ 0 \\ 1 \end{bmatrix} \ ; \ g_2 = \begin{bmatrix} \cos\theta \\ \sin\theta \\ \frac{1}{d}\tan\phi \\ 0 \end{bmatrix} \tag{10.109}$$

orthogonal to w_1 *and* w_2 *and write the corresponding control system in the form of Equation (10.101). It is left as an exercise (Problem 10-19) to show that the above constraints are nonholonomic.*

◇

Example 10.9 A Hopping Robot

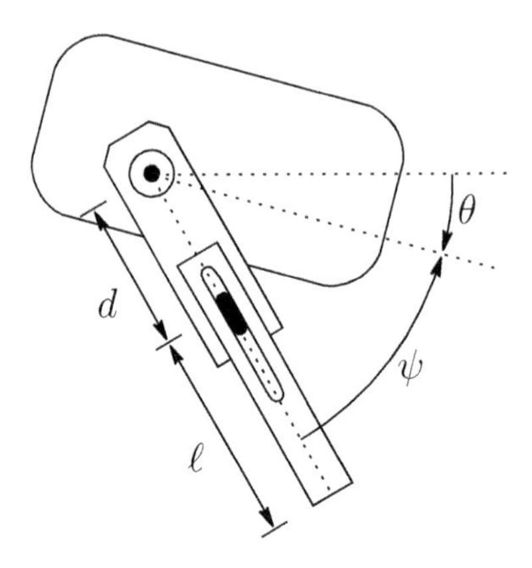

Figure 10.8: Hopping robot.

Consider the hopping robot in Figure 10.8. The configuration of this robot is defined by $q = [\psi, \ell, \theta]^T$*, where*

ψ = *the leg angle*
θ = *the body angle*
ℓ = *the leg extension*

During its flight phase the hopping robot's angular momentum is conserved. Letting I *and* m *denote the body moment of inertia and leg mass, respectively, conservation of angular momentum leads to the expression*

$$I\dot{\theta} + m(\ell + d)^2(\dot{\theta} + \dot{\psi}) = 0 \qquad (10.110)$$

assuming the initial angular momentum is zero. This constraint may be written as

$$< w, \dot{q} >= 0 \qquad (10.111)$$

where $w = \left[\begin{array}{ccc} m(\ell+d)^2 & 0 & I + m(\ell+d)^2 \end{array} \right]$*. Since the dimension of the configuration space is three and there is one constraint, we need to find two independent vectors,* g_1 *and* g_2 *spanning the annihilating distribution*

$\Delta = \Omega^{\perp}$, *where* $\Omega = \text{span}\,\{w\}$. *It is easy to see that*

$$g_1 = \begin{bmatrix} 0 \\ 1 \\ 0 \end{bmatrix} \quad \text{and} \quad g_2 = \begin{bmatrix} 1 \\ 0 \\ -\frac{m(\ell+d)^2}{I+m(\ell+d)^2} \end{bmatrix} \tag{10.112}$$

are linearly independent at each point and orthogonal to w. *Checking involutivity of* Δ *we find that*

$$[g_1, g_2] = \begin{bmatrix} 0 \\ 0 \\ \frac{-2Im(\ell+d)}{[I+m(\ell+d)^2]^2} \end{bmatrix} \tag{10.113}$$

which is not a linear combination of g_1 *and* g_2 *it follows that* Δ *is not an involutive distribution and hence the constraint is nonholonomic.*

◇

10.6 CHOW'S THEOREM

In this section we discuss the controllability properties of driftless systems of the form

$$\dot{x} = g_1(x)u_1 + \cdots + g_m(x)u_m \tag{10.114}$$

with $x \in \mathbb{R}^n$. We assume that the vector fields $g_1(x), \ldots, g_m(x)$ are smooth, complete,[5] and linearly independent at each $x \in \mathbb{R}^n$.

We have seen previously that, if the $k < n$, the Pfaffian constraints are holonomic then the trajectory of the system lies on an $m = (n-k)$-dimensional surface (an integral manifold) found by integrating the constraints. In fact, at each $x \in \mathbb{R}$ the tangent space to this manifold is spanned by the vectors $g_1(x)$, $\ldots$, $g_m(x)$. If we examine Equation (10.114) we see that any instantaneous direction in this tangent space, that is, any linear combination of $g_1, \ldots, g_m$, is achievable by a suitable choice of the control input terms u_i, $i = 1, \ldots, m$. Thus, every point on the manifold may be reached from any other point on the manifold by a suitable control input. However, points not lying on the manifold cannot be reached no matter what control input is applied. Thus, for an initial condition x_0, only points on the particular integral manifold through x_0 are reachable.

What happens if the constraints are nonholonomic? Then no such integral manifold of dimension m exists. Thus, it might be possible to reach a

[5]A complete vector field is one for which the solution of the associated differential equation exists for all time t.

space (manifold) of dimension larger than m by suitable application of the control inputs u_i. It turns out that this interesting possibility is true. In fact, by suitable combinations of two vector fields g_1 and g_2, it is possible to move in the direction defined by the Lie bracket vector field $[g_1, g_2]$. If the distribution $\Delta = \text{span}\{g_1, g_2\}$ is not involutive, then the Lie bracket vector field $[g_1, g_2]$ defines a direction not in the span of g_1 and g_2. Therefore, given vector fields $g_1, \ldots, g_m$ one may reach points not only in the span of these vector field but in the span of the distribution obtained by augmenting $g_1, \ldots, g_m$ with various Lie bracket directions.

Definition 10.11 *Let $\Delta = \text{span}\{g_1, \ldots, g_m\}$ be a distribution. The* **involutive closure** $\bar{\Delta}$ *of Δ is the smallest involutive distribution containing Δ. In other words, $\bar{\Delta}$ is an involutive distribution such that if Δ_0 is any involutive distribution satisfying $\Delta \subset \Delta_0$ then, $\bar{\Delta} \subset \Delta_0$.*

Conceptually, the involutive closure of Δ can be found by forming larger and larger distributions by repeatedly computing Lie brackets until an involutive distribution is found, that is,

$$\bar{\Delta} = \text{span}\{g_1, \ldots, g_m, [g_i, g_j], [g_k, [g_i, g_j]], \ldots\} \tag{10.115}$$

The involutive closure $\bar{\Delta}$ in Equation (10.115) is also called the **control Lie algebra** for the driftless control system (10.114). Intuitively, if $\dim\bar{\Delta} = n$, then all points in $\mathbb{R}^n$ should be reachable from x_0. This is essentially the conclusion of **Chow's theorem**.

Definition 10.12 *A driftless system of the form (10.101) is said to be* **Controllable** *if, for any x_0 and $x_1 \in \mathbb{R}^n$, there exists a time $T > 0$ and a control input $u = [u_1, \ldots, u_m]^T : [0, T] \to R^m$ such that the solution $x(t)$ of Equation (10.101) satisfies $x(0) = x_0$ and $x(T) = x_1$.*

The next result, known as **Chow's Theorem**, gives a sufficient condition for the system given by Equation (10.101) to be controllability.

Theorem 7 *The driftless system*

$$\dot{x} = g_1(x)u_1 + \cdots + g_m(x)u_m \tag{10.116}$$

is controllable if and only if rank $\bar{\Delta}(x) = n$ *at each $x \in \mathbb{R}^n$.*

The proof of Chow's Theorem is beyond the scope of this text. The condition rank $\bar{\Delta}(x) = n$ is called the **Controllability Rank Condition**.

Example 10.10

Consider the following system on $\mathbb{R}^3$

$$\begin{aligned} \begin{bmatrix} \dot{x}_1 \\ \dot{x}_2 \\ \dot{x}_3 \end{bmatrix} &= \begin{bmatrix} x_3 \\ 1 - x_3^2 \\ 0 \end{bmatrix} u_1 + \begin{bmatrix} 0 \\ 0 \\ 1 \end{bmatrix} u_2 \\ &= g_1(x)u_1 + g_2(x)u_2 \end{aligned} \tag{10.117}$$

with three states x_1, x_2, x_3, *and two control inputs,* u_1 *and* u_2. *It is easy to see that the distribution* $\Delta = \text{span}\{g_1, g_2\}$ *has rank two for all values of* $x \in \mathbb{R}^3$. *It is also easy to compute the Lie bracket* $[g_1, g_2]$ *as*

$$[g_1, g_2] = \begin{bmatrix} -1 \\ 2x_3 \\ 0 \end{bmatrix}$$

and, therefore, we have

$$\text{rank}[g_1, g_2, [g_1, g_2]] = \text{rank} \begin{bmatrix} x_3 & 0 & -1 \\ 1 - x_3^2 & 0 & 2x_3 \\ 0 & 1 & 0 \end{bmatrix} = 3$$

for all values of $x \in \mathbb{R}^3$. *Therefore, by Chow's Theorem, the system is controllable on* $\mathbb{R}^3$. $\diamond$

Example 10.11 Attitude Control of a Satellite with Reaction Wheels

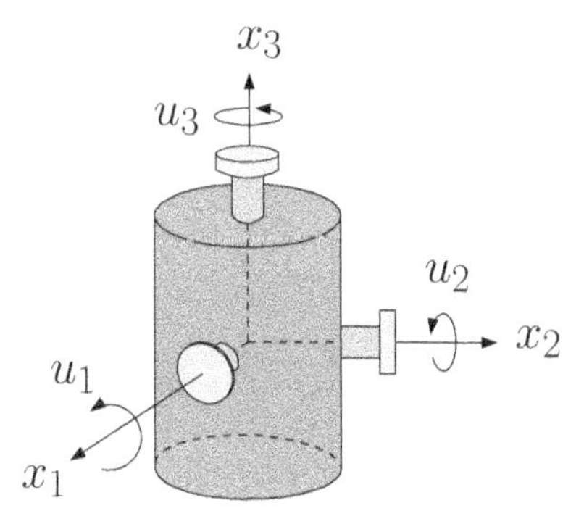

Figure 10.9: Satellite with reaction wheels.

Consider a cylindrical satellite equipped with reaction wheels for control as shown in Figure 10.9. Suppose we can control the angular velocity about *the x_1, x_2, and x_3 axes with controls u_1, u_2, and u_3, respectively. The equations of motion are then given by*

$$\dot{\omega} = \omega \times u$$

with

$$\omega = \begin{bmatrix} \omega_1 \\ \omega_2 \\ \omega_3 \end{bmatrix} \quad u = \begin{bmatrix} u_1 \\ u_2 \\ u_3 \end{bmatrix}$$

Carrying out the above calculation, it is readily shown (Problem 10-20) that

$$\begin{aligned} \dot{\omega} &= \begin{bmatrix} 0 \\ \omega_3 \\ -\omega_2 \end{bmatrix} u_1 + \begin{bmatrix} -\omega_3 \\ 0 \\ \omega_1 \end{bmatrix} u_2 + \begin{bmatrix} \omega_2 \\ -\omega_1 \\ 0 \end{bmatrix} u_3 \\ &= g_1(\omega)u_1 + g_2(\omega)u_2 + g_3(\omega)u_3 \end{aligned} \tag{10.118}$$

It is easy to show (Problem 10-21) that the distribution $\Delta = span\{g_1, g_2, g_3\}$ is involutive of rank 2 on $\mathbb{R}^3 \setminus \{0\}$. A more interesting property is that the satellite is controllable on $SO(3)$ as long as any two of the three reaction wheels are functioning. The proof of this strictly nonlinear phenomenon is left as an exercise (Problem 10-22).

◇

10.7 CONTROL OF DRIFTLESS SYSTEMS

Chow's theorem tells us when a driftless system is controllability but does not tell us how to find the control input to steer the system from a given initial state x_0 to a desired final state x_1. A detailed treatment of this topic is outside the scope of the present text. There are several approaches to designing controllers for this class of systems based on optimal control methods, Fourier methods, piecewise constant inputs, and other approaches. We will give a simple example to illustrate one approach on the unicycle example.

Example 10.12

Consider again the unicycle example but ignore the wheel orientation ϕ for simplicity. Thus, the model becomes

$$\begin{aligned} \dot{x} &= r\cos(\theta)u_1 \\ \dot{y} &= r\sin(\theta)u_1 \\ \dot{\theta} &= u_2 \end{aligned}$$

Let us change state and input variables by defining

$$\begin{aligned} x_1 &= x & v_1 &= r\cos(\theta)u_1 \\ x_2 &= \theta & v_2 &= u_2 \\ x_3 &= y \end{aligned}$$

With respect to these new variables the system is

$$\begin{aligned} \dot{x}_1 &= v_1 \\ \dot{x}_2 &= v_2 \\ \dot{x}_3 &= tan(x_2)v_1 \end{aligned}$$

For simplicity we consider first the small angle approximation $\tan(\theta) \approx \theta$ *and write*

$$\begin{aligned} \dot{x}_1 &= v_1 \\ \dot{x}_2 &= v_2 \\ \dot{x}_3 &= x_2 v_1 \end{aligned} \tag{10.119}$$

Note from these equations that x_1 *and* x_2 *can be independently controlled to any desired value. In the process however,* x_3 *may drift. One approach therefore is to first move* x_1 *and* x_2 *to their desired final values and then to execute a periodic motion of* x_1 *and* x_2 *to move* x_3 *to its desired final value.*

For example, if we set

$$\begin{aligned} v_1 &= a\sin(\omega t) \\ v_2 &= b\cos(\omega t) \end{aligned}$$

it is easily shown (Problem 10-23) that after $2\pi/\omega$ *seconds,* x_1 *and* x_2 *return to their initial values whereas the change in* x_3 *is* ab/ω^2. *Suppose that we wish to move the system from the origin* $(x_1, x_2, x_3) = (0, 0, 0)$ *to* $(x_1, x_2, x_3) = (0, 0, 10)$ *in two seconds. Using the above controls with* $a = \pi$, $\omega = \pi$, $b = 10$ *results in the response shown in Figure 10.10.*

◇

10.8 SUMMARY

This chapter introduced some basic concepts from differential geometric nonlinear control theory and serves as a foundation to explore more advanced literature.

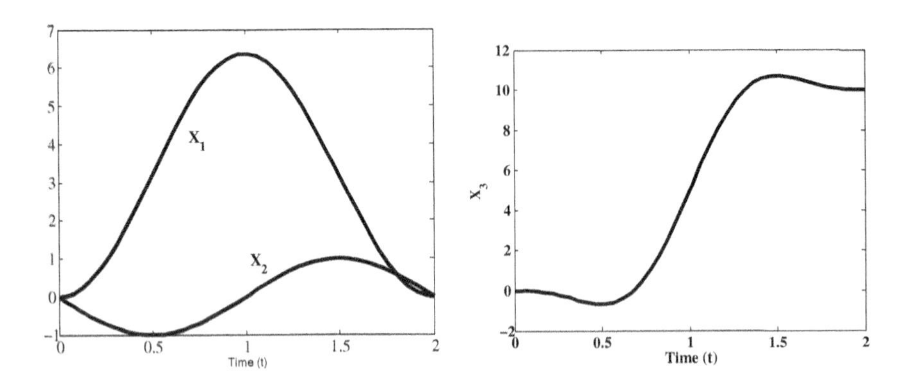

Figure 10.10: Response of x_1, x_2 and x_3. Note that x_1 and x_2 return to their original values after 2 seconds while x_3 moves from the origin to $x_3 = 10$ as desired.

Manifolds, Vector Fields, Distributions

We introduced basic definitions from differential geometry, such as a differentiable manifold, vector field, and distribution. We introduced some geometric operations such as the Lie derivative and Lie bracket and showed how they are related. We stated the Frobenius Theorem, which is an important tool for nonlinear analysis.

Feedback Linearization

We derived the necessary and sufficient conditions for feedback linearization of single-input nonlinear systems. This important result serves as a basis for controller design for a wide range of physical systems. In particular, we showed how feedback linearization can be used to design globally stable tracking controllers for flexible joint robots.

Nonholonomic Systems

We introduced the notion of nonholonomic systems, which has applications in mobile robots, hopping robots, gymnastic robots, and other systems that are subject to either rolling without slipping constraints or conservation of momentum constraints. We presented a fundamental result for so-called driftless systems known as Chow's Theorem, which tells us when a driftless system is controllable.

PROBLEMS

10-1 Complete the proof of Lemma 10.1 by direct calculation.

10-2 Show that the function $h = z - \phi(x, y)$ satisfies the system given by Equation (10.17) if ϕ is a solution of Equations (10.10) and (10.11) and X_1, X_2 are defined by Equation (10.15).

10-3 Show that if $h(x, y, z)$ satisfies Equation (10.17) and $\frac{\partial h}{\partial z} \neq 0$, then Equation (10.18) can be solved for z as $z = \phi(x, y)$ where ϕ satisfies Equations (10.10) and (10.11). Also show that $\frac{\partial h}{\partial z} = 0$ can occur only in the case of the trivial solution $h = 0$ of Equation (10.17).

10-4 Verify Equations (10.49) and (10.50).

10-5 Show that the system below is locally feedback linearizable.

$$\begin{aligned} \dot{x}_1 &= x_1^3 + x_2 \\ \dot{x}_2 &= x_2^3 + u \end{aligned}$$

Find explicitly the change of coordinates and nonlinear feedback to linearize the system.

10-6 Derive Equation (10.54) which gives the equations of motion for the single-link manipulator with joint elasticity of Figure 10.5 using Lagrange's equations.

10-7 Repeat Problem 10-6 if there is viscous friction both on the link side and on the motor side of the spring in Figure 10.5.

10-8 Perform the calculations necessary to verify Equation (10.60).

10-9 Derive the system of partial differential equations (10.63) from the condition (10.62). Also verify Equation (10.64).

10-10 Compute the change of coordinates (10.66).

10-11 Verify Equations (10.68) and (10.69).

10-12 Verify Equations (10.73).

10-13 Design and simulate a linear outer-loop control law v for the system given by Equation (10.70) so that the link angle $y_1(t)$ follows a desired trajectory $y_1^d(t) = \theta_\ell^d(t) = \sin 8t$. Use various techniques such as pole placement, linear quadratic optimal control, etc.

10-14 Consider again a single-link manipulator (either rigid or elastic joint). Add to your equations of motion the dynamics of a permanent magnet DC-motor. What can you say now about feedback linearizability of the system?

10-15 What happens to the inverse coordinate transformation given by Equation (10.73) as the joint stiffness $k \to \infty$? Give a physical interpretation. Use this to show that the system given by Equation (10.54) reduces to the equation governing the rigid joint manipulator in the limit as $k \to \infty$.

10-16 Consider the single-link manipulator with elastic joint of Figure 10.5 but suppose that the spring characteristic is nonlinear, that is, suppose that the spring force F is given by $F = \phi(q_1 - q_2)$, where ϕ is a diffeomorphism. Derive conditions under which the system is feedback linearizable and carry out the details of the feedback linearizing transformation. Specialize the result to the case of a cubic spring characteristic $\phi = k_1(q_1 - q_2) + (q_1 - q_2)^3$. The cubic spring characteristic is a more accurate description for many manipulators than is the linear spring, especially for elasticity arising from gear flexibility.

10-17 Consider again the single link flexible joint robot given by Equation (10.54) and suppose that only the link angle q_1 is measurable. Design an observer to estimate the full state vector, $x = [q_1, \dot{q}_1, q_2, \dot{q}_2]^T$.

Hint: Set $y = q_1 = Cx$ and show that the system can be written in state space as

$$\dot{x} = Ax + bu + \phi(y)$$

where $\phi(y)$ is a nonlinear function depending only on the output y. Then a **linear observer with output injection** can be designed as

$$\dot{\hat{x}} = A\hat{x} + bu + \phi(y) + L(y - C\hat{x})$$

10-18 Fill in the details in Example 10.7 showing that the constraints are nonholonomic.

10-19 Fill in the details in Example 10.8 necessary to derive the vector fields g_1 and g_2 and show that the constraints are nonholonomic.

10-20 Carry out the calculations necessary to show that the equations of motion for the satellite with reaction wheels is given by Equation (10.118).

10-21 Show that the distribution $\Delta = \text{span}(g_1, g_2, g_3)$ for the satellite model given by Equation (10.118) is involutive of rank 3.

10-22 Using Chow's theorem, show that the satellite with reaction wheels described by Equation (10.118) is controllable as long as any two of the three reaction wheels are functioning.

10-23 Complete the details of Example 10.12 showing that the control law given by Equations (10.120) and (10.120) applied to the system given by Equation (10.119) moves the state from $(0, 0, 0)$ to $(0, 0, 10)$.

NOTES AND REFERENCES

A rigorous treatment of differential geometry can be found, for example, in a number of texts, for example, [12] or [120]. A comprehensive treatment of differential geometric methods in control is [56]. For specific applications in robotics of these advanced methods, the reader is referred to [93] and [25].

Our treatment of feedback linearization for single-input, affine, nonlinear systems follows closely the pioneering result of Su [129]. The first application of the method of feedback linearization for the single-link flexible joint robot appeared in Marino and Spong [85]. The corresponding result for the case of n-link flexible joint robots is due to Spong [122]. Dynamic feedback linearization for flexible joint robots was treated in DeLuca [80].

A more complete treatment of the control of nonholonomic systems, including mobile robots, can be found in [93] and [25]. The problem of designing nonlinear observers is treated in [67] and [68].

Chapter 11

COMPUTER VISION

If a robot is to interact with its environment, then the robot must be able to sense its environment. Computer vision is one of the most powerful sensing modalities that currently exist. Therefore, in this chapter we present a number of basic concepts from the field of computer vision. It is not our intention here to cover the now vast field of computer vision. Rather, we aim to present a number of basic techniques that are applicable to the highly constrained problems that often present themselves in industrial applications. The material in this chapter, when combined with the material of previous chapters, should enable the reader to implement a rudimentary vision-based robotic manipulation system. For example, using techniques presented in this chapter, one could design a system that locates objects on a conveyor belt and determines the positions and orientations of those objects. This information could then be used in conjunction with the inverse kinematic solution for the robot to enable it to grasp these objects.

We begin by examining the geometry of the image formation process. This will provide us with the fundamental geometric relationships between objects in the world and their projections in an image. We then describe a calibration process that can be used to determine the values for the various camera parameters that appear in these relationships. We then consider image segmentation, the problem of dividing the image into distinct regions corresponding to the background and to objects in the scene. When there are multiple objects in the scene, it is often useful to deal with them individually; therefore, we present an approach to component labelling. Finally, we describe how to compute the positions and orientations of objects in the image.

11.1 THE GEOMETRY OF IMAGE FORMATION

A digital image is a two-dimensional array whose elements are called **pixels** (derived from *picture element*). In this chapter, we will denote by *Image* the array of dimension $N_{rows} \times N_{cols}$ that contains the image. The image is formed by focusing light onto a two-dimensional array of sensing elements, and each pixel's value corresponds to the intensity of the light incident on a particular sensing element. A lens with focal length λ is used to focus the light onto the sensing array, which is often composed of CCD (charge-coupled device) sensors. The lens and sensing array are packaged together in a camera, which is connected to a digitizer or frame grabber. In the case of analog cameras, the digitizer converts the analog video signal that is output by the camera into discrete values that are then transferred to the pixel array by the frame grabber. In the case of digital cameras, a frame grabber merely transfers the digital data from the camera to the pixel array.

In robotics applications, it is often sufficient to consider only the geometric aspects of image formation. Therefore, in this section we will describe only the geometry of the image formation process. We will not deal with the photometric aspects of image formation, such as issues related to depth of field, lens models, or radiometry.

We begin by assigning a coordinate frame to the imaging system. We then discuss the pinhole model of image formation, and derive the corresponding equations relating the coordinates of a point in the world to its image coordinates. Finally, we describe camera calibration, the process by which all of the relevant parameters associated with the imaging process can be determined.

11.1.1 The Camera Coordinate Frame

In order to simplify many of the equations of this chapter, it is useful to express the coordinates of objects relative to a camera centered coordinate frame. For this purpose we define the camera coordinate frame as follows. We define the image plane as the plane that contains the sensing array. The axes x_c and y_c form a basis for the image plane and are typically taken to be parallel to the horizontal and vertical axes (respectively) of the image. The axis z_c is perpendicular to the image plane and aligned with the optical axis of the lens, that is, it passes through the focal center of the lens. The origin of the camera frame is located at a distance λ behind the image plane. This point is also referred to as the **center of projection**. The point at which the optical axis intersects the image plane is known as the **principal point**. This coordinate frame is illustrated in Figure 11.1.

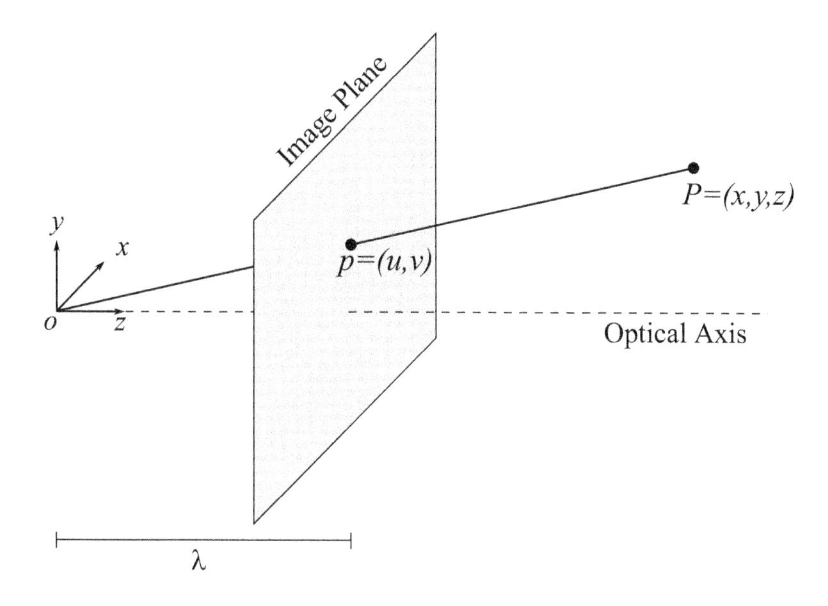

Figure 11.1: Camera coordinate frame.

With this assignment of the camera frame, any point in the image plane will have coordinates (u, v, λ). Thus, we can use (u, v) to parameterize the image plane and we will refer to (u, v) as **image plane coordinates**.

11.1.2 Perspective Projection

The image formation process is often modeled by the pinhole lens model. With this approximation, the lens is considered to be an ideal pinhole that is located at the focal center of the lens.[1] Light rays pass through this pinhole and intersect the image plane.

Let P be a point in the world with coordinates (x, y, z) relative to the camera frame. Let p denote the projection of P onto the image plane with coordinates (u, v, λ). Under the pinhole assumption, the points P, p, and the origin of the camera frame will be collinear. This is illustrated in Figure 11.1. Thus, for some unknown positive constant k we have

$$k \begin{bmatrix} x \\ y \\ z \end{bmatrix} = \begin{bmatrix} u \\ v \\ \lambda \end{bmatrix}$$

[1]Note that in our mathematical model, illustrated in Figure 11.1, we have placed the pinhole behind the image plane in order to simplify the model.

which can be rewritten as the system of equations

$$kx = u \tag{11.1}$$
$$ky = v \tag{11.2}$$
$$kz = \lambda \tag{11.3}$$

This gives $k = \lambda/z$, which can be substituted into Equations (11.1) and (11.2) to obtain

$$u = \lambda\frac{x}{z}, \qquad v = \lambda\frac{y}{z} \tag{11.4}$$

These are the well-known equations for **perspective projection**.

11.1.3 The Image Plane and the Sensor Array

As described above, the image is a discrete array of gray level values. We will denote the row and column indices for a pixel by the **pixel coordinates** (r, c). In order to relate digital images to the 3D world, we must determine the relationship between the image plane coordinates (u, v) and the pixel coordinates(r, c).

We typically define the origin of the pixel array to be located at a corner of the image rather than at the center of the image. Let the pixel array coordinates of the pixel that contains the principal point be given by (o_r, o_c). In general, the sensing elements in the camera will not be of unit size, nor will they necessarily be square. Denote by s_x and s_y the horizontal and vertical dimensions, respectively, of a pixel. Finally, it is often the case that the horizontal and vertical axes of the pixel array coordinate system point in opposite directions from the horizontal and vertical axes of the camera coordinate frame.[2] Combining these, we obtain the following relationship between image plane coordinates and pixel array coordinates

$$-\frac{u}{s_x} = (r - o_r), \qquad -\frac{v}{s_y} = (c - o_c) \tag{11.5}$$

Note that the coordinates (r, c) will be integers, since they are the discrete indices into an array that is stored in computer memory. Therefore, this relationship is only an approximation. In practice, the value of (r, c) can be obtained by truncating or rounding the ratio on the left-hand side of these equations.

[2]This is an artifact of our choice to place the center of projection behind the image plane. The directions of the pixel array axes may vary, depending on the frame grabber.

11.2 CAMERA CALIBRATION

The objective of camera calibration is to determine all of the parameters that are necessary to relate the pixel coordinates (r, c) to the (x, y, z) coordinates of a point in the camera's field of view. In other words, given the coordinates of P relative to the world coordinate frame, after we have calibrated the camera we will be able to predict (r, c), the image pixel coordinates for the projection of this point.

11.2.1 Extrinsic Camera Parameters

To this point in our derivations of the equations for perspective projection, we have dealt only with coordinates expressed relative to the camera frame. In typical robotics applications, tasks are expressed in terms of the world coordinate frame. If we know the position and orientation of the camera frame relative to the world coordinate frame we have

$$x^w = R^w_c x^c + O^w_c$$

or, if we know x^w and wish to solve for x^c,

$$x^c = R^c_w (x^w - O^w_c)$$

In the remainder of this section, to simplify notation, we will define

$$R = R^c_w, \quad T = -R^c_w O^w_c$$

and we write

$$x^c = Rx^w + T$$

Together, R and T are called the **extrinsic camera parameters.**

Cameras are typically mounted on tripods or on mechanical positioning units. In the latter case, a popular configuration is the pan/tilt head. A pan/tilt head has two degrees of freedom: a rotation about the world z axis and a rotation about the pan/tilt head's x axis. These two degrees of freedom are analogous to those of a human head, which can easily look up or down, and can turn from side to side. In this case, the rotation matrix R is given by

$$R = R_{z,\theta} R_{x,\alpha}$$

where θ is the pan angle and α is the tilt angle. More precisely, θ is the angle between the world x-axis and the camera x-axis, about the world z-axis, while α is the angle between the world z-axis and the camera z-axis, about the camera x-axis.

11.2.2 Intrinsic Camera Parameters

The mapping from 3D world coordinates to pixel coordinates is obtained by combining Equations (11.4) and (11.5) to obtain

$$r = -\frac{\lambda}{s_x}\frac{x}{z} + o_r, \qquad c = -\frac{\lambda}{s_y}\frac{y}{z} + o_c \tag{11.6}$$

Thus, once we know the values of the parameters $\lambda, s_x, o_r, s_y, o_c$ we can determine (r, c) from (x, y, z), where (x, y, z) are coordinates relative to the camera frame. In fact, we don't need to know all of λ, s_x, s_y; it is sufficient to know the ratios

$$f_x = \frac{\lambda}{s_x} \qquad f_y = \frac{\lambda}{s_y}$$

These parameters f_x, o_r, f_y, o_c are known as the intrinsic parameters of the camera. They are constant for a given camera and do not change when the camera moves.

11.2.3 Determining the Camera Parameters

We will first determine the parameters associated with the image center and then solve for the remaining parameters.

Of all the camera parameters, o_r, o_c (the image pixel coordinates of the principal point) are the easiest to determine. This can be done by using the idea of vanishing points. Although a full treatment of vanishing points is beyond the scope of this text, the idea is simple: a set of parallel lines in the world will project onto image lines that intersect at a single point, and this intersection point is known as a **vanishing point**. The vanishing points for three mutually orthogonal sets of parallel lines in the world will define a triangle in the image. The orthocenter of this triangle (that is, the point at which the three altitudes intersect) is the image principal point (Problem 11-9). Thus, a simple way to compute the principal point is to position a cube in the workspace, find the edges of the cube in the image (this will produce the three sets of mutually orthogonal parallel lines), compute the intersections of the image lines that correspond to each set of parallel lines in the world, and determine the orthocenter for the resulting triangle.

Once we know the principal point, we proceed to determine the remaining camera parameters. This is done by constructing a linear system of equations in terms of the known coordinates of points in the world and the pixel coordinates of their projections in the image. The unknowns in this system are the camera parameters. The first step is to acquire a data set of the form $\{r_i, c_i, x_i, y_i, z_i\}$ for $i = 1 \cdots N$, in which r_i, c_i are the image pixel coordinates of the projection of a point in the world with coordinates x_i, y_i, z_i relative to the world coordinate frame. This acquisition is often done manually, for example, by placing a small bright light at known (x, y, z) coordinates in the world and then hand selecting the corresponding image point.

Once we have acquired the data set, we proceed to set up the linear system of equations. The extrinsic parameters of the camera are given by

$$R = \begin{bmatrix} r_{11} & r_{12} & r_{13} \\ r_{21} & r_{22} & r_{23} \\ r_{31} & r_{32} & r_{33} \end{bmatrix}, \quad T = \begin{bmatrix} T_x \\ T_y \\ T_z \end{bmatrix}$$

With respect to the camera frame, the coordinates of a point in the world are thus given by

$$\begin{aligned} x^c &= r_{11}x + r_{12}y + r_{13}z + T_x \\ y^c &= r_{21}x + r_{22}y + r_{23}z + T_y \\ z^c &= r_{31}x + r_{32}y + r_{33}z + T_z \end{aligned}$$

Combining these three equations with Equation (11.6) we obtain

$$r - o_r = -f_x \frac{x^c}{z^c} = -f_x \frac{r_{11}x + r_{12}y + r_{13}z + T_x}{r_{31}x + r_{32}y + r_{33}z + T_z} \tag{11.7}$$

$$c - o_c = -f_y \frac{y^c}{z^c} = -f_y \frac{r_{21}x + r_{22}y + r_{23}z + T_y}{r_{31}x + r_{32}y + r_{33}z + T_z} \tag{11.8}$$

Since we know the coordinates of the principal point, we can simplify these equations by using the coordinate transformation

$$r \leftarrow r - o_r, \qquad c \leftarrow c - o_c$$

We now write the two transformed projection equations as functions of the unknown variables $r_{ij}, T_x, T_y, T_z, f_x, f_y$. This is done by solving Equations (11.7) and (11.8) for z^c, and setting the resulting equations to be equal to one another. In particular, for the data points r_i, c_i, x_i, y_i, z_i we have

$$r_i f_y (r_{21}x_i + r_{22}y_i + r_{23}z_i + T_y) = c_i f_x (r_{11}x_i + r_{12}y_i + r_{13}z_i + T_x)$$

Defining $\alpha = f_x / f_y$, we can rewrite this as

$$r_i r_{21} x_i + r_i r_{22} y_i + r_i r_{23} z_i + r_i T_y - \alpha c_i r_{11} x_i - \alpha c_i r_{12} y_i - \alpha c_i r_{13} z_i - \alpha c_i T_x = 0$$

We can combine the N such equations into the matrix equation

$$Ax = 0 \tag{11.9}$$

in which

$$A = \begin{bmatrix} r_1x_1 & r_1y_1 & r_1z_1 & r_1 & -c_1x_1 & -c_1y_1 & -c_1z_1 & -c_1 \\ r_2x_2 & r_2y_2 & r_2z_2 & r_2 & -c_2x_2 & -c_2y_2 & -c_2z_2 & -c_2 \\ \vdots & \vdots & \vdots & \vdots & \vdots & \vdots & \vdots & \vdots \\ r_Nx_N & r_Ny_N & r_Nz_N & r_N & -c_Nx_N & -c_Ny_N & -c_Nz_N & -c_N \end{bmatrix}$$

and

$$x = \begin{bmatrix} r_{21} \\ r_{22} \\ r_{23} \\ T_y \\ \alpha r_{11} \\ \alpha r_{12} \\ \alpha r_{13} \\ \alpha T_x \end{bmatrix}$$

If $\bar{x} = [\bar{x}_1, \ldots, \bar{x}_8]^T$ is a solution for Equation (11.9) we only know that this solution is some scalar multiple of the desired solution x, namely,

$$\bar{x} = k[r_{21}, r_{22}, r_{23}, T_y, \alpha r_{11}, \alpha r_{12}, \alpha r_{13}, \alpha T_x]^T$$

in which k is an unknown scale factor.

In order to solve for the true values of the camera parameters, we can exploit constraints that arise from the fact that R is a rotation matrix. In particular,

$$(\bar{x}_1^2 + \bar{x}_2^2 + \bar{x}_3^2)^{\frac{1}{2}} = (k^2(r_{21}^2 + r_{22}^2 + r_{23}^2))^{\frac{1}{2}} = |k|$$

and likewise

$$(\bar{x}_5^2 + \bar{x}_6^2 + \bar{x}_7^2)^{\frac{1}{2}} = (\alpha^2 k^2(r_{21}^2 + r_{22}^2 + r_{23}^2))^{\frac{1}{2}} = \alpha|k|$$

Note that by definition, $\alpha > 0$.

Our next task is to determine the sign of k. Using Equation (11.6) we see that $rx^c < 0$ (recall that we have used the coordinate transformation $r \leftarrow r - o_r$). Therefore, we choose k such that $r(r_{11}x + r_{12}y + r_{13}z + T_x) < 0$.

At this point we know the values for $k, \alpha, r_{21}, r_{22}, r_{23}, r_{11}, r_{12}, r_{13}, T_x, T_y$, and all that remains is to determine T_z, f_x, f_y, since the third column of

R can be determined as the vector cross product of its first two columns. Since $\alpha = f_x/f_y$, we need only determine T_z and f_x. Returning again to the projection equations, we can write

$$r = -f_x \frac{x^c}{z^c} = -f_x \frac{r_{11}x + r_{12}y + r_{13}z + T_x}{r_{31}x + r_{32}y + r_{33}z + T_z}$$

Using an approach similar to that used above to solve for the first eight parameters, we can write this as the linear system

$$r(r_{31}x + r_{32}y + r_{33}z + T_z) = -f_x(r_{11}x + r_{12}y + r_{13}z + T_x)$$

which can easily be solved for T_z and f_x.

11.3 SEGMENTATION BY THRESHOLDING

Segmentation is the process by which an image is divided into meaningful components. Segmentation has been the topic of computer vision research since its earliest days, and the approaches to segmentation are far too numerous to survey here. These approaches are sometimes concerned with finding **features** in an image, such as edges, and sometimes concerned with partitioning the image into homogeneous regions (region-based segmentation). In many practical applications the goal of segmentation is merely to divide the image into two regions: one region that corresponds to an object in the scene and one region that corresponds to the background. The resulting image is called a **binary image** since each pixel belongs to one of two classes: object or background. In many industrial applications this segmentation can be accomplished by a straightforward thresholding approach. For light objects against a dark background, pixels whose gray levels are greater than the threshold are considered to belong to the object and pixels whose gray level is less than or equal to the threshold are considered to belong to the background. For dark objects against a light background, these categories are reversed.

In this section we will describe an algorithm that automatically selects a threshold. The basic idea behind the algorithm is that the pixels should be divided into two groups, background and object, and that the intensities of the pixels in a particular group should all be fairly similar. To quantify this idea, we will use some standard techniques from statistics. Thus, we begin the section with a quick review of the necessary concepts from statistics and then proceed to describe the threshold selection algorithm.

11.3.1 A Brief Statistics Review

Many approaches to segmentation exploit statistical information contained in the image. The basic premise for most of these statistical concepts is that the gray level value associated with a pixel in an image is a random variable that takes on values in the set $\{0, 1, \ldots N-1\}$. Let $P(z)$ denote the probability that a pixel has gray level value z. In general, we will not know this probability, but we can estimate it with the use of a **histogram**. A histogram is an array H that encodes the number of occurrences of each gray level value in the image. In particular, the entry $H[z]$ is the number of times gray level value z occurs in the image. Thus, $0 \leq H[z] \leq N_{rows} \times N_{cols}$ for all z. A simple algorithm to compute the histogram for an image is as follows.

```
1. FOR i = 0 TO N - 1
2.        H[i] ← 0
3. FOR r = 0 TO N_rows - 1
4.        FOR c = 0 TO N_cols - 1
5.               Index ← Image[r, c]
6.               H[Index] ← H[Index] + 1
```

Given the histogram for the image, we estimate the probability that a pixel will have gray level z by

$$P(z) = \frac{H[z]}{N_{rows} \times N_{cols}} \tag{11.10}$$

Thus, the image histogram is a scaled version of our approximation of P.

Given P we can compute the average or **mean** value of the gray level values in the image. We denote the mean by μ and compute it as

$$\mu = \sum_{z=0}^{N-1} z P(z) \tag{11.11}$$

In many applications the image will consist of one or more objects against some background. In such applications, it is often useful to compute the mean for each object in the image and also for the background. This computation can be effected by constructing individual histogram arrays for each object and for the background in the image. If we denote by H_i the histogram for the ith object in the image, where $i = 0$ denotes the background, the mean for the ith object is given by

$$\mu_i = \sum_{z=0}^{N-1} z \frac{H_i[z]}{\sum_{z=0}^{N-1} H_i[z]} \tag{11.12}$$

which is a straightforward generalization of Equation (11.11). The term

$$\frac{H_i[z]}{\sum_{z=0}^{N-1} H_i[z]}$$

is in fact an estimate of the probability that a pixel will have gray level value z given that the pixel is a part of object i in the image. For this reason, μ_i is sometimes called a **conditional mean**.

The mean conveys useful but limited information about the distribution of gray level values in an image. For example, if half of the pixels have gray value 127 and the remaining half have gray value 128, the mean will be $\mu = 127.5$. Likewise, if half or the pixels have gray value 255 and the remaining half have gray value 0, the mean will be $\mu = 127.5$. Clearly, these two images are very different, but this difference is not reflected by the mean. One way to capture this difference is to compute the average deviation of gray values from the mean. This average would be small for the first example and large for the second. We could, for example, use the average value of $|z - \mu|$. It is more convenient mathematically to use the square of this value instead. The resulting quantity is known as the **variance**, which is denoted by σ^2 and is defined by

$$\sigma^2 = \sum_{z=0}^{N-1} (z - \mu)^2 P(z) \tag{11.13}$$

As with the mean, we can also compute the conditional variance σ_i^2 for each object in the image as

$$\sigma_i^2 = \sum_{z=0}^{N-1} (z - \mu_i)^2 \frac{H_i[z]}{\sum_{z=0}^{N-1} H_i[z]}$$

11.3.2 Automatic Threshold Selection

We are now prepared to develop an automatic threshold selection algorithm. We will assume that the image consists of an object and a background and that the background pixels have gray level values less than or equal to some threshold while the object pixels are above the threshold. Thus, for a given threshold value z_t we divide the image pixels into two groups: those pixels with gray level value $z \leq z_t$ and those pixels with gray level value $z > z_t$.

We can compute the mean and variance for each of these groups using the equations of Section 11.3.1. Clearly, the conditional means and variances depend on the choice of z_t, since it is the choice of z_t that determines which pixels will belong to each of the two groups. The approach that we take in this section is to determine the value for z_t that minimizes a function of the variances of these two groups of pixels.

It is convenient to rewrite the conditional means and variances in terms of the pixels in the two groups. To do this, we define $q_i(z_t)$ as the probability that a pixel in the image will belong to group i for a particular choice of threshold z_t. Since all pixels in the background have gray value less than or equal to z_t and all pixels in the object have gray value greater than z_t, we can define $q_i(z_t)$ for $i = 0, 1$ by

$$q_0(z_t) = \frac{\sum_{z=0}^{z_t} H[z]}{(N_{rows} \times N_{cols})}, \quad q_1(z_t) = \frac{\sum_{z=z_t+1}^{N-1} H[z]}{(N_{rows} \times N_{cols})}$$

We now rewrite Equation (11.12) as

$$\mu_i = \sum_{z=0}^{N-1} z \frac{H_i[z]}{\sum_{z=0}^{N-1} H_i[z]} = \sum_{z=0}^{N-1} z \frac{H_i[z]/(N_{rows} \times N_{cols})}{\sum_{z=0}^{N-1} H_i[z]/(N_{rows} \times N_{cols})}$$

Using again the fact that the two pixel groups are defined by the threshold z_t, we have

$$H_0[z] = \begin{cases} (N_{rows} \times N_{cols})P(z) & z \leq z_t \\ 0 & \text{otherwise} \end{cases}$$

and

$$H_1[z] = \begin{cases} 0 & z \leq z_t \\ (N_{rows} \times N_{cols})P(z) & \text{otherwise} \end{cases}$$

Thus, we can write the conditional means for the two groups as

$$\mu_0(z_t) = \sum_{z=0}^{z_t} z \frac{P(z)}{q_0(z_t)}, \quad \mu_1(z_t) = \sum_{z=z_t+1}^{N-1} z \frac{P(z)}{q_1(z_t)} \tag{11.14}$$

Similarly, we can write the equations for the conditional variances by

$$\begin{aligned} \sigma_0^2(z_t) &= \sum_{z=0}^{z_t} (z - \mu_0(z_t))^2 \frac{P(z)}{q_0(z_t)} \\ \sigma_1^2(z_t) &= \sum_{z=z_t+1}^{N} (z - \mu_1(z_t))^2 \frac{P(z)}{q_1(z_t)} \end{aligned}$$

We now turn to the selection of z_t. If nothing is known about the true values of μ_i or σ_i^2, how can we determine the optimal value of z_t? To answer this question, recall that the variance is a measure of the average deviation of pixel intensities from the mean. Thus, if we make a good choice for z_t, we would expect that the variances $\sigma_i^2(z_t)$ would be small. This reflects the assumption that pixels belonging to the object will have intensity values that are clustered closely about μ_1 and that pixels belonging to the background will have intensity values that are clustered closely about μ_0. We could, therefore, select the value of z_t that minimizes the sum of these two variances. However, it is unlikely that the object and background will occupy the same number of pixels in the image; merely adding the variances gives both regions equal importance. A more reasonable approach is to weight the variances σ_i^2 by the probability that a pixel will belong to the corresponding region

$$\sigma_w^2(z_t) = q_0(z_t)\sigma_0^2(z_t) + q_1(z_t)\sigma_1^2(z_t)$$

The value σ_w^2 is known as the **within-group variance**. The approach that we will take minimizes this within-group variance, giving a threshold that divides the image into two groups,

At this point we could implement a threshold selection algorithm. The naive approach would be to simply iterate over all possible values of z_t and select the one for which $\sigma_w^2(z_t)$ is smallest. Such an algorithm performs an enormous amount of calculation, much of which is identical for successive candidate values of the threshold. As we will see, most of the calculations required to compute $\sigma_w^2(z_t)$ are also required to compute $\sigma_w^2(z_t + 1)$; the summations that are required change only slightly from one iteration to the next.

To develop a more efficient algorithm, we take two steps. First, we will derive the **between-group variance** σ_b^2, which depends on the within-group variance and the variance over the entire image. The between-group variance is a bit simpler to deal with than the within-group variance, and we will show that maximizing the between-group variance is equivalent to minimizing the within-group variance. Then, we will derive a recursive formulation for the between-group variance that lends itself to an efficient implementation.

To derive the between-group variance, we begin by expanding the equation for the total variance of the image and then simplifying and grouping terms. The total variance for the image σ^2 is a constant, and does not depend on the choice of threshold value. The total variance of the gray level values in the image is given by Equation (11.13), which can be rewritten as two summations, one for the background pixels, and one for the object pixels

$$\begin{aligned}
\sigma^2 &= \sum_{z=0}^{N-1} (z-\mu)^2 P(z) \\
&= \sum_{z=0}^{z_t} (z-\mu_0+\mu_0-\mu)^2 P(z) + \sum_{z=z_t+1}^{N-1} (z-\mu_1+\mu_1-\mu)^2 P(z) \\
&= \sum_{z=0}^{z_t} [(z-\mu_0)^2 + 2(z-\mu_0)(\mu_0-\mu) + (\mu_0-\mu)^2] P(z) \\
&\quad + \sum_{z=z_t+1}^{N-1} [(z-\mu_1)^2 + 2(z-\mu_1)(\mu_1-\mu) + (\mu_1-\mu)^2] P(z)
\end{aligned}$$

We have not explicitly noted the dependence on z_t here. In the remainder of this section, to simplify notation, we will refer to the group probabilities and conditional means and variances as q_i, μ_i, and σ_i^2, without explicitly noting the dependence on z_t. The final expression in the derivation above can be further simplified by examining the cross-terms

$$\begin{aligned}
&\sum (z-\mu_i)(\mu_i-\mu)P(z) \\
&\quad = \sum z\mu_i P(z) - \sum z\mu P(z) - \sum \mu_i^2 P(z) + \sum \mu_i \mu P(z) \\
&\quad = \mu_i \sum zP(z) - \mu \sum zP(z) - \mu_i^2 \sum P(z) + \mu_i\mu \sum P(z) \\
&\quad = \mu_i(\mu_i q_i) - \mu(\mu_i q_i) - \mu_i^2 q_i + \mu_i \mu q_i \\
&\quad = 0
\end{aligned}$$

in which the summations are taken for z from 0 to z_t for the background pixels (that is, $i=0$) and z from z_t+1 to $N-1$ for the object pixels (that is, $i=1$). Therefore, we can simplify our expression for σ^2 to obtain

$$\begin{aligned}
\sigma^2 &= \sum_{z=0}^{z_t} [(z-\mu_0)^2 + (\mu_0-\mu)^2] P(z) + \sum_{z=z_t+1}^{N-1} [(z-\mu_1)^2 + (\mu_1-\mu)^2] P(z) \\
&= q_0\sigma_0^2 + q_0(\mu_0-\mu)^2 + q_1\sigma_1^2 + q_1(\mu_1-\mu)^2 \\
&= \{q_0\sigma_0^2 + q_1\sigma_1^2\} + \{q_0(\mu_0-\mu)^2 + q_1(\mu_1-\mu)^2\} \\
&= \sigma_w^2 + \sigma_b^2
\end{aligned}$$

in which

$$\sigma_b^2 = q_0(\mu_0-\mu)^2 + q_1(\mu_1-\mu)^2 \tag{11.15}$$

Since σ^2 does not depend on the threshold value, minimizing σ_w^2 is equivalent to maximizing σ_b^2. This is preferable because σ_b^2 is a function only of the q_i and μ_i, and is thus simpler to compute than σ_w^2, which depends also on the σ_i^2. In fact, by expanding the squares in Equation (11.15), using the facts that $q_1 = 1 - q_0$ and $\mu = q_1\mu_0 + q_1\mu_1$, we obtain

$$\sigma_b^2 = q_0(1 - q_0)(\mu_0 - \mu_1)^2 \tag{11.16}$$

The simplest algorithm to maximize σ_b^2 is to iterate over all possible threshold values, and select the one that maximizes σ_b^2. However, as discussed above, such an algorithm performs many redundant calculations, since most of the calculations required to compute $\sigma_b^2(z_t)$ are also required to compute $\sigma_b^2(z_t + 1)$. A more efficient algorithm would reuse the computations needed for $\sigma_b^2(z_t)$ when computing $\sigma_b^2(z_t + 1)$. In particular, we will derive expressions for the necessary terms at iteration $z_t + 1$ in terms of expressions that were computed at iteration z_t. We begin with the group probabilities, and determine the recursive expression for q_0 as

$$q_0(z_t + 1) = \sum_{z=0}^{z_t+1} P(z) = P(z_t + 1) + \sum_{z=0}^{z_t} P(z) = P(z_t + 1) + q_0(z_t) \tag{11.17}$$

In this expression, $P(z_t + 1)$ can be obtained directly from the histogram array, and $q_0(z_t)$ is directly available because it was computed on the previous iteration of the algorithm. Thus, given the results from iteration z_t, very little computation is required to compute the value for q_0 at iteration $z_t + 1$.

For the conditional mean $\mu_0(z_t)$ we have

$$\begin{aligned}
\mu_0(z_t + 1) &= \sum_{z=0}^{z_t+1} z\frac{P(z)}{q_0(z_t + 1)} \\
&= \frac{(z_t + 1)P(z_t + 1)}{q_0(z_t + 1)} + \sum_{z=0}^{z_t} z\frac{P(z)}{q_0(z_t + 1)} \\
&= \frac{(z_t + 1)P(z_t + 1)}{q_0(z_t + 1)} + \frac{q_0(z_t)}{q_0(z_t + 1)} \sum_{z=0}^{z_t} z\frac{P(z)}{q_0(z_t)} \\
&= \frac{(z_t + 1)P(z_t + 1)}{q_0(z_t + 1)} + \frac{q_0(z_t)}{q_0(z_t + 1)}\mu_0(z_t)
\end{aligned} \tag{11.18}$$

Again, all of the quantities in this expression are available either from the histogram, or as the results of calculations performed at iteration z_t of the algorithm.

To compute $\mu_1(z_t+1)$, we use the relationship $\mu = q_0\mu_0 + q_1\mu_1$, which can be easily obtained using Equations (11.11) and (11.14). Thus, we have

$$\begin{aligned}\mu_1(z_t+1) &= \frac{\mu - q_0(z_t+1)\mu_0(z_t+1)}{q_1(z_t+1)} \\ &= \frac{\mu - q_0(z_t+1)\mu_0(z_t+1)}{1 - q_0(z_t+1)}\end{aligned} \tag{11.19}$$

We can now construct an efficient algorithm to automatically select a threshold that minimizes the within-group variance. This algorithm simply iterates from 0 to $N-1$ (where N is the total number of gray level values), computing q_0, μ_0, μ_1 and σ_b^2 at each iteration using the recursive formulations given in Equations (11.16), (11.17), (11.18), and (11.19). The algorithm returns the value of z_t for which σ_b^2 is largest. Figure 11.2 shows a gray level image and the binary thresholded image that results from the application of this algorithm, along with the histogram and within-group variance for the gray level image.

11.4 CONNECTED COMPONENTS

It is often the case that multiple objects will be present in a single image. When this occurs, there will be multiple connected components with gray level values that are above the threshold. In this section, we will first make precise the notion of a **connected component** and then describe an algorithm that assigns a unique label to each connected component, that is, all pixels within a single connected component have the same label, but pixels in different connected components have different labels.

In order to define what is meant by a connected component, it is first necessary to define what is meant by connectivity. For our purposes, it is sufficient to say that a pixel with image pixel coordinates (r, c) is adjacent to four pixels, those with image pixel coordinates $(r-1, c)$, $(r+1, c)$, $(r, c+1)$, and $(r, c-1)$. In other words, each image pixel, except those at the edges of the image, has four neighbors: the pixel directly above, directly below, directly to the right and directly to the left of the pixel. This relationship is sometimes referred to as **4-connectivity**. Two pixels are 4-connected if they are adjacent by this definition. If we expand the definition of adjacency to include those pixels that are diagonally adjacent, that is, the pixels with coordinates $(r-1, c-1)$, $(r-1, c+1)$, $(r+1, c-1)$, and $(r+1, c+1)$, then we say that adjacent pixels are **8-connected**. In this text, we will consider only the case of 4-connectivity.

A connected component is a set of pixels, S, such that for any two pixels, say P and P' in S, there is a 4-connected path between them and this path

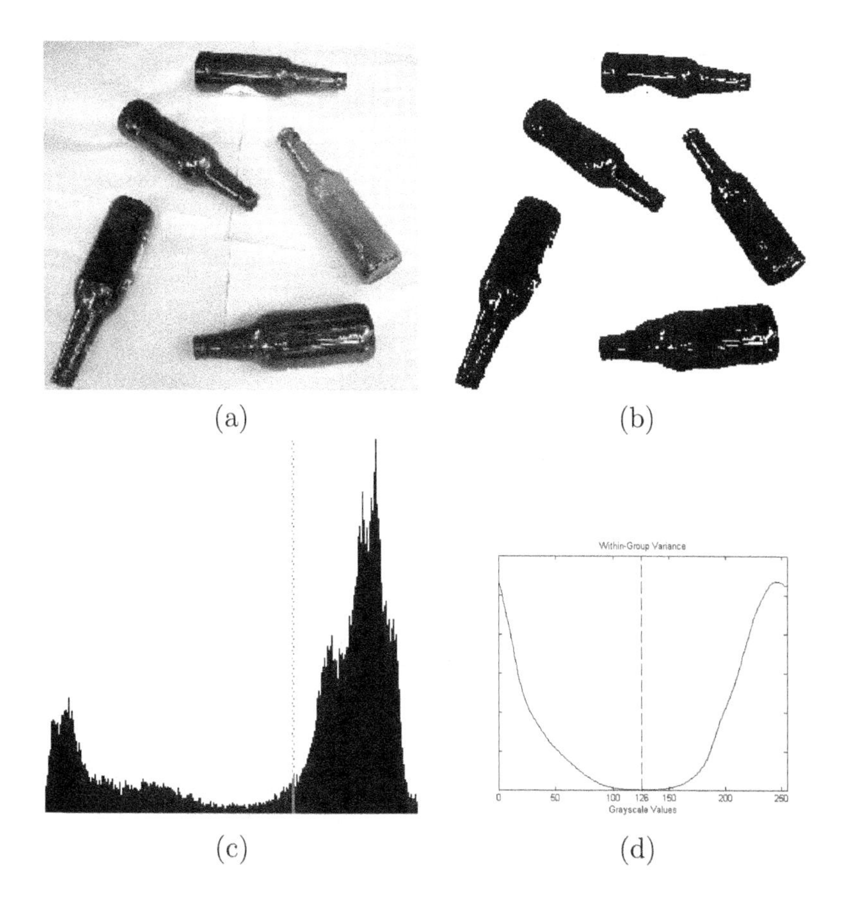

Figure 11.2: (a) An image with 256 gray levels. (b) Thresholded version of the image. (c) Histogram for the image. (d) Within-group variance for the image shown.

is contained in S. Intuitively, this definition means that it is possible to move from P to P' by "taking steps" only to adjacent pixels without ever leaving the region S. The purpose of a component labeling algorithm is to assign a unique label to each such S.

There are many component labeling algorithms that have been developed over the years. Here, we describe a simple algorithm that requires two passes over the image. This algorithm performs two raster scans of the image. A raster scan visits each pixel in the image by traversing from left to right and top to bottom, in the same way that one reads a page of text. On the first raster scan, when an object pixel P, that is, a pixel whose gray level is above the threshold value, is encountered, its previously visited neighbors, the pixel immediately above and the pixel immediately to the left of P, are examined. If they have gray value that is below the threshold, so that they are background pixels, a new label is given to P. This is done by using a global counter that is initialized to zero and is incremented each time a new label is needed. If either of these two neighbors have already received labels, then P is given the smaller of these, and in the case when both of the neighbors have received labels, an equivalence is noted between those two labels. For example, in Figure 11.3, after the first raster scan labels $(2, 3, 4)$ are noted as equivalent. In the second raster scan, each pixel's label is replaced by the smallest label to which it is equivalent. Thus, in the example of Figure 11.3, at the end of the second raster scan labels 3 and 4 have been replaced by the label 2.

After this algorithm has assigned labels to the components in the image, it is not necessarily the case that the labels will be the consecutive integers $(1, 2, \dots)$. Therefore, a second stage of processing is sometimes used to relabel the components to achieve this. In other cases, it is desirable to give each component a label that is very different from the labels of the other components. For example, if the component labeled image is to be displayed, it is useful to increase the contrast, so that distinct components will actually appear distinct in the image. A component with the label 2 will appear almost indistinguishable from a component with label 3 if the component labels are used as pixel gray values in the displayed component labeled image. The results of applying this process to the image in Figure 11.2 are shown in Figure 11.4.

When there are multiple connected object components, it is often useful to process each component individually. For example, we might like to compute the sizes of the various components. For this purpose, it is useful to introduce the **indicator function** for a component. The indicator function for component i, denoted by $\mathcal{I}_i$, is a function that takes on the value 1 for pixels that are contained in component i and the value 0 for all other pixels:

$$\mathcal{I}_i(r, c) = \begin{cases} 1 & : \text{ pixel } r, c \text{ is contained in component } i \\ 0 & : \text{ otherwise} \end{cases}$$

We will make use of the indicator function below, when we discuss computing statistics associated with the various objects in the image.

11.5 POSITION AND ORIENTATION

The ultimate goal of a robotic system is to manipulate objects in the world. In order to achieve this, it is necessary to know the positions and orientations

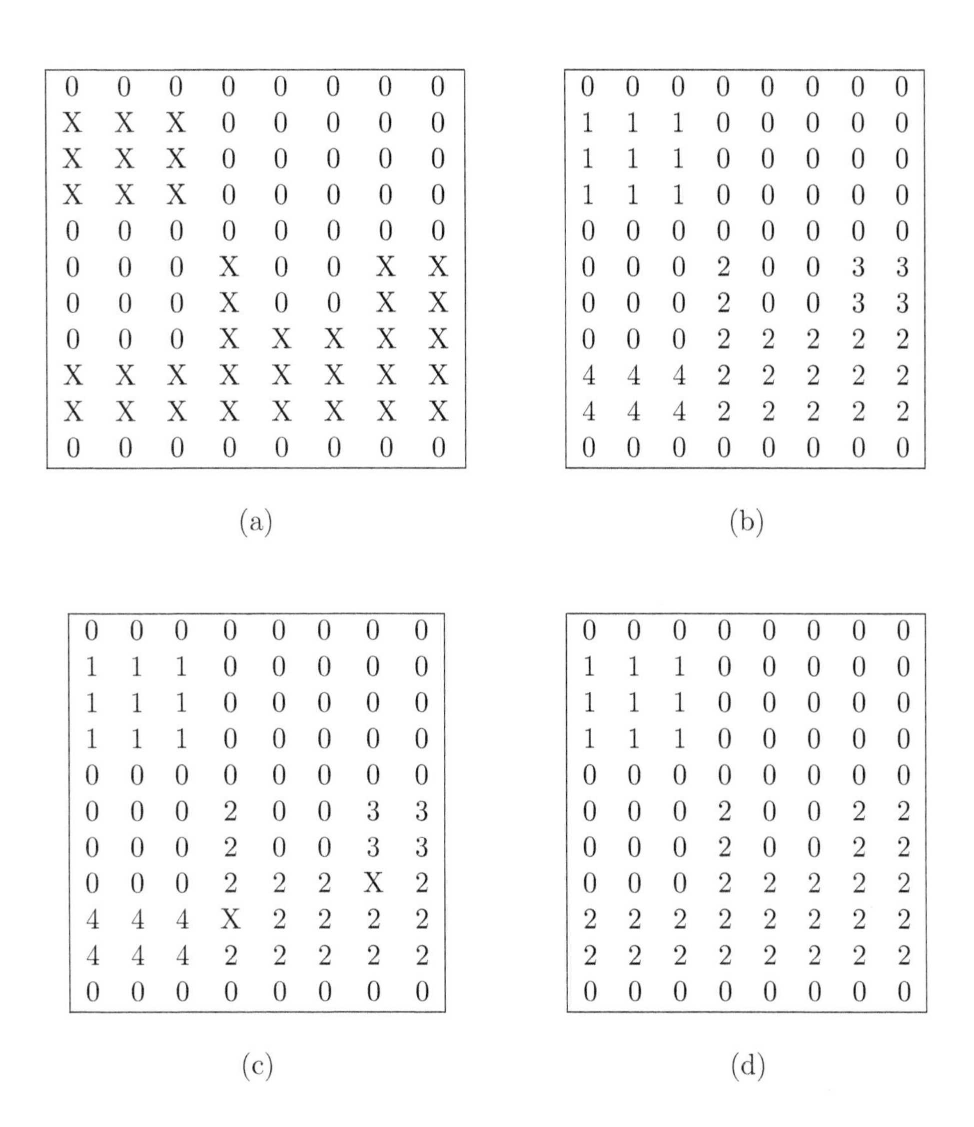

(a)

0	0	0	0	0	0	0	0
X	X	X	0	0	0	0	0
X	X	X	0	0	0	0	0
X	X	X	0	0	0	0	0
0	0	0	0	0	0	0	0
0	0	0	X	0	0	X	X
0	0	0	X	0	0	X	X
0	0	0	X	X	X	X	X
X	X	X	X	X	X	X	X
X	X	X	X	X	X	X	X
0	0	0	0	0	0	0	0

(b)

0	0	0	0	0	0	0	0
1	1	1	0	0	0	0	0
1	1	1	0	0	0	0	0
1	1	1	0	0	0	0	0
0	0	0	0	0	0	0	0
0	0	0	2	0	0	3	3
0	0	0	2	0	0	3	3
0	0	0	2	2	2	2	2
4	4	4	2	2	2	2	2
4	4	4	2	2	2	2	2
0	0	0	0	0	0	0	0

(c)

0	0	0	0	0	0	0	0
1	1	1	0	0	0	0	0
1	1	1	0	0	0	0	0
1	1	1	0	0	0	0	0
0	0	0	0	0	0	0	0
0	0	0	2	0	0	3	3
0	0	0	2	0	0	3	3
0	0	0	2	2	2	X	2
4	4	4	X	2	2	2	2
4	4	4	2	2	2	2	2
0	0	0	0	0	0	0	0

(d)

0	0	0	0	0	0	0	0
1	1	1	0	0	0	0	0
1	1	1	0	0	0	0	0
1	1	1	0	0	0	0	0
0	0	0	0	0	0	0	0
0	0	0	2	0	0	2	2
0	0	0	2	0	0	2	2
0	0	0	2	2	2	2	2
2	2	2	2	2	2	2	2
2	2	2	2	2	2	2	2
0	0	0	0	0	0	0	0

Figure 11.3: The image in (a) is a simple binary image. Background pixels are denoted by 0 and object pixels are denoted by X. Image (b) shows the assigned labels after the first raster scan. In image (c) an X denotes those pixels at which an equivalence is noted during the first raster scan. Image (d) shows the final component labeled image.

Figure 11.4: The image of Figure 11.2 after connected components have been labeled.

of the objects that are to be manipulated. In this section, we address the problem of determining the position and orientation of objects *in the image.* Once the camera has been calibrated, it is then possible to use these image positions and orientations to infer the 3D positions and orientations of the objects. In general, this problem of inferring the 3D position and orientation from image measurements can be a difficult problem; however, for many cases that are faced by industrial robots we can obtain adequate solutions. For example, when grasping parts from a conveyor belt, the depth z is fixed and the perspective projection equations can be inverted if z is known.

We begin the section with a general discussion of moments, which will be used in the computation of both position and orientation of objects in the image.

11.5.1 Moments

Moments are functions defined on the image that can be used to summarize various aspects of the shape and size of objects in the image. The i, j moment for the kth object, denoted by $m_{ij}(k)$, is defined by

$$m_{ij}(k) = \sum_{r,c} r^i c^j \mathcal{I}_k(r, c)$$

From this definition, it is evident that m_{00} is merely the number of pixels in the object. The order of a moment is defined to be the sum $i + j$. The first order moments are of particular interest when computing the centroid of an object, and they are given by

$$m_{10}(k) = \sum_{r,c} r\mathcal{I}_k(r, c), \qquad m_{01}(k) = \sum_{r,c} c\mathcal{I}_k(r, c)$$

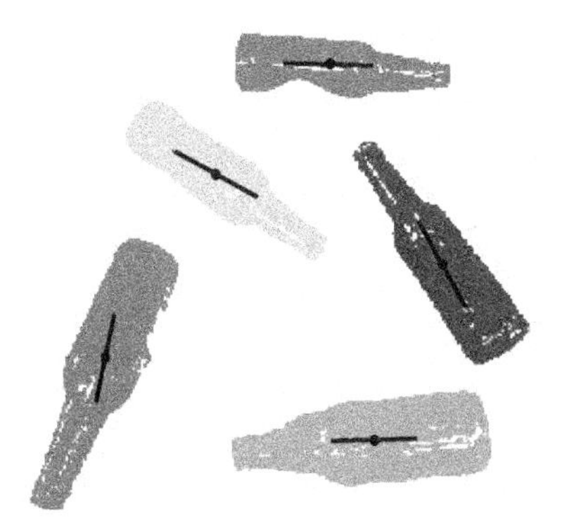

Figure 11.5: The segmented component-labeled image of Figure 11.2 showing the centroids and orientation of each component.

11.5.2 The Centroid of an Object and Central Moments

It is convenient to define the position of an object to be the object's center of mass or **centroid**. By definition, the center of mass of an object is that point $(\bar{r}, \bar{c})$ such that, if all of the object's mass were concentrated at $(\bar{r}, \bar{c})$, the first moments would not change. Thus, we have

$$\sum_{r,c} \bar{r}_i \mathcal{I}_i(r,c) = \sum_{r,c} r \mathcal{I}_i(r,c) \quad \Rightarrow \quad \bar{r}_i = \frac{\sum_{r,c} r \mathcal{I}_i(r,c)}{\sum_{r,c} \mathcal{I}_i(r,c)} = \frac{m_{10}(i)}{m_{00}(i)}$$

$$\sum_{r,c} \bar{c}_i \mathcal{I}_i(r,c) = \sum_{r,c} c \mathcal{I}_i(r,c) \quad \Rightarrow \quad \bar{c}_i = \frac{\sum_{r,c} c \mathcal{I}_i(r,c)}{\sum_{r,c} \mathcal{I}_i(r,c)} = \frac{m_{01}(i)}{m_{00}(i)}$$

Figure 11.5 shows the centroids for the connected components of the image of Figure 11.2.

It is often useful to compute moments with respect to the object center of mass. By doing so, we obtain characteristics that are invariant with respect to translation of the object. These moments are called **central moments**. The i, j central moment for the kth object is defined by

$$C_{ij}(k) = \sum_{r,c} (r - \bar{r}_k)^i (c - \bar{c}_k)^j \mathcal{I}_k(r,c) \tag{11.20}$$

in which $(\bar{r}_k, \bar{c}_k)$ are the coordinates for the centroid of the kth object.

11.5.3 The Orientation of an Object

We will define the orientation of an object in the image to be the orientation of an axis passing through the object such that the second moment of the object about that axis is minimal. This axis is merely the two-dimensional equivalent of the axis of least inertia.

For a given line in the image, the second moment of the object about that line is given by

$$\mathcal{L} = \sum_{r,c} d^2(r,c)\mathcal{I}(r,c)$$

in which $d(r,c)$ is the minimum distance from the pixel with coordinates (r,c) to the line. Our task is to minimize $\mathcal{L}$ with respect to all possible lines in the image plane. To do this, we will use the ρ, θ parameterization of lines and compute the partial derivatives of $\mathcal{L}$ with respect to ρ and θ. We find the minimum by setting these partial derivatives to zero.

With the ρ, θ parameterization, a line consists of all those points x, y that satisfy

$$x\cos\theta + y\sin\theta - \rho = 0$$

Thus, $(\cos\theta, \sin\theta)$ gives the unit normal to the line and ρ gives the perpendicular distance to the line from the origin. This parameterization is illustrated in Figure 11.6. Under this parameterization the distance from the line to the point with coordinates (r,c) is given by

$$d(r,c) = r\cos\theta + c\sin\theta - \rho$$

Thus, the value $\mathcal{L}^\star$ that minimizes $\mathcal{L}$ is given by

$$\mathcal{L}^\star = \min_{\rho,\theta} \sum_{r,c} (r\cos\theta + c\sin\theta - \rho)^2 \mathcal{I}(r,c)$$

We compute the partial derivative of $\mathcal{L}$ with respect to ρ as

$$\begin{aligned}
\frac{\partial}{\partial\rho}\mathcal{L} &= \frac{\partial}{\partial\rho}\sum_{r,c}(r\cos\theta + c\sin\theta - \rho)^2\mathcal{I}(r,c) \\
&= -2\cos\theta\sum_{r,c} r\mathcal{I}(r,c) - 2\sin\theta\sum_{r,c} c\mathcal{I}(r,c) + 2\rho\sum_{r,c}\mathcal{I}(r,c) \\
&= -2m_{00}(\bar{r}\cos\theta + \bar{c}\sin\theta - \rho) \qquad (11.21)
\end{aligned}$$

Now, setting this to zero we obtain

$$\bar{r}\cos\theta + \bar{c}\sin\theta - \rho = 0$$

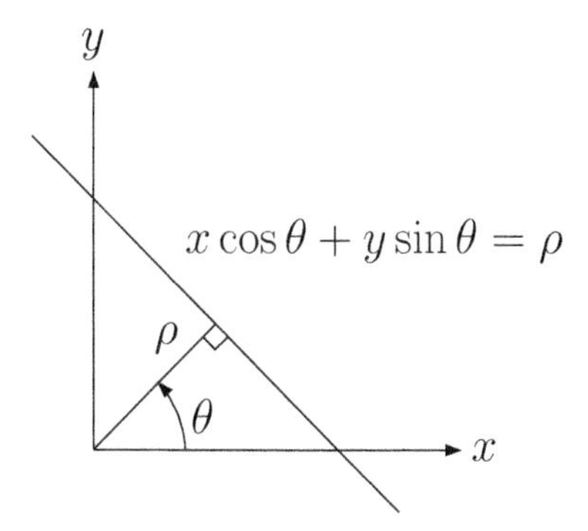

Figure 11.6: The ρ, θ parameterization of a line

But this is just the equation of a line that passes through the point $(\bar{r}, \bar{c})$, and therefore we conclude that the inertia is minimized by a line that passes through the center of mass. We can use this knowledge to simplify the remaining computations. In particular, define the new coordinates (r', c') as

$$r' = r - \bar{r}, \qquad c' = c - \bar{c}$$

The line that minimizes $\mathcal{L}$ passes through the point $r' = 0, c' = 0$, and therefore its equation can be written as

$$r' \cos\theta + c' \sin\theta = 0$$

Before computing the partial derivative of $\mathcal{L}$ (expressed in the new coordinate system) with respect to θ, it is useful to perform some simplifications.

$$\begin{aligned}
\mathcal{L} &= \sum_{r,c} (r' \cos\theta + c' \sin\theta)^2 \mathcal{I}(r, c) \\
&= \cos^2\theta \sum_{r,c} (r')^2 \mathcal{I}(r, c) + 2\cos\theta \sin\theta \sum_{r,c} (r'c') \mathcal{I}(r, c) + \sin^2\theta \sum_{r,c} (c')^2 \mathcal{I}(r, c) \\
&= C_{20} \cos^2\theta + 2C_{11} \cos\theta \sin\theta + C_{02} \sin^2\theta
\end{aligned}$$

in which the C_{ij} are the central moments given in Equation (11.20). Note that the central moments depend on neither ρ nor θ.

The final set of simplifications that we will make all rely on the double angle identities:

$$\begin{aligned}
cos^2\theta &= \frac{1}{2} + \frac{1}{2}\cos 2\theta \\
sin^2\theta &= \frac{1}{2} - \frac{1}{2}\cos 2\theta \\
cos\theta \sin\theta &= \frac{1}{2}\sin 2\theta
\end{aligned}$$

Substituting these into our expression for $\mathcal{L}$ we obtain

$$\mathcal{L} = \frac{1}{2}(C_{20} + C_{02}) + \frac{1}{2}(C_{20} - C_{02})\cos 2\theta + C_{11}\sin 2\theta$$

It is now easy to compute the partial derivative with respect to θ:

$$\frac{\partial}{\partial \theta}\mathcal{L} = -(C_{20} - C_{02})\sin 2\theta + 2C_{11}\cos 2\theta$$

and setting this to zero we obtain

$$\tan 2\theta = \frac{2C_{11}}{C_{20} - C_{02}}$$

Figure 11.5 shows the orientations for the connected components of the image of Figure 11.2.

11.6 SUMMARY

In this chapter we studied basic image formation and processing. We began with the geometry of the image formation process, which is typically modeled using perspective projection. In this case, the projection onto the image plane of a point with coordinates (x, y, z) is given by the perspective projection equations

$$u = \lambda\frac{x}{z}, \qquad v = \lambda\frac{y}{z}$$

The actual discrete image array coordinates are related to the u, v coordinates by

$$u = -s_x(r - o_r), \quad v = -s_y(c - o_c)$$

in which the principal point has coordinates (o_r, o_c) and s_x and s_y are the horizontal and vertical dimensions, respectively, of a pixel. The parameters o_r, o_c, s_x and s_y, along with the focal length of the camera imaging system, are known as intrinsic camera parameters. The position and orientation of the camera frame with respect to the world coordinate frame comprise the extrinsic parameters. All of these parameters can be estimated by the process of camera calibration.

Segmentation is the process of partitioning an image into foreground and background components. We described a threshold-based segmentation method, in which the threshold is automatically selected by maximizing the between-group variance given by

$$\sigma_b^2 = q_0(1 - q_0)(\mu_0 - \mu_1)^2$$

in which q_0 is the fraction of background pixels, and μ_0 and μ_1 are the mean gray level values of the background and foreground pixels, respectively. This maximization can be achieved by an efficient recursive scheme.

Once the image has been segmented into foreground and background, pixels can be grouped together using a connected components algorithm. If the segmentation algorithm is effective, this gives an image in which each individual object has a unique label. It is then possible to compute properties of the objects using moments. In this chapter, we demonstrated how first and second order moments can be used to determine position and orientation of a 2D object in an image. The moments for the kth object are given by

$$m_{ij}(k) = \sum_{r,c} r^i c^j \mathcal{I}_k(r, c)$$

The centroid of an object has coordinates given by

$$\bar{r}_i = \frac{m_{10}(i)}{m_{00}(i)}, \qquad \bar{c}_i = \frac{m_{01}(i)}{m_{00}(i)}$$

The i, j central moments for an object are defined with respect to the object's centroid. For the kth object the central moments are defined by

$$C_{ij}(k) = \sum_{r,c} (r - \bar{r}_k)^i (c - \bar{c}_k)^j \mathcal{I}_k(r, c)$$

The orientation of an object can be determined from its central moments as

$$\tan 2\theta = \frac{2C_{11}}{C_{20} - C_{02}}$$

PROBLEMS

11-1 For a camera with focal length $\lambda = 10$, find the image plane coordinates for the 3D points whose coordinates with respect to the camera frame are given below. Indicate if any of these points will not be visible to a physical camera.

1. $(25, 25, 50)$
2. $(-25, -25, 50)$
3. $(20, 5, -50)$
4. $(15, 10, 25)$
5. $(0, 0, 50)$

6. $(0, 0, 100)$

11-2 Repeat problem 11-1 for the case when the coordinates of the points are given with respect to the world frame. Suppose the optical axis of the camera is aligned with the world x-axis, the camera x-axis is parallel to the world y-axis, and the center of projection has coordinates $(0, 0, 100)$.

11-3 A stereo camera system consists of two cameras that share a common field of view. By using two cameras, stereo vision methods can be used to compute 3D properties of the scene. Consider stereo cameras with coordinate frames $o_1x_1y_1z_1$ and $o_2x_2y_2z_2$ such that

$$H_2^1 = \begin{bmatrix} 1 & 0 & 0 & B \\ 0 & 1 & 0 & 0 \\ 0 & 0 & 1 & 0 \\ 0 & 0 & 0 & 1 \end{bmatrix}$$

Here, B is called the **baseline distance** between the two cameras. Suppose that a 3D point P projects onto these two images with image plane coordinates (u_1, v_1) in the first camera and (u_2, v_2) in the second camera. Determine the depth of the point P.

11-4 Show that the projection of a 3D line is a line in the image.

11-5 Consider two parallel lines in 3D, given parametrically by

$$\begin{bmatrix} x \\ y \\ z \end{bmatrix} = \begin{bmatrix} x_i \\ y_i \\ z_i \end{bmatrix} + \gamma u_i$$

in which $\gamma \in \mathbb{R}$, u_i is a unit vector and (x_i, y_i, z_i) is a point on the line. Show that if two lines are parallel, that is, if $u_1 = u_2$, then the projections of these two lines in an image intersect at a single point. This point is called the **vanishing point**.

11-6 Show that the vanishing points for all 3D horizontal lines must lie on the line $v = 0$ of the image plane.

11-7 Suppose the vanishing point for two parallel lines has the image coordinates (u_∞, v_∞). Show that the direction vector for the 3D line is given by

$$u = \frac{1}{\sqrt{u_\infty^2 + v_\infty^2 + \lambda^2}} \begin{bmatrix} u_\infty \\ v_\infty \\ \lambda \end{bmatrix}$$

in which λ is the focal length of the imaging system.

11-8 Two parallel lines define a plane. Consider a set of pairs of parallel lines such that the corresponding planes are all parallel. Show that the vanishing points for the images of these lines are collinear. Hint: let n be the normal vector for the parallel planes and exploit the fact that $u_i \cdot n = 0$ for the direction vector u_i associated to the ith line.

11-9 A cube has twelve edges, each of which defines a line in three space. We can group these lines into three groups, such that in each of the groups there are four parallel lines. Let (a_1, a_2, a_3), (b_1, b_2, b_3), and (c_1, c_2, c_3) be the direction vectors for these three sets of parallel lines. Each set of parallel lines gives rise to a vanishing point in the image. Let the three vanishing points be $V_a = (u_a, v_a)$, $V_b = (u_b, v_b)$, and $V_c = (u_c, c_c)$, respectively.

1. If C is the optical center of the camera, show that the three angles $\angle V_a C V_b$, $\angle V_a C V_c$ and $\angle V_b C V_c$ are each equal to $\frac{\pi}{2}$. Hint: In the world coordinate frame, the image plane is the plane $z = \lambda$.
2. Let h_a be the altitude from V_a to the line defined by V_b and V_c. Show that the plane containing both h_a and the line through points C and V_a is orthogonal to the line defined by V_b and V_c.
3. Let h_a be the altitude from V_a to the line defined by V_b and V_c, h_b the altitude from V_b to the line defined by V_a and V_c, and h_c the altitude from V_c to the line defined by V_a and V_b. We define the following three planes:
   - P_a is the plane containing both h_a and the line through points C and V_a.
   - P_b is the plane containing both h_b and the line through points C and V_b.
   - P_c is the plane containing both h_c and the line through points C and V_c.

   Show that each of these planes is orthogonal to the image plane (it is sufficient to show that P_i is orthogonal to the image plane for a specific value of i).
4. The three vanishing points V_a, V_b, V_c define a triangle, and the three altitudes h_a, h_b, h_c intersect at the orthocenter of this triangle. For this special case, where the three direction vectors are mutually orthogonal, what is the significance of this point?

11-10 Use your results on vanishing points to draw a nice cartoon scene with a road or two, some houses and maybe a roadrunner and coyote.

11-11 Suppose that a circle lies in a plane parallel to the image plane. Show that the perspective projection of the circle is a circle in the image plane and determine its radius.

11-12 Show that

$$\sum_{i=1}^{N}(X-\mu)^2P(X) = \left[\sum_{i=1}^{N} X^2P(X)\right] - \mu^2$$

In other words, show that the variance of X is equal to the difference between the expected value of X^2 and the square of the mean.

11-13 Verify Equation (11.21).

11-14 Suppose that an image consists of a light object on a dark background. Further, suppose that the image is hand segmented, giving histograms for both the object and background. Thus, it is a simple matter to compute $P_0(z)$ (the probability that a pixel with intensity value z belongs to the background) and $P_1(z)$ (the probability that a pixel with intensity value z belongs to the object). Give an expression for the probability that a pixel will be misclassified if the threshold value of t is selected.

11-15 Suppose again that an image consists of a light object on a dark background and that the image has been hand segmented, giving $P_0(z)$ and $P_1(z)$. Give an algorithm that determines t^*, the optimal threshold value, that is, the threshold value that minimizes the probability of misclassification of an image pixel. Your algorithm should employ a recursive formulation whenever possible.

NOTES AND REFERENCES

Computer vision research dates back to the early sixties. In the early eighties several computer vision texts appeared. These books approached computer vision from the perspective of cognitive modeling of human vision [87], image processing [109], and applied robotic vision [54]. A comprehensive review of computer vision techniques through the early nineties (including the segmentation method described in this chapter) can be found in [50], and an introductory treatment of methods in 3D vision can be found in [133]. Detailed treatments of the geometric aspects of computer vision can be found in [33] and [83]. A comprehensive review of the state of the art in computer vision at the turn of the century can be found in [38].

Chapter 12

VISION-BASED CONTROL

In Chapter 9 we described how feedback from a force sensor can be used to control the forces and torques applied by the manipulator. In the case of force control, the quantities to be controlled, that is, forces and torques, are measured directly by a sensor. Indeed, the output of a typical force sensor comprises six electric voltages that are proportional to the forces and torques experienced by the sensor. Force control is very similar to state-feedback control in this regard.

In this chapter we consider the problem of vision-based control. Unlike force control, with vision-based control the quantities to be controlled cannot always be measured directly from the sensor. For example, if the task is to grasp an object, the quantities to be controlled are pose variables, while the vision sensor, as we have seen in Chapter 11, provides a two-dimensional array of intensity values. There is, of course, a relationship between this array of intensity values and the geometry of the robot's workspace, but the task of inferring this geometry from an image is a difficult one that has been at the heart of computer vision research for many years. The problem faced in vision-based control is that of extracting a relevant and robust set of parameters from an image and using these parameters to control the motion of the manipulator in real time.

Over the years, a variety of approaches have been developed for the problem of vision-based control. These vary based on how the image data are used, the relative configuration of camera and manipulator, choices of coordinate systems, etc. Here, we focus primarily on one specific approach, namely, **image-based, visual servo control** for **eye-in-hand camera systems**. We begin the chapter with a brief description of this approach, contrasting it with other options. Next, we develop the specific mathematical tools needed for this approach, both design and analysis.

12.1 DESIGN CONSIDERATIONS

A number of questions confront the designer of a vision-based control system. What kind of camera should be used? Should a zoom lens or a lens with fixed focal length be used? How many cameras should be used? Where should the cameras be placed? What image features should be used? Should the features be used to derive a three-dimensional description of the scene, or should two-dimensional image data be used? For the questions of camera and lens selection, in this chapter we will consider only systems that use a single camera with a fixed focal length lens. We briefly discuss the remaining questions below.

12.1.1 Camera Configuration

Perhaps the first decision to be made when constructing a vision-based control system is where to place the camera. There are essentially two options: the camera can be mounted in a fixed location in the workspace or it can be attached to the robot. These are often referred to as **fixed camera** and **eye-in-hand** configurations, respectively.

With a fixed camera configuration, the camera is positioned so that it can observe the manipulator and any objects to be manipulated. There are several advantages to this approach. Since the camera position is fixed, the field of view does not change as the manipulator moves. The geometric relationship between the camera and the workspace is fixed, and can be calibrated offline. A disadvantage to this approach is that as the manipulator moves through the workspace, it can occlude the camera's field of view. This can be particularly important for tasks that require high precision. For example, if an insertion task is to be performed, it may be difficult to find a position from which the camera can view the entire insertion task without occlusion from the end effector.

With an eye-in-hand system, the camera is often attached to the manipulator above the wrist so that the motion of the wrist does not affect the camera motion. In this way, the camera can observe the motion of the end effector at a fixed resolution and without occlusion as the manipulator moves through the workspace. One difficulty that confronts the eye-in-hand configuration is that the geometric relationship between the camera and the workspace changes as the manipulator moves. The field of view can change drastically for even small motion of the manipulator, particularly if the link to which the camera is attached experiences a change in orientation. For example, a camera attached to link three of an elbow manipulator (such as the one shown in Figure 3.1) will experience a significant change in field of view when joint 3 moves.

For either the fixed camera or eye-in-hand configuration, motion of the manipulator will produce changes in the images obtained by the camera (assuming that the manipulator is in the field of view of the fixed camera system). The analysis of the relationships between manipulator motion and changes in the image for the two cases are similar, and in this text we will consider only the case of eye-in-hand systems.

12.1.2 Image-Based vs. Position-Based Approaches

There are two basic ways to approach the problem of vision-based control, and these are distinguished by the way in which the data provided by the vision system are used. These two approaches can also be combined in various ways to yield what are known as partitioned control schemes.

The first approach to vision-based control is known as **position-based visual servo control**. With this approach, the vision data are used to build a partial 3D representation of the world. For example, if the task is to grasp an object, the perspective projection equations from Chapter 11 can be solved to determine the 3D coordinates of the grasp points relative to the camera coordinate frame. If these 3D coordinates can be obtained in real time, then they can be provided as set points to the robot controller. The main difficulties with position-based methods are related to the difficulty of building the 3D representation in real time. In particular, these methods tend not to be robust with respect to errors in camera calibration. Furthermore, with position-based methods, there is no direct control over the image itself. Therefore, a common problem with position-based methods is that camera motion can cause the object of interest to leave the camera field of view.

A second method known as **image-based, visual servo control** uses the image data directly to control the robot motion. An error function is defined in terms of quantities that can be directly measured in an image (for example, image coordinates of points or the orientations of lines in an image) and a control law is constructed that maps this error directly to robot motion. To date, the most common approach has been to use easily detected points on an object as feature points. The error function is then the vector difference between the desired and measured locations of these points in the image. Typically, relatively simple control laws are used to map the image error to robot motion. We will describe image-based control in some detail in this chapter.

It is possible to combine multiple approaches, using different control algorithms to control different degrees of freedom of the robot motion. Such methods essentially partition the degrees of freedom into disjoint sets, and are thus known as **partitioned methods**. We briefly describe one particular partitioned method in Section 12.6.

12.2 CAMERA MOTION AND THE INTERACTION MATRIX

As mentioned above, image-based methods map an image error function directly to robot motion without solving the 3D reconstruction problem. Recall the inverse velocity problem discussed in Chapter 4. Even though the inverse kinematics problem is difficult to solve and often ill posed, the inverse velocity problem is typically fairly easy to solve: one merely inverts the manipulator Jacobian matrix, assuming the Jacobian is nonsingular. This can be understood mathematically by noting that while the inverse kinematic equations represent a nonlinear mapping between possibly complicated geometric spaces (for example, even for the simple two-link planar arm the mapping is from $\mathbb{R}^2$ to the torus), the mapping of velocities is a linear map between linear subspaces (in the two-link example, a mapping from $\mathbb{R}^2$ to a plane that is tangent to the torus). Likewise, the relationship between vectors defined in terms of image features and camera velocities is a linear mapping between linear subspaces. We will now give a more rigorous explanation of this basic idea.

Let $s(t)$ denote a vector of feature values that can be measured in an image. Its derivative $\dot{s}(t)$ is referred to as an **image feature velocity**. For example, if a single image point is used as a feature, we would have

$$s(t) = \begin{bmatrix} u(t) \\ v(t) \end{bmatrix}$$

In this case $\dot{s}(t)$ would be the image plane velocity of the image point.

The image feature velocity is linearly related to the camera velocity. Let the camera velocity ξ consist of linear velocity v and angular velocity ω

$$\xi = \begin{bmatrix} v \\ \omega \end{bmatrix} \tag{12.1}$$

so that the origin of the camera frame is moving with linear velocity v and the camera frame is rotating about the axis ω, which passes through the origin of the camera frame. There is no difference between ξ as used here and as used in Chapter 4; in each case, ξ encodes the linear and angular velocity of a moving frame. In Chapter 4 the frame was attached to the end effector while here it is attached to the moving camera.

The relationship between $\dot{s}$ and ξ is given by

$$\dot{s} = L(s, q)\xi \tag{12.2}$$

Here, the matrix $L(s, q)$ is known as the **interaction matrix**. The interaction matrix is a function of both the configuration of the robot, as was also true for the manipulator Jacobian described in Chatper 4, and of the image feature values s.

The interaction matrix L is also called the **image Jacobian matrix**. This is due, at least in part, to the analogy that can be drawn between the manipulator Jacobian discussed in Chapter 4 and the interaction matrix. In each case, a velocity ξ is related to the variation in a set of parameters, either joint angles or image feature velocities, by a linear transformation. Strictly speaking, the interaction matrix is not a Jacobian matrix, since ξ is not actually the derivative of some set of pose parameters. However, using techniques analogous to those used to develop the analytic Jacobian in Section 4.8, it is straightforward to construct an actual Jacobian matrix that represents a linear transformation from the derivatives of a set of pose parameters to the image feature velocities, which are derivatives of the image feature values.

The specific form of the interaction matrix depends on the features that are used to define s. The simplest features are coordinates of points in the image, and we will focus our attention on this case.

12.3 THE INTERACTION MATRIX FOR POINT FEATURES

In this section we derive the interaction matrix for the case of a moving camera observing a point that is fixed in space. This scenario is useful for postioning a camera relative to some object that is to be manipulated. For example, a camera can be attached to a manipulator arm that is to grasp a stationary object. Vision-based control can then be used to bring the manipulator to a grasping configuration that may be defined in terms of image features. In Section 12.3.4 we extend the development to the case of multiple feature points.

At time t, the orientation of the camera frame is given by a rotation matrix $R_c^0 = R(t)$, which specifies the orientation of the camera frame relative to the fixed frame. We denote by $o(t)$ the position of the origin of the camera frame relative to the fixed frame. We denote by P the fixed point in the workspace, and by $s = [u, v]^T$ the feature vector corresponding to the projection of P in the image. This is illustrated in Figure 12.1.

Our goal is to derive the interaction matrix L that relates the velocity of the camera ξ to the derivatives of the coordinates of the projection of the

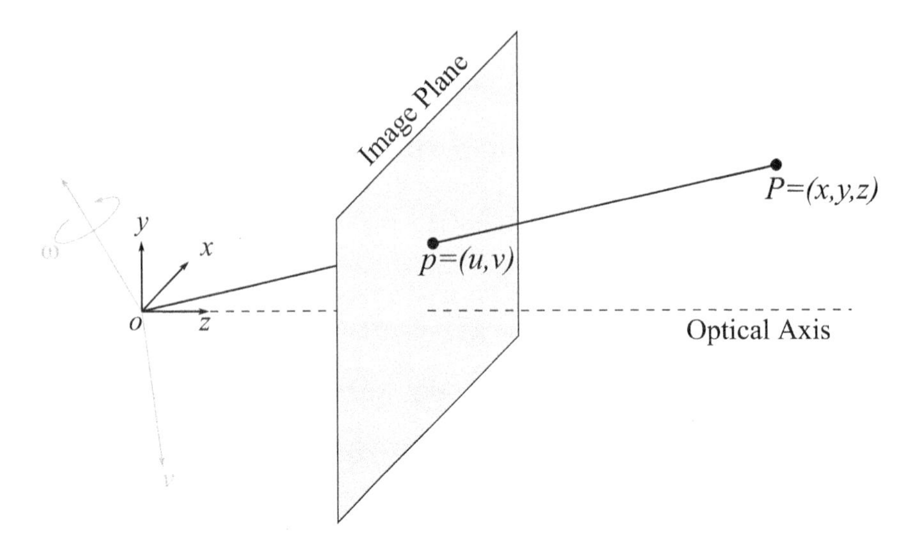

Figure 12.1: The point P is fixed with respect to the world coordinate frame, and the camera frame moves with angular velocity ω and linear velocity v.

point in the image $\dot{s}$. We begin by finding an expression for the velocity of the point P relative to the moving camera. We then use the perspective projection equations to relate this velocity to the image velocity $\dot{s}$. Finally, after a bit of algebraic manipulation we arrive to the interaction matrix that satisfies $\dot{s} = L\xi$.

12.3.1 Velocity of a Fixed Point Relative to a Moving Camera

We denote by p^0 the coordinates of P relative to the world frame. Note that p^0 does not vary with time, since P is fixed with respect to the world frame. If we denote by $p^c(t)$ the coordinates of P relative to the moving camera frame at time t, using Equation (2.51) we have

$$p^0 = R(t)p^c(t) + o(t)$$

Thus, at time t we can solve for the coordinates of P relative to the camera frame by

$$p^c(t) = R^T(t)\left[p^0 - o(t)\right] \tag{12.3}$$

Now, to find the velocity of the point P relative to the moving camera frame we merely differentiate this equation, as was done in Chapter 4. We will drop the explicit reference to time in these equations to simplify notation, but the reader is advised to bear in mind that both the rotation matrix

R and the location of the origin of the camera frame o are time varying quantities. Using the product rule for differentiation we obtain

$$\frac{d}{dt}p^c(t) = \dot{R}^T\left(p^0 - o\right) - R^T\dot{o} \tag{12.4}$$

From Equation (4.19) we have $\dot{R} = S(\omega)R$, and thus $\dot{R}^T = R^T S(\omega)^T = R^T S(-\omega)$. This allows us to write Equation (12.4) as

$$\begin{aligned}
\dot{R}^T\left(p^0 - o\right) - R^T\dot{o} &= R^T S(-\omega)\left(p^0 - o\right) - R^T\dot{o} \\
&= R^T S(-\omega) R R^T\left(p^0 - o\right) - R^T\dot{o} \\
&= -R^T\omega \times R^T\left(p^0 - o\right) - R^T\dot{o}
\end{aligned}$$

In this equation the rotation matrix R^T is applied to three vectors, yielding three new vectors whose coordinates are expressed with respect to the camera frame. From Equation (12.3) we see that $R^T(p^0 - o) = p^c$. The vector ω gives the angular velocity vector for the moving frame in coordinates expressed with respect to the fixed frame, that is, $\omega = \omega^0$. Therefore, $R^T\omega = R_0^c\omega^0 = \omega^c$ gives the angular velocity vector for the moving frame in coordinates expressed with respect to the moving frame. Finally, note that $R^T\dot{o} = \dot{o}^c$. Using these conventions we can immediately write the equation for the velocity of P relative to the moving camera frame

$$\dot{p}^c = -\omega^c \times p^c - \dot{o}^c \tag{12.5}$$

Relating this to the velocity ξ we see that ω^c is the angular velocity of the camera frame expressed relative to the moving camera frame, and $\dot{o}^c$ is the linear velocity v of the camera frame, also expressed relative to the moving camera frame. It is interesting to note that the velocity of a fixed point relative to a moving frame is merely -1 times the velocity of a moving point relative to a fixed frame.

Example 12.1 Camera Motion in the Plane

Consider a camera whose optical axis is parallel and opposite to the world z axis. If the camera motion is constrained to rotation about the optical axis and translation parallel to the x-y plane, we have

$$R = \begin{bmatrix} \cos\theta & -\sin\theta & 0 \\ \sin\theta & \cos\theta & 0 \\ 0 & 0 & 1 \end{bmatrix}, \quad o(t) = \begin{bmatrix} x_c \\ y_c \\ z_0 \end{bmatrix}$$

in which z_0 is the fixed height of the camera frame relative to the world frame. This gives

$$\omega^c = \begin{bmatrix} 0 \\ 0 \\ \dot{\theta} \end{bmatrix}, \qquad \dot{o}^c = \begin{bmatrix} v_x \\ v_y \\ 0 \end{bmatrix}$$

If the point P has coordinates (x, y, z) relative to the camera frame, we have

$$\dot{p}^c = -\omega^c \times p^c - \dot{o}^c = \begin{bmatrix} y\dot{\theta} - v_x \\ -x\dot{\theta} - v_y \\ 0 \end{bmatrix}$$

◇

12.3.2 Constructing the Interaction Matrix

Using Equation (12.5) and the equations of perspective projection, it is not difficult to derive the interaction matrix for point features. To simplify notation, we define the coordinates for P relative to the camera frame as $p^c = [x, y, z]^T$. By this convention, the velocity of P relative to the moving camera frame is merely the vector $\dot{p}^c = [\dot{x}, \dot{y}, \dot{z}]^T$. We will denote the coordinates for the angular velocity vector by $\omega^c = [\omega_x, \omega_y, \omega_z]^T = R^T\omega$. To further simplify notation, we assign coordinates $R^T\dot{o} = [v_x, v_y, v_z]^T = \dot{o}^c$. Using these conventions, we can write Equation (12.5) as

$$\begin{bmatrix} \dot{x} \\ \dot{y} \\ \dot{z} \end{bmatrix} = -\begin{bmatrix} \omega_x \\ \omega_y \\ \omega_z \end{bmatrix} \times \begin{bmatrix} x \\ y \\ z \end{bmatrix} - \begin{bmatrix} v_x \\ v_y \\ v_z \end{bmatrix}$$

which can be written as the system of three equations

$$\dot{x} = y\omega_z - z\omega_y - v_x \quad (12.6)$$

$$\dot{y} = z\omega_x - x\omega_z - v_y \quad (12.7)$$

$$\dot{z} = x\omega_y - y\omega_x - v_z \quad (12.8)$$

Since u and v are the the image coordinates of the projection of P in the image, using Equation (11.4), we can express x and y as

$$x = \frac{uz}{\lambda}, \qquad y = \frac{vz}{\lambda}$$

Substituting these into Equations (12.6)–(12.8) we obtain

$$\dot{x} = \frac{vz}{\lambda}\omega_z - z\omega_y - v_x \quad (12.9)$$

$$\dot{y} = z\omega_x - \frac{uz}{\lambda}\omega_z - v_y \quad (12.10)$$

$$\dot{z} = \frac{uz}{\lambda}\omega_y - \frac{vz}{\lambda}\omega_x - v_z \quad (12.11)$$

These equations express the velocity $\dot{p}^c$ in terms of the image coordinates u, v, the depth z of the point P, and the angular and linear velocity of the camera. We will now find expressions for $\dot{u}$ and $\dot{v}$ and then combine these with Equations (12.9)–(12.11).

Using the quotient rule for differentiation with the equations of perspective projection we obtain

$$\dot{u} = \frac{d}{dt}\frac{\lambda x}{z} = \lambda \frac{z\dot{x} - x\dot{z}}{z^2}$$

Substituting Equations (12.9) and (12.11) into this expression gives

$$\begin{aligned} \dot{u} &= \frac{\lambda}{z^2}\left(z\left[\frac{vz}{\lambda}\omega_z - z\omega_y - v_x\right] - \frac{uz}{\lambda}\left[\frac{uz}{\lambda}\omega_y - \frac{vz}{\lambda}\omega_x - v_z\right]\right) \\ &= -\frac{\lambda}{z}v_x + \frac{u}{z}v_z + \frac{uv}{\lambda}\omega_x - \frac{\lambda^2 + u^2}{\lambda}\omega_y + v\omega_z \end{aligned} \tag{12.12}$$

We can apply the same technique for $\dot{v}$

$$\dot{v} = \frac{d}{dt}\frac{\lambda y}{z} = \lambda \frac{z\dot{y} - y\dot{z}}{z^2}$$

and substituting Equations (12.10) and (12.11) into this expression gives

$$\begin{aligned} \dot{v} &= \frac{\lambda}{z^2}\left(z\left[-\frac{uz}{\lambda}\omega_z + z\omega_x - v_y\right] - \frac{vz}{\lambda}\left[\frac{uz}{\lambda}\omega_y - \frac{vz}{\lambda}\omega_x - v_z\right]\right) \\ &= -\frac{\lambda}{z}v_y + \frac{v}{z}v_z + \frac{\lambda^2 + v^2}{\lambda}\omega_x - \frac{uv}{\lambda}\omega_y - u\omega_z \end{aligned} \tag{12.13}$$

Equations (12.12) and (12.13) can be combined and written in matrix form as

$$\begin{bmatrix} \dot{u} \\ \dot{v} \end{bmatrix} = \begin{bmatrix} -\frac{\lambda}{z} & 0 & \frac{u}{z} & \frac{uv}{\lambda} & -\frac{\lambda^2 + u^2}{\lambda} & v \\ 0 & -\frac{\lambda}{z} & \frac{v}{z} & \frac{\lambda^2 + v^2}{\lambda} & -\frac{uv}{\lambda} & -u \end{bmatrix} \begin{bmatrix} v_x \\ v_y \\ v_z \\ \omega_x \\ \omega_y \\ \omega_z \end{bmatrix} \tag{12.14}$$

The matrix in this equation is the interaction matrix for a point. To make explicit its dependence on u, v, and z, this equation is often written as

$$\dot{s} = L_p(u, v, z)\xi \tag{12.15}$$

Example 12.2 Camera Motion in the Plane (cont.)
Consider the situation described in Example 12.1. Suppose that the point P has coordinates $p^0 = [x_p, y_p, 0]^T$ relative to the world frame. Relative to the camera frame, P has coordinates given by

$$\begin{aligned} p = R^T(p^0 - o) &= \begin{bmatrix} \cos\theta & \sin\theta & 0 \\ -\sin\theta & \cos\theta & 0 \\ 0 & 0 & 1 \end{bmatrix} \left(\begin{bmatrix} x_p \\ y_p \\ 0 \end{bmatrix} - \begin{bmatrix} x_c \\ y_c \\ z_0 \end{bmatrix} \right) \\ &= \\ &= \begin{bmatrix} \cos\theta(x_p - x_c) + \sin\theta(y_p - y_c) \\ -\sin\theta(x_p - x_c) + \cos\theta(y_p - y_c) \\ -z_0 \end{bmatrix} \end{aligned}$$

The image coordinates for P are thus given by

$$\begin{aligned} u &= -\lambda \frac{\cos\theta(x_p - x_c) + \sin\theta(y_p - y_c)}{z_0} \\ v &= -\lambda \frac{-\sin\theta(x_p - x_c) + \cos\theta(y_p - y_c)}{z_0} \end{aligned}$$

These can be substituted into Equation (12.15) to yield

$$\begin{aligned} \begin{bmatrix} \dot{u} \\ \dot{v} \end{bmatrix} &= \begin{bmatrix} -\frac{\lambda}{z} & 0 & \frac{u}{z} & \frac{uv}{\lambda} & -\frac{\lambda^2 + u^2}{\lambda} & v \\ 0 & -\frac{\lambda}{z} & \frac{v}{z} & \frac{\lambda^2 + v^2}{\lambda} & -\frac{uv}{\lambda} & -u \end{bmatrix} \begin{bmatrix} v_x \\ v_y \\ 0 \\ 0 \\ 0 \\ \dot{\theta} \end{bmatrix} \\ &= \frac{\lambda}{z_0} \begin{bmatrix} v_x + (\sin\theta(x_p - x_c) - \cos\theta(y_p - y_c))\dot{\theta} \\ v_y + (\cos\theta(x_p - x_c) + \sin\theta(y_p - y_c))\dot{\theta} \end{bmatrix} \end{aligned}$$

◇

12.3.3 Properties of the Interaction Matrix for Points

Equation (12.15) can be decomposed as

$$\dot{s} = L_v(u, v, z)v + L_\omega(u, v)\omega \tag{12.16}$$

in which $L_v(u, v, z)$ contains the first three columns of the interaction matrix, and is a function of both the image coordinates of the point and its depth, while $L_\omega(u, v)$ contains the last three columns of the interaction matrix, and is a function of only the image coordinates of the point, that is, it does not depend on depth. This can be particularly beneficial in real-world situations when the exact value of z may not be known. In this case, errors in the value of z merely cause a scaling of the matrix $L_v(u, v, z)$, and this kind of scaling effect can be compensated for by using fairly simple control methods. This kind of decomposition is at the heart of the partitioned method that we discuss in Section 12.6.

The camera velocity ξ has six degrees of freedom, while only the two values u and v are observed in the image. Thus, one would expect that not all camera motions cause observable changes in the image. More precisely, $L \in \mathbb{R}^{2\times 6}$ and therefore has a null space of dimension 4. Therefore, the system

$$0 = L(s, q)\xi$$

has solution vectors ξ that lie in a four-dimensional subspace of $\mathbb{R}^6$. For the case of a single point, it can be shown that the null space of the interaction matrix given in Equation (12.14) is spanned by the four vectors

$$\begin{bmatrix} u \\ v \\ \lambda \\ 0 \\ 0 \\ 0 \end{bmatrix}, \begin{bmatrix} 0 \\ 0 \\ 0 \\ u \\ v \\ \lambda \end{bmatrix}, \begin{bmatrix} uvz \\ -(u^2+\lambda^2)z \\ \lambda vz \\ -\lambda^2 \\ 0 \\ u\lambda \end{bmatrix}, \begin{bmatrix} \lambda(u^2+v^2+\lambda^2)z \\ 0 \\ -u(u^2+v^2+\lambda^2)z \\ uv\lambda \\ -(u^2+\lambda^2)z \\ u\lambda^2 \end{bmatrix}$$

The first two of these vectors have particularly intuitive interpretations. The first corresponds to motion of the camera frame along the projection ray that contains the point P, and the second corresponds to rotation of the camera frame about a projection ray that contains P.

12.3.4 The Interaction Matrix for Multiple Points

It is straightforward to generalize the development above to the case in which several points are used to define the image feature vector. Consider the case for which the feature vector consists of the coordinates of n image points. Here, the ith feature point has an associated depth z_i and we define the feature vector s and the vector of depth values z by

$$s = \begin{bmatrix} u_1 \\ v_1 \\ \vdots \\ u_n \\ v_n \end{bmatrix}, \quad z = \begin{bmatrix} z_1 \\ \vdots \\ z_n \end{bmatrix}$$

For this case, the composite interaction matrix L_c that relates camera velocity to image feature velocity is a function of the image coordinates of the n points and also of the n depth values,

$$\dot{s} = L_c(s, z)\xi$$

This interaction matrix is obtained by stacking the n interaction matrices for the individual feature points,

$$\begin{aligned} L_c(s, z) &= \begin{bmatrix} L_1(u_1, v_1, z_1) \\ \vdots \\ L_n(u_n, v_n, z_n) \end{bmatrix} \\ &= \begin{bmatrix} -\frac{\lambda}{z_1} & 0 & \frac{u_1}{z_1} & \frac{u_1 v_1}{\lambda} & -\frac{\lambda^2 + u_1^2}{\lambda} & v_1 \\ 0 & -\frac{\lambda}{z_1} & \frac{v_1}{z_1} & \frac{\lambda^2 + v_1^2}{\lambda} & -\frac{u_1 v_1}{\lambda} & -u_1 \\ & & & & & \\ \vdots & \vdots & \vdots & \vdots & \vdots & \vdots \\ & & & & & \\ -\frac{\lambda}{z_n} & 0 & \frac{u_n}{z_n} & \frac{u_n v_n}{\lambda} & -\frac{\lambda^2 + u_n^2}{\lambda} & v_n \\ 0 & -\frac{\lambda}{z_n} & \frac{v_n}{z_n} & \frac{\lambda^2 + v_n^2}{\lambda} & -\frac{u_n v_n}{\lambda} & -u_n \end{bmatrix} \end{aligned}$$

Thus, we have $L_c \in \mathbb{R}^{2n \times 6}$ and therefore three points are sufficient to solve for ξ given the image measurements $\dot{s}$.

12.4 IMAGE-BASED CONTROL LAWS

With image-based control, the goal configuration is defined by a desired configuration of image features, denoted by s^d. The image error function is then given by

$$e(t) = s(t) - s^d$$

The image-based control problem is to find a mapping from this error function to a commanded camera motion. As we have seen in previous chapters, there are a number of control approaches that can be used to determine the joint-level inputs to achieve a desired trajectory. Therefore, in this chapter we will treat the manipulator as a kinematic positioning device, that is, we will ignore manipulator dynamics and develop controllers that compute desired end effector trajectories. The underlying assumption is that these trajectories can then be tracked by a lower level manipulator controller.

The most common approach to image-based control is to compute a desired camera velocity ξ and use this as the control input. Relating image feature velocities to the camera velocity ξ is typically done by solving Equation (12.2), which gives the camera velocity that will produce a desired value for $\dot{s}$. In some cases, this can be done simply by inverting the interaction matrix, but in other cases the pseudoinverse must be used. Below we describe various pseudoinverses of the interaction matrix and then explain how these can be used to construct an image-based control law.

12.4.1 Computing Camera Motion

For the case of k feature values and m components of the camera body velocity ξ, we have $L \in \mathbb{R}^{k \times m}$. In general we will have $m = 6$, but in some cases we may have $m < 6$, for example if the camera is attached to a SCARA arm used to manipulate objects on a moving conveyor. When L is full rank ($\text{rank}(L) = \min(k, m)$), it can be used to compute ξ from $\dot{s}$. There are three cases that must be considered: $k = m$, $k > m$, and $k < m$. We now discuss each of these.

When $k = m$ and L is full rank, we have $\xi = L^{-1}\dot{s}$.

When $k < m$, L^{-1} does not exist, and the system is underconstrained. In the visual servo application, this implies that we are not observing enough feature velocities to uniquely determine the camera motion ξ, that is, there are certain components of the camera motion that cannot be observed. In this case we can compute a solution given by

$$\xi = L^+\dot{s} + (I_m - L^+L)b$$

where L^+ is the pseudoinverse for L given by

$$L^+ = L^T(LL^T)^{-1}$$

I_m is the $m \times m$ identity matrix, and $b \in \mathbb{R}^m$ is an arbitrary vector. Note the similarity between this equation and Equation (4.113) which gives the solution for the inverse velocity problem (that is, solving for joint velocities to achieve a desired end-effector velocity) for redundant manipulators.

In general, for $k < m$, $(I - LL^+) \neq 0$, and all vectors of the form $(I - LL^+)b$ lie in the null space of L, which implies that those components of the camera velocity that are unobservable lie in the null space of L. If we let $b = 0$, we obtain the value for ξ that minimizes the norm

$$\|\dot{s} - L\xi\|$$

When $k > m$ and L is full rank, we will typically have an inconsistent system, especially when the feature values s are obtained from measured image data. In the visual servo application, this implies that we are observing more feature velocities than are required to uniquely determine the camera motion ξ. In this case the rank of the null space of L is zero, since the dimension of the column space of L equals rank(L). In this situation, we can use the least squares solution

$$\xi = L^{+}\dot{s} \tag{12.17}$$

in which the pseudoinverse is given by

$$L^{+} = (L^T L)^{-1} L^T \tag{12.18}$$

12.4.2 Proportional Control Schemes

Lyapunov theory (see Appendix D) can be used to analyze the stability of dynamic systems, but it can also be used to aid in the design of stable control systems. For the system given by Equation (12.2) with error defined by Equation (12.1), consider the candidate Lyapunov function

$$V(t) = \frac{1}{2}\|e(t)\|^2 = \frac{1}{2}e^T e$$

The derivative of this function is

$$\dot{V} = \frac{d}{dt}\frac{1}{2}e^T e = e^T \dot{e}$$

Thus, if we could design a controller such that

$$\dot{e} = -\lambda e \tag{12.19}$$

with $\lambda > 0$ we would have

$$\dot{V} = -\lambda e^T e < 0$$

and this would ensure asymptotic stability of the closed-loop system. In fact, if we could design such a controller, we would have exponential stability, which ensures that the closed-loop system is asympotically stable even under small perturbations, for example, small errors in camera calibration.

For the case of visual servo control, it is often possible to design such a controller. The derivative of the error function is given by

$$\dot{e}(t) = \frac{d}{dt}(s(t) - s^d) = \dot{s}(t) = L\xi$$

and substituting this into Equation (12.19) we obtain

$$-\lambda e(t) = L\xi$$

If $k = m$ and L has full rank, then L^{-1} exists, and we have

$$\xi = -\lambda L^{-1} e(t)$$

and the system is exponentially stable.

When $k > m$ we obtain the control

$$\xi = -\lambda L^{+} e(t)$$

with $L^+ = (L^T L)^{-1} L$. Unfortunately, in this case we do not obtain exponential stability. To see this, consider again the Lyapunov function given above. We have

$$\begin{aligned} \dot{V} &= e^T \dot{e} \\ &= e^T L\xi \\ &= -\lambda e^T L L^+ e \end{aligned}$$

But in this case, the matrix LL^+ is only positive semidefinite, not positive definite, and therefore we can not demonstrate asymptotic stability by Lyapunov theory. This follows because $L^+ \in \mathbb{R}^{m \times k}$, and since $k > m$, it has a nonzero nullspace. Therefore, $eLL^+e = 0$ for certain values of e, and we can demonstrate only stability, not asymptotic stability.

In practice, we will not know the exact value of L or L^+ since these depend on knowledge of depth information that must be estimated by the computer vision system. In this case, we will have an estimate for the interaction matrix $\widehat{L}^+$ and we can use the control $\xi = -\widehat{L}^+ e(t)$. It is easy to show, by a proof analogous to the one above, that the resulting visual servo system will be stable when $L\widehat{L}^+$ is positive definite. This helps explain the robustness of image-based control methods to calibration errors in the computer vision system.

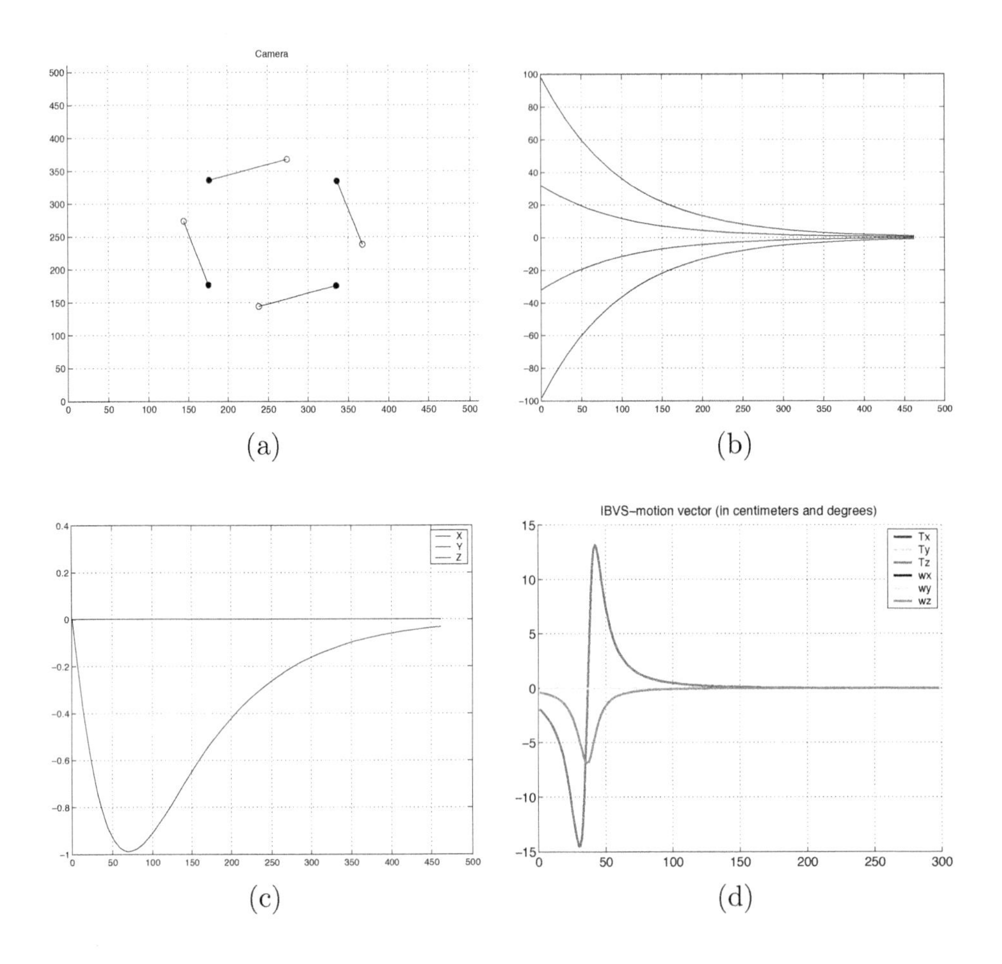

Figure 12.2: The required camera motion is a small rotation (approximately 30°) about the camera z-axis. As can be seen in (a), the feature points move on straight-line trajectories in the image to the desired positions, while the image error decreases exponentially to zero, as shown in part (b). Unfortunately, the required camera motion includes a significant retreat along the camera z-axis, as illustrated in parts (c) and (d).

12.4.3 Performance of IBVS systems

While the image-based control law described above performs well with respect to the image error, it can sometimes induce large camera motions that cause task failure, for example, if the required camera motion exceeds the physical range of the manipulator. Such a case is illustrated in Figure 12.2. In this example, the desired camera motion is a rotation about the camera's z-axis. Figure 12.2(a) shows the image feature trajectories for the four feature points. As can be seen in the figure, the feature points move on straight lines to their goal positions in the image. Figure 12.2(b) shows the image feature errors for the four points; the errors converge exponentially to zero. Unfortunately, as can be seen in Figure 12.2(c), to achieve this performance the camera retreats by a full one meter along its z-axis. Such a large motion is not possible for most manipulators. Figure 12.2(d) shows the corresponding camera velocities. The velocities along and about the camera x- and y-axes are very small, but the linear velocity along camera z-axis varies significantly.

The most extreme version of this problem occurs then the required camera motion is a rotation by π about the camera's optical axis. This case is shown in Figure 12.3. In Figure 12.3(a) the feature points again move on straight line trajectories in the image. However, in this case, these trajectories pass through the image center. This occurs only when the camera has retreated infinitely far along its z-axis. The corresponding camera position is shown in Figure 12.3(d).

These two examples are special cases that illustrate one of the key problems that confront image-based visual servo systems. Such systems explicitly control the error in the image, but exert no explicit control over the trajectory of the camera. Thus, it is possible that the required camera motions will exceed the capabilities of the robot manipulator. Partitioned methods provide one way to cope with these problems, and we describe one such method in Section 12.6.

12.5 END EFFECTOR AND CAMERA MOTIONS

The output of a visual servo controller is a camera velocity ξ_c, typically expressed in coordinates relative to the camera frame. If the camera frame were coincident with the end-effector frame, we could use the manipulator Jacobian to determine the joint velocities that would achieve the desired camera motion as described in Section 4.11. In most applications, the camera frame is not coincident with the end effector frame, but is rigidly attached to it. Suppose the two frames are related by the constant homogeneous transfor-

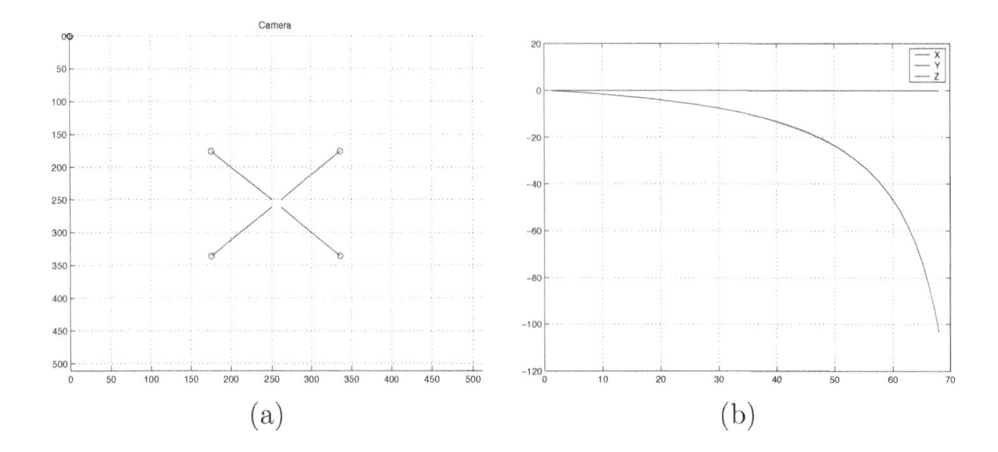

Figure 12.3: The required camera motion is a rotation by π about the camera z-axis. In (a) the feature points move along straight line trajectories in the image, but in (b) this requires the camera to reatreat to $z = -\infty$.

mation

$$T_c^6 = \begin{bmatrix} R & d \\ 0 & 1 \end{bmatrix} \tag{12.20}$$

In this case, we can use Equation (4.82) to determine the required velocity of the end effector to achieve the desired camera velocity. This gives

$$\xi_6^6 = \begin{bmatrix} R & S(d)R \\ 0_{3\times3} & R \end{bmatrix} \xi_c^c$$

If we wish to express the end effector velocity with respect to the base frame, we merely apply a rotational transformation to the two free vectors v_6 and ω_6, and this can be written as the matrix equation

$$\xi_6^0 = \begin{bmatrix} R_6^0 & 0_{3\times3} \\ 0_{3\times3} & R_6^0 \end{bmatrix} \xi_6^6$$

Example 12.3 Eye-in-hand System with SCARA Arm

Consider the camera system described in Example 12.2. Recall that in this example the camera motion was restricted to three degrees of freedom $\xi_c = [v_x, v_y, 0, 0, 0, \dot{\theta}]^T$. Suppose that this camera is attached to the end effector of a SCARA manipulator, such that the optical axis of the camera is aligned with the z-axis of the end effector frame. In this case, we can express *the orientation of the camera frame relative to the end effector frame by*

$$R_c^6 = \begin{bmatrix} \cos\alpha & -\sin\alpha & 0 \\ \sin\alpha & \cos\alpha & 0 \\ 0 & 0 & 1 \end{bmatrix}$$

in which α gives the angle from x_6 to x_c. Let the origin of the camera frame relative to the end effector frame be given by $d_c^6 = [10, 5, 0]^T$. The relationship between end effector and camera velocities is then given by

$$\begin{aligned}
\xi_6^6 &= \begin{bmatrix} \cos\alpha & -\sin\alpha & 0 & 0 & 0 & 5 \\ \sin\alpha & \cos\alpha & 0 & 0 & 0 & -10 \\ 0 & 0 & 1 & -5 & 10 & 0 \\ 0 & 0 & 0 & \cos\alpha & -\sin\alpha & 0 \\ 0 & 0 & 0 & \sin\alpha & \cos\alpha & 0 \\ 0 & 0 & 0 & 0 & 0 & 1 \end{bmatrix} \begin{bmatrix} v_x \\ v_y \\ 0 \\ 0 \\ 0 \\ \dot\theta \end{bmatrix} \\
&= \begin{bmatrix} v_x\cos\alpha - v_y\sin\alpha + 5\dot\theta \\ v_x\sin\alpha + v_y\cos\alpha - 10\dot\theta \\ 0 \\ 0 \\ 0 \\ \dot\theta \end{bmatrix}
\end{aligned}$$

This can be used with the Jacobian matrix of the SCARA arm (derived in Chapter 4) to solve for the joint velocities required to achieve the desired camera motion.

◇

12.6 PARTITIONED APPROACHES

Although image-based methods are versatile and robust to calibration and sensing errors, they sometimes fail when the required camera motion is large. Consider, for example, the case when the required camera motion is a large rotation about the optical axis. If point features are used, a pure rotation of the camera about the optical axis would cause each feature point to trace a trajectory in the image that lies on a circle. Image-based methods, in contrast, would cause each feature point to move in a straight line from its current image position to its desired position. The induced camera motion would be a retreat along the optical axis, and for a required rotation of π the camera would retreat to $z = -\infty$, at which point $\det L = 0$ and the controller would fail. This problem is a consequence of the fact that image-based control does not explicitly take camera motion into account. Instead, image-based control determines a desired trajectory in the image feature space, and maps this trajectory, using the interaction matrix, to a camera velocity.

One way to combat this problem is to use a **partitioned method**. Partitioned methods use the interaction matrix to control only a subset of the camera degrees of freedom, and use other methods to control the remaining degrees of freedom. Consider Equation (12.14). We can write this equation as

$$\begin{aligned} \dot{s} &= \begin{bmatrix} L_{v_z} & L_{v_y} & L_{v_z} & L_{\omega_x} & L_{\omega_y} & L_{\omega_z} \end{bmatrix} \xi \\ &= \begin{bmatrix} L_{v_x} & L_{v_y} & L_{\omega_x} & L_{\omega_y} \end{bmatrix} \begin{bmatrix} v_x \\ v_y \\ \omega_x \\ \omega_y \end{bmatrix} + \begin{bmatrix} L_{v_z} & L_{\omega_z} \end{bmatrix} \begin{bmatrix} v_z \\ \omega_z \end{bmatrix} \\ &= L_{xy}\xi_{xy} + L_z\xi_z \end{aligned} \tag{12.21}$$

Here, $\dot{s}_z = L_z\xi_z$ gives the component of $\dot{s}$ due to the camera motion along and rotation about the optical axis, while $\dot{s}_{xy} = L_{xy}\xi_{xy}$ gives the component of $\dot{s}$ due to velocity along and rotation about the camera x and y axes.

Equation (12.21) allows us to partition the control into two components, ξ_{xy} and ξ_z. Suppose that we have established a control scheme to determine the value $\xi_z = u_z$. Using an image-based method to find ξ_{xy} we would solve Equation (12.21) as

$$\xi_{xy} = L_{xy}^{+} \left\{ \dot{s} - L_z\xi_z \right\} \tag{12.22}$$

This equation has an intuitive explanation. $-L_{xy}^{+}L_z\xi_z$ is the required value of ξ_{xy} to cancel the feature motion $\dot{s}_z$. The control $u_{xy} = \xi_{xy} = L_{xy}^{+}\dot{s}$ gives the velocity along and rotation about the camera x and y axes that produce the desired $\dot{s}$ once image feature motion due to ξ_z has been accounted for.

If we use the Lyapunov design method described above, we set $\dot{e} = -\lambda e$, and obtain

$$-\lambda e = \dot{e} = \dot{s} = L_{xy}\xi_{xy} + L_z\xi_z$$

which leads to

$$\xi_{xy} = -L_{xy}^{+} \left(\lambda e(t) + L_z\xi_z \right)$$

We can consider $(\lambda e(t) + L_z\xi_z)$ as a modified error that incorporates the original image feature error while taking into account the feature error that will be induced by the translation along and rotation about the optical axis due to ξ_z.

The only remaining task is to construct a control law to determine the value of ξ_z. To determine ω_z, we can use the angle θ_{ij} from the horizontal axis of the image plane to the directed line segment joining two feature points. For numerical conditioning it is advantageous to select the longest line segment that can be constructed from the feature points, allowing this choice to change during the motion as the feature point configuration changes. The value for ω_z is given by

$$\omega_z = \gamma_{\omega_z}(\theta_{ij}^d - \theta_{ij})$$

in which θ_{ij}^d is the desired value, and γ_{ω_z} is a scalar gain coefficient.

We can use the apparent size of an object in the image to determine v_z. Let σ^2 denote the area of some polygon in the image. We define v_z as

$$v_z = \gamma_{v_z} \ln\left(\frac{\sigma^d}{\sigma}\right)$$

The advantages to using the apparent size as a feature are that (1) it is a scalar; (2) it is rotation invariant, thus decoupling camera rotation from z-axis translation; (3) it can be easily computed.

Figure 12.4 illustrates the performance of this partitioned controller for the case of desired rotation by π about the optical axis. Note that the camera does not retreat (σ is constant), the angle θ monotonically decreases, and the feature points move in a circle. The feature coordinate error is initially increasing, unlike the classical image-based methods, in which feature error is monotonically decreasing.

12.7 MOTION PERCEPTIBILITY

Recall the that notion of manipulability described in Section 4.12 gave a quantitative measure of the scaling from joint velocities to end-effector velocities. **Motion perceptibility** is an analogous concept that relates camera velocity to the velocity of features in the image. Intuitively, motion perceptibility quantifies the magnitude of changes to image features that result from motion of the camera.

Consider the set of all robot tool velocities ξ such that

$$||\xi||^2 = (\xi_1^2 + \xi_2^2 + \ldots \xi_m^2) \leq 1.$$

Suppose that there are redundant image features, that is, $k > m$. We may use Equation (12.17) to obtain

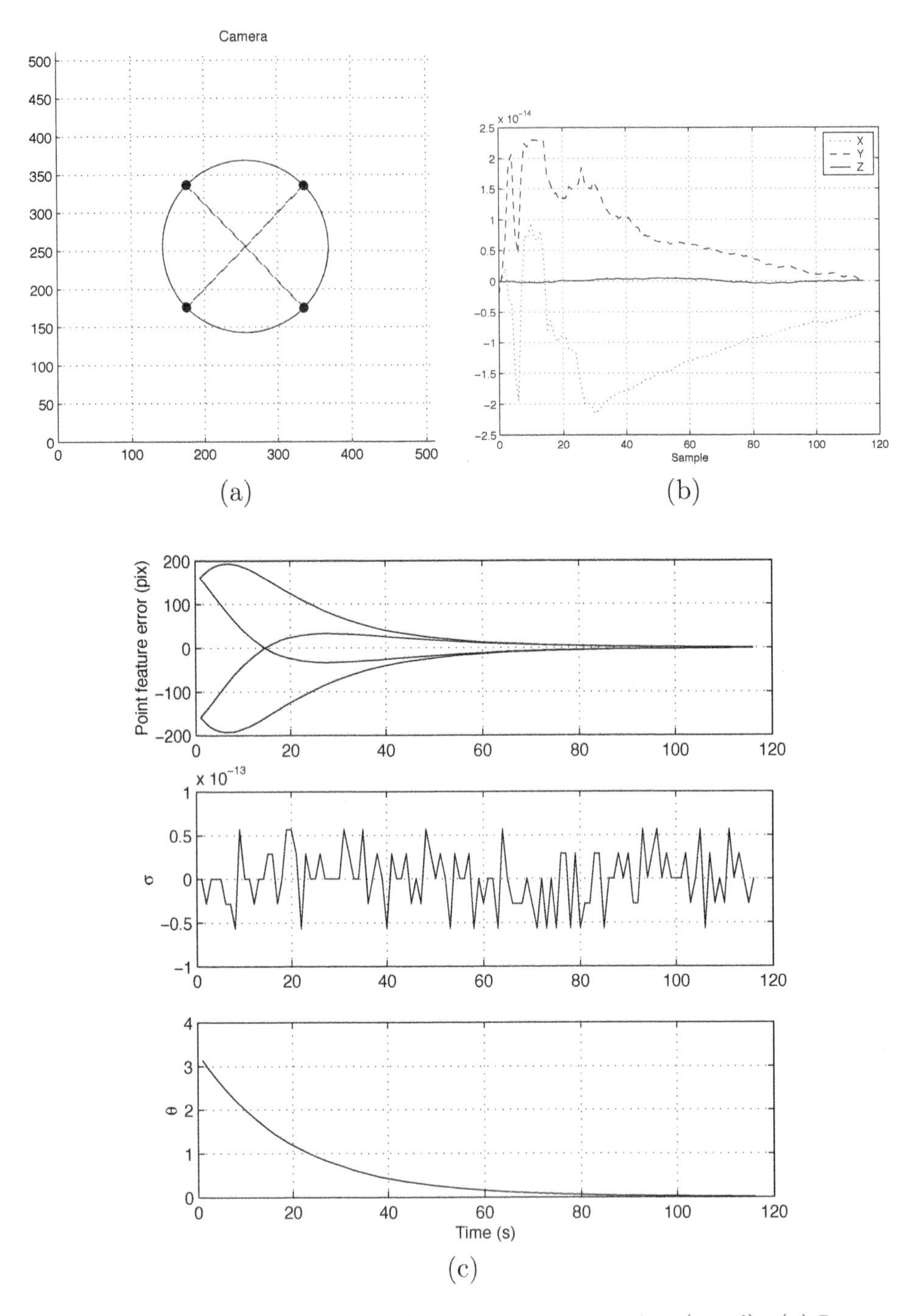

Figure 12.4: Partitioned method for pure target rotation (πrad). (a) Image-plane feature motion with initial location denoted by $\circ$ and desired location denoted by $\bullet$. (b) Cartesian translation trajectory. (c) Feature error trajectory.

$$
\begin{aligned}
||\xi||^2 &= \xi^T \xi \\
&= (L^+ \dot{s})^T (L^+ \dot{s}) \\
&= \dot{s}^T (L^{+^T} L^+) \dot{s} \leq 1
\end{aligned}
\tag{12.23}
$$

Now, consider the singular value decomposition of L (see Appendix B) given by

$$
L = U \Sigma V^T. \tag{12.24}
$$

in which

$$
U = [u_1 u_2 \dots u_k], \; V = [v_1 v_2 \dots v_m]
$$

are orthogonal matrices and $\Sigma \in \mathbb{R}^{k \times m}$ with

$$
\Sigma = \begin{bmatrix} \sigma_1 & & & & \\ & \sigma_2 & & & \\ & & \cdot & & \\ & & & \cdot & \\ & & & & \sigma_m \\ & & 0 & & \end{bmatrix}
$$

and the σ_i are the singular values of L and $\sigma_1 \geq \sigma_2 \dots \geq \sigma_m$.

For this case, the pseudoinverse of the interaction matrix L^+ is given by Equation (12.18). Using this with Equations (12.23) and (12.24) we obtain

$$
\dot{s}^T U \begin{bmatrix} \sigma_1^{-2} & & & & \\ & \sigma_2^{-2} & & & \\ & & \cdot & & \\ & & & \cdot & \\ & & & & \sigma_m^{-2} \\ & & 0 & & \end{bmatrix} U^T \dot{s} \leq 1 \tag{12.25}
$$

Consider the orthogonal transformation of $\dot{s}$ given by

$$
\tilde{\dot{s}} = U^T \dot{s}
$$

Substituting this into Equation (12.25) we obtain

$$
\sum_{i=1}^{m} \frac{1}{\sigma_i^2} \tilde{\dot{s}}_i \leq 1 \tag{12.26}
$$

Equation (12.26) defines an ellipsoid in an m-dimensional space. We shall refer to this ellipsoid as the **motion perceptibility ellipsoid**. We may use the volume of the m-dimensional ellipsoid given in Equation (12.26) as a quantitative measure of the perceptibility of motion. The volume of the motion perceptibility ellipsoid is given by

$$K\sqrt{\det(L^T L)}$$

in which K is a scaling constant that depends on the dimension of the ellipsoid m. Because the constant K depends only on m, it is not relevant for the purpose of evaluating motion perceptibility, since m will be fixed for any particular problem. Therefore, we define the motion perceptibility, which we shall denote by ρ, as

$$\rho = \sqrt{\det(L^T L)} = \sigma_1 \sigma_2 \cdots \sigma_m$$

The motion perceptibility measure ρ has the following properties, which are direct analogs of properties derived for manipulability:

- In general, $\rho = 0$ holds if and only if $\text{rank}(L) < \min(k, m)$, that is, when L is not full rank.

- Suppose that there is some error in the measured visual feature velocity $\Delta\dot{s}$. We can bound the corresponding error in the computed camera velocity $\Delta\xi$ by

$$(\sigma_1)^{-1} \leq \frac{||\Delta\xi||}{||\Delta\dot{s}||} \leq (\sigma_m)^{-1}.$$

There are other quantitative methods that could be used to evaluate the perceptibility of motion. For example, in the context of feature selection the condition number for the interaction matrix, given by $||L||||L^{-1}||$, could be used to select image features.

12.8 SUMMARY

Image-based visual servo control is a method for using an error measured in the image to directly control the motion of a robot. The key relationship exploited by all image-based methods is given by

$$\dot{s} = L(s, q)\xi$$

in which $L(s, q)$ is the interaction matrix and s is a vector of measured image feature values. When a single image point is used as the feature, this relationship is given by

$$\begin{bmatrix} \dot{u} \\ \dot{v} \end{bmatrix} = \begin{bmatrix} -\frac{\lambda}{z} & 0 & \frac{u}{z} & \frac{uv}{\lambda} & -\frac{\lambda^2+u^2}{\lambda} & v \\ 0 & -\frac{\lambda}{z} & \frac{v}{z} & \frac{\lambda^2+v^2}{\lambda} & -\frac{uv}{\lambda} & -u \end{bmatrix} \begin{bmatrix} v_x \\ v_y \\ v_z \\ \omega_x \\ \omega_y \\ \omega_z \end{bmatrix}$$

In image-based control the image error is defined by

$$e(t) = s(t) - s^d$$

and by using the square of the error norm as a candidate Lyapunov function, we derive the control law

$$\xi = -\lambda L^{-1} e(t)$$

when the interation matrix is square and nonsingular or

$$\xi = -\lambda L^{+} e(t)$$

with $L^+ = (L^T L)^{-1} L^T$ when $L \in \mathbb{R}^{k \times m}$ and $k > m$.

In general, the camera coordinate frame and the end effector frame of the robot are not coincident. In this case, it is necessary to relate the camera velocity to the end effector velocity. This relationship is given by

$$\xi_6^6 = \begin{bmatrix} R_c^6 & S(d_c^6) R_c^6 \\ 0_{3\times 3} & R_c^6 \end{bmatrix} \xi_c^c$$

in which R_c^6 and d_c^6 specify the fixed relative orientation and position of the camera frame with respect to the end effector frame, and ξ_6 and ξ_c denote the end effector and camera velocities, respectively.

In some cases, it is advantageous to use different control laws for the different degrees of freedom. In this chapter, we described one way to partition the control system using the relationship

$$\dot{s} = L_{xy} \xi_{xy} + L_z \xi_z$$

After defining two new image features, we controlled the z-axis translation and rotation using

$$\begin{aligned} \omega_z &= \gamma_{\omega_z} (\theta_{ij}^d - \theta_{ij}) \\ v_z &= \gamma_{v_z} \ln \left(\frac{\sigma^d}{\sigma} \right) \end{aligned}$$

in which θ_{ij} is the angle between the horizontal axis of the image plane and the directed line segment joining two feature points, σ^2 is the area of a polygon in the image, and γ_{ω_z} and γ_{v_z} are scalar gain coefficients.

Finally, we defined motion perceptibility as a property of visual servo systems that is analogous to the manipulability measure for manipulators. For $k > m$ motion percpetibility is defined by

$$\rho = \sqrt{\det(L^T L)} = \sigma_1 \sigma_2 \cdots \sigma_m$$

in which σ_i are the singular values of the interaction matrix.

PROBLEMS

12-1 Give an expression for the interaction matrix for two points p_1 and p_2 that satisfies

$$\begin{bmatrix} u_1 \\ v_1 \\ u_2 \\ v_2 \end{bmatrix} = L\xi$$

in which (u_1, v_1) and (u_2, v_2) are the image coordinates of the projection of p_1 and p_2, respectively, and ξ is the velocity of the moving camera.

12-2 What is the dimension of the null space for the interaction matrix for two points? Give a basis for this null space.

12-3 Consider a stereo camera system attached to a robot manipulator. Derive the interaction matrix L that satisfies

$$\begin{bmatrix} u_l \\ v_l \\ u_r \\ v_r \end{bmatrix} = L\xi$$

in which (u_l, v_l) and (u_r, v_r) are the image coordinates of the projection of p in the left and right images, respectively, and ξ is the velocity of the moving stereo camera system.

12-4 Consider a stereo camera system mounted to a fixed tripod observing the manipulator end effector. If the end effector velocity is given by

ξ, derive the interaction matrix L that satisfies

$$\begin{bmatrix} u_l \\ v_l \\ u_r \\ v_r \end{bmatrix} = L\xi$$

in which (u_l, v_l) and (u_r, v_r) are the image coordinates of the projection of p in the left and right images, respectively.

12-5 Consider a fixed camera that observes the motion of a robot arm. Derive the interaction matrix that relates the velocity of the end effector frame to the image coordinates of the projection of the origin of the end effector frame.

12-6 Consider a camera mounted above a conveyor belt such that the optical axis is parallel to the world z axis. The camera can translate and rotate about its optical axis, so in this case we have $\xi = [v_x, v_y, v_z, \dot{\theta}]^T$. Suppose the camera observes a planar object whose moments are given by

$$m_{ij} = \sum_{r,c} r^i c^j \mathcal{I}(r, c).$$

Derive the interaction matrix that satisfies

$$\begin{bmatrix} \dot{m}_{00} \\ \dot{m}_{10} \\ \dot{m}_{01} \end{bmatrix} = L\xi$$

12-7 Write a simulator for an image-based visual servo controller that uses the coordinates of four image points as features. Find initial and goal images for which the classical image-based control scheme will stop in a local minimum, that is, where the error lies in null space of the interaction matrix. Find initial and goal images for which the system diverges.

12-8 Use the simulator you implemented for Problem 12-7 to demonstrate the robustness of image-based methods to errors in depth estimation.

12-9 Use the simulator you implemented for Problem 12-7 to demonstrate the robustness of image-based methods to errors in the estimation of camera orientation. For example, use Euler angles to construct a rotation matrix that can be used to perturb the world coordinates of the four points used as features.

12-10 It is often acceptible to use a fixed value of depth $z = z^d$ when constructing the interaction matrix. Here, z^d denotes the depth when the camera is in its goal position. Use the simulator you implemented for Problem 12-7 to compare the performance of image-based control using the true value of depth vs. using the fixed value z^d.

12-11 Write a simulator for a partitioned controller that uses the coordinates of four image points as features. Find initial and goal images for which the controller fails. Compare the performance of this controller to the one you implemented for Problem 12-7.

NOTES AND REFERENCES

Vision-based control of robotic systems dates back to the 1960's and the robot known as Shakey that was built at SRI. However, the vision system for Shakey was much too slow for real-time control applications. Some of the earliest results of real-time, vision-based control were reported in [4] and [5], which described a robot that played ping pong.

The interaction matrix was first introduced in [111], where it was referred to as the *feature sensitivity matrix*. In [35] it was refered to as a Jacobian matrix, subsequently referred to in the literature as the image Jacobian, and in [31] it was given the name interaction matrix, the term we use in this text.

The performance problems associated with image-based methods were first rigorously investigated in [17]. This paper charted a course for the next several years of research in visual servo control.

The first partitioned method for visual servo was introduced in [84], which describes a system in which the three rotational degrees of freedom are controlled using position-based methods and the three translational degrees of freedom are controlled using image-based methods. Other partitioned methods have been reported in [26] and [92]. The method described in this chapter was reported in [20].

Motion perceptibility was introduced in [115] and [114]. The notion of **resolvability**, introduced in [94] and [95], is similar. In [34] the condition number of the interaction matrix is used for the purpose of feature selection.

Appendix A

TRIGONOMETRY

A.1 THE TWO-ARGUMENT ARCTANGENT FUNCTION

The usual inverse tangent function returns an angle in the range $(-\pi/2, \pi/2)$. In order to express the full range of angles we will find it useful to define the so-called **two-argument arctangent function**, $\text{Atan2}(x, y)$, which is defined for all $(x, y) \neq (0, 0)$ and equals the unique angle θ such that

$$\cos\theta = \frac{x}{(x^2 + y^2)^{\frac{1}{2}}}, \qquad \sin\theta = \frac{y}{(x^2 + y^2)^{\frac{1}{2}}} \tag{A.1}$$

This function uses the signs of x and y to select the appropriate quadrant for the angle θ. For example, $\text{Atan2}(1, -1) = -\frac{\pi}{4}$, while $\text{Atan2}(-1, 1) = +\frac{3\pi}{4}$. Note that if both x and y are zero, then Atan2 is undefined.

A.2 USEFUL TRIGONOMETRIC FORMULAS

Reduction Formulas

$$\sin(-\theta) = -\sin\theta \qquad \sin(\tfrac{\pi}{2} + \theta) = \cos\theta$$

$$\cos(-\theta) = \cos\theta \qquad \tan(\tfrac{\pi}{2} + \theta) = -\cot\theta$$

$$\tan(-\theta) = -\tan\theta \qquad \tan(\theta - \pi) = \tan\theta$$

Double Angle Identities

$$\begin{aligned}
\sin(x \pm y) &= \sin x \cos y \pm \cos x \sin y \\
\cos(x \pm y) &= \cos x \cos y \mp \sin x \sin y \\
\tan(x \pm y) &= \frac{\tan(x) \pm \tan(y)}{1 \mp \tan x \tan y}
\end{aligned}$$

Law of Cosines

If a triangle has sides of length a, b, and c, and θ is the angle opposite the side of length c, then

$$c^2 = a^2 + cb^2 - 2ab \cos \theta$$

Appendix B

LINEAR ALGEBRA

In this book we assume that the reader has some familiarity with basic properties of vectors and matrices, such as matrix addition, subtraction, multiplication, matrix transpose, and determinants. For additional background see [8].

B.1 VECTORS

The symbol $\mathbb{R}$ will denote the set of real numbers and $\mathbb{R}^n$ will denote the usual vector space of n-tuples over $\mathbb{R}$. We use lower case letters a, b, c, x, y, etc., to denote scalars in $\mathbb{R}$ and vectors in $\mathbb{R}^n$. Uppercase letters A, B, C, R, etc., denote matrices. Unless otherwise stated, vectors will be defined as column vectors. Thus, the statement $x \in \mathbb{R}^n$ means that

$$x = \begin{bmatrix} x_1 \\ \vdots \\ x_n \end{bmatrix}, \text{ with } x_i \in \mathbb{R} \; i = 1, \ldots, n \tag{B.1}$$

The vector x is thus an n-tuple, arranged in a column with real-valued components $x_1, \ldots, x_n$. We will frequently denote this as

$$x = [x_1, \ldots, x_n]^T \tag{B.2}$$

where the superscript T denotes transpose.

The **scalar product** of vectors x and y belonging to $\mathbb{R}^n$, denoted $\langle x, y \rangle$ or $x^T y$, is a real number defined by

$$\langle x, y \rangle = x^T y = x_1 y_1 + \cdots + x_n y_n \tag{B.3}$$

The scalar product of vectors is commutative, that is,

$$x^T y = y^T x \tag{B.4}$$

The length or **norm** of a vector $x \in \mathbb{R}^n$ is

$$\|x\| = \langle x, x \rangle^{\frac{1}{2}} = (x_1^2 + \cdots + x_n^2)^{\frac{1}{2}} \tag{B.5}$$

We also have the useful inequalities,

$$|x^T y| \leq \|x\| \; \|y\| \qquad \text{(Cauchy-Schwartz inequality)} \tag{B.6}$$

$$\|x + y\| \leq \|x\| \; + \|y\| \qquad \text{(triangle inequality)} \tag{B.7}$$

For vectors in $\mathbb{R}^2$ or $\mathbb{R}^3$ the scalar product can be expressed as

$$x^T y = \|x\| \; \|y\| \cos(\theta) \tag{B.8}$$

where θ is the angle between the vectors x and y.

The **outer product** of two vectors x and y belonging to $\mathbb{R}^n$ is an $n \times n$ matrix defined by

$$xy^T = \begin{bmatrix} x_1 y_1 & \cdot & \cdot & x_1 y_n \\ x_2 y_1 & \cdot & \cdot & x_2 y_n \\ \cdot & \cdot & \cdot & \cdot \\ x_n y_1 & \cdot & \cdot & x_n y_n \end{bmatrix} \tag{B.9}$$

From Equation (B.9) we can see that the scalar product and the outer product are related by

$$x^T y = Tr(xy^T) \tag{B.10}$$

where the function $Tr(\cdot)$ denotes the **trace** of a matrix, that is, the sum of the diagonal elements of the matrix.

We will sometimes use i, j, and k to denote the standard unit vectors in $\mathbb{R}^3$

$$i = \begin{bmatrix} 1 \\ 0 \\ 0 \end{bmatrix}, \quad j = \begin{bmatrix} 0 \\ 1 \\ 0 \end{bmatrix}, \quad k = \begin{bmatrix} 0 \\ 0 \\ 1 \end{bmatrix} \tag{B.11}$$

Using this notation a vector $x = [x_1, x_2, x_3]^T$ may be written as

$$x = x_1 i + x_2 j + x_3 k \tag{B.12}$$

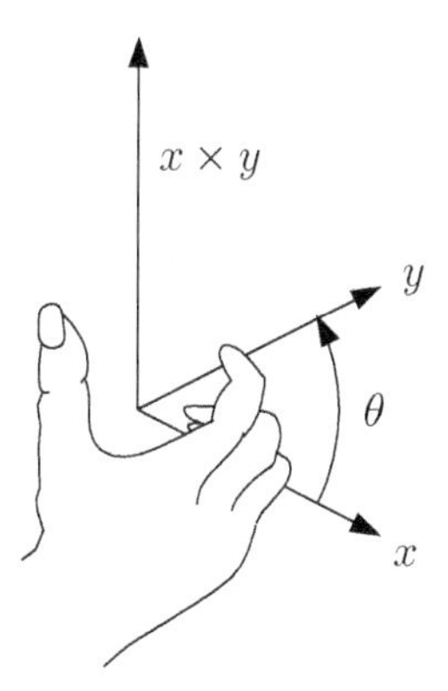

Figure B.1: The right hand rule.

The **vector product** or **cross product** $x \times y$ of two vectors x and y belonging to $\mathbb{R}^3$ is a vector c defined by

$$\begin{aligned} c &= x \times y = \det \begin{bmatrix} i & j & k \\ x_1 & x_2 & x_3 \\ y_1 & y_2 & y_3 \end{bmatrix} \\ &= (x_2y_3 - x_3y_2)i + (x_3y_1 - x_1y_3)j + (x_1y_2 - x_2y_1)k \end{aligned} \quad \text{(B.13)}$$

The cross product is a vector whose magnitude is

$$\|c\| = \|x\| \, \|y\| \cdot \|\sin(\theta)\| \quad \text{(B.14)}$$

where θ is the angle between x and y and whose direction is given by the right hand rule shown in Figure B.1.

A right-handed coordinate frame x-y-z is a coordinate frame with axes mutually perpendicular and that also satisfies the right hand rule in the sense that $k = i \times j$, where i, j, and k are unit vectors along the x, y, and z axes, respectively.

We can remember the right hand rule as being the direction of advancement of a right-handed screw rotated from the positive x axis into the positive y axis through the smallest angle between the axes. The cross product has the properties

$$\begin{aligned} x \times y &= -y \times x \\ x \times (y + z) &= x \times y + x \times z \quad \text{(B.15)} \\ \alpha(x \times y) &= (\alpha x) \times y = x \times (\alpha y) \quad \text{(B.16)} \end{aligned}$$

B.2 DIFFERENTIATION OF VECTORS

Suppose that the vector $x(t) = [x_1(t), \ldots, x_n(t)]^T$ is a function of time. Then the time derivative $\dot{x}$ of x is the vector

$$\dot{x} = [\dot{x}_1(t), \ldots, \dot{x}_n(t)]^T \tag{B.17}$$

Similarly, the derivative dA/dt of a matrix $A = (a_{ij})$ is the matrix $(\dot{a}_{ij})$. Similar statements hold for integration of vectors and matrices. The scalar and vector products satisfy the following product rules for differentiation similar to the product rule for differentiation of ordinary functions.

$$\frac{d}{dt}\langle x, y\rangle = \langle \frac{dx}{dt}, y\rangle + \langle x, \frac{dy}{dt}\rangle \tag{B.18}$$

$$\frac{d}{dt}(x \times y) = \frac{dx}{dt} \times y + x \times \frac{dy}{dt} \tag{B.19}$$

B.3 LINEAR INDEPENDENCE

A set of vectors $\{x_1, \ldots, x_n\}$ is said to be **linearly independent** if and only if

$$\sum_{i=1}^{n} \alpha_i x_i = 0 \text{ implies } \alpha_i = 0 \text{ for all } i \tag{B.20}$$

A **basis** of a vector space X is a linearly independent set of vectors $(e_1, \ldots, e_n)$ such that every vector $x \in X$ can be written as a linear combination

$$x = x_1 e_1 + \ldots x_n e_n$$

The representation $x = [x_1, \ldots, x_n]^T$ is uniquely determined by the particular basis $(e_1, \ldots, e_n)$. The dimension of the vector space X is the number of basis vectors.

B.4 MATRICES

An $n \times m$ matrix $A = (a_{ij})$ is an ordered array of real numbers with n row vectors $(a_{i1}, \ldots, a_{im})$ for $i = 1, \ldots, n$ (likewise m column vectors $[a_{1j}, \ldots, a_{nj}]^T, j = 1, \ldots, m$).

The **rank** of a matrix A is the largest number of linearly independent rows (or columns) of A. Thus, the rank of an $n \times m$ matrix can be no greater than the minimum of n and m.

The **Transpose** of a matrix A is denoted A^T and is formed by interchanging rows and columns of A. Some properties of the matrix transpose are

$$\begin{aligned} (A^T)^T &= A \\ (AB)^T &= B^T A^T, \text{ where } A \text{ and } B \text{ have compatible dimensions} \\ (A+B)^T &= A^T + B^T \end{aligned}$$

A square $n \times n$ matrix A is said to be

- **symmetric** if and only if $A^T = A$
- **skew symmetric** if and only if $A^T = -A$
- **orthogonal** if and only if $A^T A = AA^T = I$

The **inverse** of a square matrix $A \in \mathbb{R}^{n\times n}$ is a matrix $B \in \mathbb{R}^{n\times n}$ satisfying

$$AB = BA = I$$

where I is the $n \times n$ identity matrix. We denote the inverse of A by A^{-1}. The inverse of a matrix A exists and is unique if and only if A has rank n, equivalently if and only if the determinant $\det(A)$ is nonzero. The inverse of a square matrix satisfies

1. $(A^{-1})^{-1} = A$
2. $(AB)^{-1} = B^{-1}A^{-1}$, where B is square of the same dimension as A.

If A is an orthogonal matrix, $A^T = A^{-1}$, the inverse of A.

The **Null Space** $\mathcal{N}$ of a matrix A is defined as

$$\mathcal{N}(A) = \{x \in \mathbb{R}^n : Ax = 0\}$$

The null space of a matrix is a subspace of $\mathbb{R}^n$, that is, a subset of $\mathbb{R}^n$ that is also a vector space in its own right. An important property of the null space is that, if A is an $n \times n$ matrix, then

$$\text{rank}(A) + \dim\mathcal{N}(A) = n$$

Thus, a matrix is invertible if and only if the nullspace consists of only the zero vector, that is,

$$Ax = 0 \quad \text{implies} \quad x = 0$$

The **norm of a matrix** $A \in \mathbb{R}^{n\times n}$ is defined as

$$||A|| = \sup_{||x||\neq 0} \frac{||Ax||}{||x||}$$

B.5 CHANGE OF COORDINATES

An $n \times n$ matrix A represents a linear transformation from $\mathbb{R}^n$ to $\mathbb{R}^n$ in the sense that it takes a vector x to a new vector y according to

$$y = Ax \tag{B.21}$$

The vector y is called the **image** of x under the transformation A. The numerical values of the entries of A are determined by the particular basis used to represent vectors in $\mathbb{R}^n$. If the vectors x and y are represented in terms of the standard unit vectors

$$e_1 = [1, 0, \ldots, 0]^T, \ldots, e_n = [0, 0, \ldots, 1]^T \tag{B.22}$$

then the column vectors of A represent the images of the basis vectors $e_1, \ldots, e_n$. Often it is desired to represent vectors with respect to a second coordinate frame with different basis vectors $f_1, \ldots, f_n$. In this case the matrix representing the same linear transformation as A, but relative to this new basis, is given by

$$A' = T^{-1}AT \tag{B.23}$$

where T is a nonsingular matrix with column vectors $f_1, \ldots, f_n$. The transformation $T^{-1}AT$ is called a **similarity transformation** of the matrix A.

B.6 EIGENVALUES AND EIGENVECTORS

The **eigenvalues** of a matrix A are the solutions in s of the equation

$$\det(sI - A) = 0 \tag{B.24}$$

The function $\det(sI - A)$ is a polynomial in s called the **characteristic polynomial** of A. If s_e is an eigenvalue of A, an eigenvector of A corresponding to s_e is a nonzero vector x satisfying the system of linear equations

$$(s_e I - A)x = 0 \tag{B.25}$$

If the eigenvalues $s_1, \ldots, s_n$ of A are distinct, then there exists a similarity transformation $A' = T^{-1}AT$, such that A' is a diagonal matrix with the eigenvalues $s_1, \ldots, s_n$ on the main diagonal, that is,

$$A' = \mathrm{diag}[s_1, \ldots, s_n] \tag{B.26}$$

B.7 SINGULAR VALUE DECOMPOSITION (SVD)

For square matrices, we can use tools such as the determinant, eigenvalues, and eigenvectors to analyze their properties. However, for nonsquare matrices these tools simply do not apply. Their generalizations are captured by the **singular value decomposition (SVD)** .

As we described above, for $A \in \mathbb{R}^{m\times n}$, we have $AA^T \in \mathbb{R}^{m\times m}$. This square matrix has eigenvalues and eigenvectors that satisfy

$$(AA^T - \lambda_i I)u_i = 0 \tag{B.27}$$

which implies that the matrix $(AA^T - \lambda_i I)$ is singular, and we can express this in terms of its determinant as

$$\det(AA^T - \lambda_i I) = 0 \tag{B.28}$$

We can use Equation (B.28) to find the eigenvalues $\lambda_1 \geq \lambda_2 \cdots \geq \lambda_m \geq 0$ for AA^T. The **singular values** for the matrix A are given by the square roots of the eigenvalues of AA^T,

$$\sigma_i = \sqrt{\lambda_i} \tag{B.29}$$

The **singular value decomposition** (SVD) of the matrix A is then given by

$$A = U\Sigma V^T \tag{B.30}$$

in which

$$U = [u_1, u_2, \ldots, u_m], \; V = [v_1, v_2, \ldots, v_n] \tag{B.31}$$

are orthogonal matrices, and $\Sigma \in R^{m\times n}$ is given by

$$\Sigma = \left[\begin{array}{ccccc|c} \sigma_1 & & & & & \\ & \sigma_2 & & & & \\ & & \cdot & & & 0 \\ & & & \cdot & & \\ & & & & \sigma_m & \end{array}\right] \tag{B.32}$$

We can compute the SVD of A as follows. We begin by finding the singular values σ_i of A using Equations (B.28) and (B.29). These singular values can then be used to find eigenvectors $u_1, \cdots, u_m$ that satisfy

$$AA^T u_i = \sigma_i^2 u_i \tag{B.33}$$

These eigenvectors comprise the matrix $U = [u_1, u_2, \ldots, u_m]$. The system of equations (B.33) can be written as

$$AA^TU = U\Sigma_m^2 \tag{B.34}$$

where the matrix Σ_m is defined as

$$\Sigma_m = \begin{bmatrix} \sigma_1 & & & & \\ & \sigma_2 & & & \\ & & \cdot & & \\ & & & \cdot & \\ & & & & \sigma_m \end{bmatrix}$$

Now, define

$$V_m = A^TU\Sigma_m^{-1} \tag{B.35}$$

and let V be any orthogonal matrix that satisfies $V = [V_m \mid V_{n-m}]$ (note that here V_{n-m} contains just enough columns so that the matrix V is an $n \times n$ matrix). It is a simple matter to combine the above equations to verify Equation (B.30):

$$\begin{aligned}
U\Sigma V^T &= U\left[\Sigma_m \mid 0\right]\begin{bmatrix} V_m^T \\ V_{n-m}^T \end{bmatrix} && \text{(B.36)}\\
&= U\Sigma_m V_m^T && \text{(B.37)}\\
&= U\Sigma_m \left(A^TU\Sigma_m^{-1}\right)^T && \text{(B.38)}\\
&= U\Sigma_m(\Sigma_m^{-1})^TU^TA && \text{(B.39)}\\
&= U\Sigma_m\Sigma_m^{-1}U^TA && \text{(B.40)}\\
&= UU^TA && \text{(B.41)}\\
&= A && \text{(B.42)}
\end{aligned}$$

Here, Equation (B.36) follows immediately from our construction of the matrices U, V, and Σ_m. Equation (B.38) is obtained by substituting Equation (B.35) into Equation (B.37). Equation (B.40) follows because Σ_m^{-1} is a diagonal matrix, and thus symmetric. Finally, Equation (B.42) is obtained using the fact that $U^T = U^{-1}$, since U is orthogonal.

Appendix C

DYNAMICAL SYSTEMS

Here we give a brief introduction to some concepts in the state space theory of linear and nonlinear systems.

Definition C.1 *A vector field f is a continuous function $f : \mathbb{R}^n \to \mathbb{R}^n$.*

We can think of a differential equation

$$\dot{x}(t) \quad = \quad f(x(t)) \tag{C.1}$$

as being defined by a vector field f on $\mathbb{R}^n$. A solution $t \to x(t)$ of Equation (C.1) with $x(t_0) = x_0$ is then a curve C in $\mathbb{R}^n$, beginning at x_0 parametrized by t, such that at each point of C, the vector field $f(x(t))$ is tangent to C. $\mathbb{R}^n$ is then called the **state space** of the system given by Equation (C.1). For two-dimensional systems, we can represent

$$t \quad \to \quad \begin{bmatrix} x_1(t) \\ x_2(t) \end{bmatrix} \tag{C.2}$$

by a curve C in the plane.

Example C.1

Consider the two-dimensional system

$$\dot{x}_1 = x_2 \qquad x_1(0) = x_{10} \tag{C.3}$$

$$\dot{x}_2 = -x_1 \qquad x_2(0) = x_{20} \tag{C.4}$$

In the plane the solutions of this equation are circles of radius

$$r \quad = \quad x_{10}^2 + x_{20}^2 \tag{C.5}$$

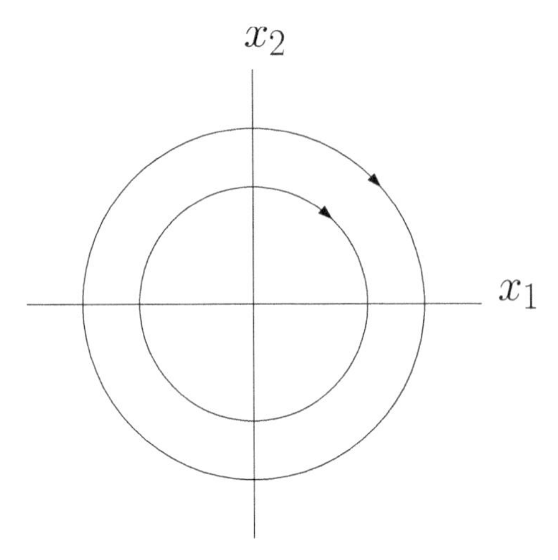

Figure C.1: Phase portrait for Example C.1.

To see this consider the equation

$$x_1^2(t) + x_2^2(t) = r \tag{C.6}$$

Clearly the initial conditions satisfy this equation. If we differentiate Equation (C.6) in the direction of the vector field $f = [x_2, -x_1]^T$ that defines Equations (C.3) and (C.4) we obtain

$$2x_1\dot{x}_1 + 2x_2\dot{x}_2 = 2x_1x_2 - 2x_2x_1 = 0 \tag{C.7}$$

Thus, f is tangent to the circle. The graph of such curves C in the $x_1 - x_2$ plane for different initial conditions are shown in Figure C.1.

◇

The x_1 - x_2 plane is called the **phase plane** and the trajectories of the system given by Equations (C.3) and (C.4) form what is called the **phase portrait**. For linear systems of the form

$$\dot{x} = Ax \tag{C.8}$$

in $\mathbb{R}^2$, the phase portrait is determined by the eigenvalues and eigenvectors of A . For example, consider the system

$$\dot{x}_1 = x_2 \tag{C.9}$$

$$\dot{x}_2 = x_1 \tag{C.10}$$

In this case

$$A = \begin{bmatrix} 0 & 1 \\ 1 & 0 \end{bmatrix} \tag{C.11}$$

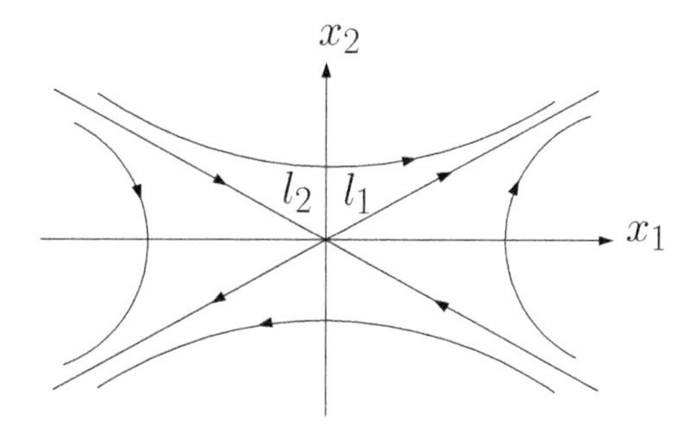

Figure C.2: Phase portrait for Example C.2.

The phase portrait is shown in Figure C.2. The lines ℓ_1 and ℓ_2 are in the direction of the eigenvectors of A and are called **eigensubspaces** of A.

State Space Representation of Linear Systems

Consider a single-input/single-output linear control system with input u and output y of the form

$$a_n \frac{d^n y}{dt^n} + a_{n-1} \frac{d^{n-1} y}{dt^{n-1}} + \cdots + a_1 \frac{dy}{dt} + a_0 y \quad = \quad u \tag{C.12}$$

The characteristic polynomial, whose roots are the open-loop poles, is given as

$$p(s) \quad = \quad a_n s^n + a_{n-1} s^{n-1} + \cdots + a_0 \tag{C.13}$$

For simplicity we suppose that $p(s)$ is monic, that is, $a_n = 1$. The standard way of representing Equation (C.12) in state space is to define n state variables $x_1, x_2, \ldots, x_n$ as

$$\begin{aligned} x_1 &= y \\ x_2 &= \dot{y} = \dot{x}_1 \\ x_3 &= \ddot{y} = \dot{x}_2 \\ &\vdots \\ x_n &= \frac{d^{n-1}y}{dt^{n-1}} = \dot{x}_{n-1} \end{aligned} \tag{C.14}$$

and express Equation (C.12) as the system of first order differential equations

$$\begin{aligned}\dot{x}_1 &= x_2 \\ \dot{x}_2 &= x_3 \\ \dot{x}_{n-1} &= x_n \\ \dot{x}_n &= \frac{d^n y}{dt^n} = -a_0 y - a_1 \frac{dy}{dt} - \cdots - a_{n-1}\frac{d^{n-1}y}{dt^{n-1}} + u \\ &= -a_0 x_1 - a_1 x_2 - \cdots - a_{n-1} x_n + u\end{aligned} \tag{C.15}$$

In matrix form this system of equations is written as

$$\begin{bmatrix} \dot{x}_1 \\ \vdots \\ \dot{x}_n \end{bmatrix} = \begin{bmatrix} 0 & 1 & \cdot & \cdot & 0 \\ 0 & 0 & 1 & \cdot & 0 \\ & & & \cdot & \cdot \\ & & & & 1 \\ -a_0 & \cdot & \cdot & \cdot & -a_{n-1} \end{bmatrix} \begin{bmatrix} x_1 \\ \vdots \\ x_n \end{bmatrix} + \begin{bmatrix} 0 \\ 0 \\ \\ 0 \\ 1 \end{bmatrix} \tag{C.16}$$

or

$$\dot{x} = Ax + bu \ , \ x \in \mathbb{R}^n$$

The output y can be expressed as

$$\begin{aligned} y &= [1, 0, \ldots, 0]x \\ &= c^T x \end{aligned} \tag{C.17}$$

It is easy to show that

$$\det(sI - A) = s^n + a_{n-1}s^{n-1} + \cdots + a_1 s + a_0 \tag{C.18}$$

and so the last row of the matrix A consists of precisely the negative of the coefficients of the characteristic polynomial of the system. Furthermore the eigenvalues of A are the open-loop poles of the system.

In the Laplace domain, the transfer function $\frac{Y(s)}{U(s)}$ is equivalent to

$$\frac{Y(s)}{U(s)} = c^T(sI - A)^{-1}b \tag{C.19}$$

Appendix D

LYAPUNOV STABILITY

We give here some basic definitions of stability and Lyapunov functions and present a sufficient condition for showing stability of a class of nonlinear systems. For simplicity we treat only time-invariant systems. For a more general treatment of the subject the reader is referred to [137].

Definition D.1 *Consider a nonlinear system on* $\mathbb{R}^n$

$$\dot{x} = f(x) \tag{D.1}$$

where $f(x)$ *is a vector field on* $\mathbb{R}^n$, *and suppose that* $f(0) = 0$. *Then the origin in* $\mathbb{R}^n$ *is said to be an* **equilibrium point** *for Equation (D.1).*

If initially the system given by Equation (D.1) satisfies $x(t_0) = 0$, then the function $x(t) \equiv 0$ for $t > t_0$ is a solution of Equation (D.1), called the **null** or **equilibrium** solution. In other words, if the system represented by Equation (D.1) starts initially at the equilibrium, then it remains at the equilibrium thereafter. The question of stability deals with the solutions of Equation (D.1) for initial conditions away from the equilibrium point. Intuitively, the null solution should be called stable if, for initial conditions close to the equilibrium, the solution remains close thereafter in some sense. We can formalize this notion into the following.

Definition D.2 *The null solution* $x(t) = 0$ *is* **stable** *if and only if, for any* $\epsilon > 0$ *there exist* $\delta(\epsilon) > 0$ *such that*

$$\|x(t_0)\| < \delta \textit{ implies } \|x(t)\| < \epsilon \textit{ for all } t > t_0 \tag{D.2}$$

The null solution $x(t) = 0$ *is* **unstable** *if it is not stable.*

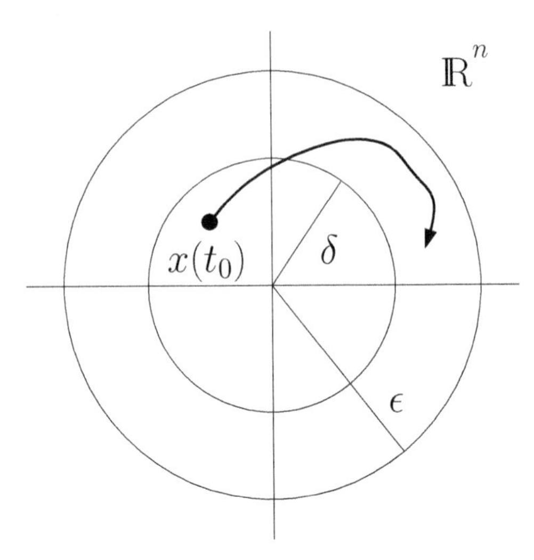

Figure D.1: Illustrating the definition of stability.

This situation is illustrated by Figure D.1 and says that the system is stable if the solution remains within a ball of radius ϵ around the equilibrium, so long as the initial condition lies in a ball of radius δ around the equilibrium. Notice that the required δ will depend on the given ϵ. To put it another way, a system is stable if "small" perturbations in the initial conditions result in "small" perturbations from the null solution.

If the equilibrium is unstable, then, for any neighborhood of the equilibrium point, no matter how small, there always exists at least one initial condition, $x(t_0) = \bar{x}$ in this neighborhood such that the trajectory of (D.1) eventually leaves this neighborhood as $t \to \infty$.

Definition D.3 *The null solution $x(t) = 0$ is asymptotically stable if and only if,*

1. *it is stable, and*

2. *there exists $\delta > 0$ such that*

$$\|x(t_0)\| \;<\; \delta \text{ implies } \|x(t)\| \to 0 \text{ as } t \to \infty. \tag{D.3}$$

In other words, asymptotic stability means that if the system is perturbed away from the equilibrium it will return asymptotically to the equilibrium. The above notions of stability are local in nature, that is, they may hold for initial conditions "sufficiently near" the equilibrium point but may fail for initial conditions farther away from the equilibrium. Stability (respectively, asymptotic stability) is said to be **global** if it holds for arbitrary initial conditions.

For a linear system

$$\dot{x} = Ax \tag{D.4}$$

the null solution is globally asymptotically stable if and only if all eigenvalues of the matrix A lie in the open left half of the complex plane. Such a matrix is called a **Hurwitz matrix**. For nonlinear systems global stability cannot be so easily determined. However, local stability of the null solution of a nonlinear system $\dot{x} = f(x)$ can sometimes be determined by examining the eigenvalues of the Jacobian of the vector field $f(x)$.

Given the system (D.1) and suppose $x = 0$ is an equilibrium point. Let A be the $n \times n$ Jacobian matrix of $f(x)$, evaluated at $x = 0$. In other words

$$A = (a_{ij}), \text{ where } a_{ij} = \frac{\partial f_i}{\partial x_j}|_{x=0} \tag{D.5}$$

The system

$$\dot{x} = Ax \tag{D.6}$$

is called the **linear approximation** about the equilibrium of the nonlinear system (D.1).

Theorem 8

1. *Suppose A in (D.6) is a Hurwitz matrix so that $x = 0$ is a globally asymptotically stable equilibrium point. Then $x = 0$ is locally asymptotically stable for the nonlinear system (D.1).*

2. *Suppose A has one or more unstable eigenvalues, that is, one or more eigenvalues in the open right half plane so that $x = 0$ is an unstable equilibrium point for the linear system (D.6). Then $x = 0$ is unstable for the nonlinear system (D.6).*

3. *Suppose A has no eigenvalues in the open right half plane but one or more eigenvalues on the $j\omega$-axis. Then the stability properties of the equilibrium $x = 0$ for the nonlinear system (D.1) cannot be determined.*

Eigenvalues on the $j\omega$-axis are called **critical eigenvalues**. Examining the eigenvalues of the linear approximation of a nonlinear system in order to determine its stability properties is referred to as **Lyapunov's first method**. We see that local stability of the equilibrium of the nonlinear system (D.1) can be determined provided the matrix A of the linear approximation has no critical eigenvalues. If A has critical eigenvalues, or if one desires to determine the global stability properties of the nonlinear system (D.1) then Lyapunov's first method is inconclusive and other methods must be used. **Lyapunov's second method**, introduced below, addresses these latter issues.

Another important notion related to stability is the notion of **uniform ultimate boundedness** of solutions.

Definition D.4 *A solution $x(t) : [t_0, \infty) \rightarrow \mathbb{R}^n$ for Equation (D.1) with initial condition $x(t_0) = x_0$ is said to be* **uniformly ultimately bounded** *(u.u.b.) with respect to a set S if there is a nonnegative constant $T(x_0, S)$ such that*

$$x(t) \in S \text{ for all } t \geq t_0 + T.$$

Uniform ultimate boundedness says that the solution trajectory of Equation (D.1) beginning at x_0 at time t_0 will ultimately enter and remain within the set S. If the set S is a small region about the equilibrium, then uniform ultimate boundedness is a practical notion of stability that is useful in control system design.

D.1 QUADRATIC FORMS AND LYAPUNOV FUNCTIONS

We next discuss the so-called **second method of Lyapunov** or Lyapunov's second method.

Definition D.5 *Given a symmetric matrix $P = (p_{ij})$ the scalar function*

$$V(x) = x^T P x = \sum_{i,j=1}^{n} p_{ij} x_i x_i \qquad \text{(D.7)}$$

is said to be a **quadratic form**. *$V(x)$, equivalently the quadratic form, is said to be* **positive definite** *if and only if*

$$V(x) > 0 \qquad \text{(D.8)}$$

for $x \neq 0$.

Note that $V(0) = 0$. $V(x)$ will be positive definite if and only if the symmetric matrix P is a positive definite matrix, that is, has all eigenvalues positive.

The level surfaces of V, given as solutions of $V(x) = constant$, are ellipsoids in $\mathbb{R}^n$. A positive definite quadratic form defines a norm on $\mathbb{R}^n$. In fact, given the usual norm $\|x\|$ on R^n, the function V given as

$$V(x) = x^T x = \|x\|^2 \tag{D.9}$$

is a positive definite quadratic form corresponding to the choice $P = I$, the $n \times n$ identity matrix.

Definition D.6 *Let $V(x) : \mathbb{R}^n \to \mathbb{R}$ be a continuous function with continuous first partial derivatives in a neighborhood of the origin in $\mathbb{R}^n$. Furthermore, suppose that V is positive definite, that is, $V(0) = 0$ and $V > 0$ for $x \neq 0$. Then V is called a* **Lyapunov function candidate** *for the system given by Equation (D.1.*

For the most part we will be utilizing Lyapunov function candidates that are quadratic forms, but the power of Lyapunov stability theory comes from the fact that any function may be used in an attempt to show stability of a given system provided it is a Lyapunov function candidate according to the above definition.

By the derivative of V along trajectories of Equation (D.1), or the derivative of V in the direction of the vector field defining Equation (D.1), we mean

$$\dot{V}(t) = \frac{\partial V}{\partial x} f(x) = \frac{\partial V}{\partial x_1} f_1(x) + \cdots + \frac{\partial V}{\partial x_n} f_n(x) \tag{D.10}$$

where

$$\frac{\partial V}{\partial x} = \left[\frac{\partial V}{\partial x_1}, \ldots, \frac{\partial V}{\partial x_n} \right] \tag{D.11}$$

denotes the **gradient** of $V(x)$.

Suppose that we evaluate the Lyapunov function candidate V at points along a solution trajectory $x(t)$ of Equation (D.1) and find that $V(t)$ is decreasing for increasing t. Intuitively, since V acts like a norm, this must mean that the given solution trajectory must be converging toward the origin. This is the idea of Lyapunov stability theory.

D.2 LYAPUNOV STABILITY

Theorem 9 *The null solution of Equation (D.1) is stable if there exists a Lyapunov function candidate V such that $\dot{V}$ is negative semi-definite along solution trajectories of Equation (D.1), that is, if*

$$\dot{V} = \frac{\partial V}{\partial x} f(x) \leq 0 \tag{D.12}$$

The inequality (D.12) says that the derivative of V computed along solutions of Equation (D.1) is nonpositive, which says that V itself is nonincreasing along solutions. Since V is a measure of how far the solution is from the origin, (D.12) says that the solution must remain near the origin. If a Lyapunov function candidate V can be found satisfying (D.12) then V is called a **Lyapunov function** for the system given by Equation (D.1). Note that Theorem 9 gives only a sufficient condition for stability of Equation (D.1). If one is unable to find a Lyapunov function satisfying the inequality (D.12) it does not mean that the system is unstable. However, an easy sufficient condition for instability of Equation (D.1) is for there to exist a Lyapunov function candidate V such that $\dot{V} > 0$ along at least one solution of the system.

Theorem 10 *The null solution of Equation (D.1) is asymptotically stable if there exists a Lyapunov function candidate V such that $\dot{V}$ is strictly negative definite along solutions of Equation (D.1), that is,*

$$\dot{V}(x) \; < \; 0. \tag{D.13}$$

The strict inequality in Equation (D.13) means that V is actually decreasing along solution trajectories of Equation (D.1) and hence, the trajectories must be converging to the equilibrium point.

Corollary D.1 *Let V be a Lyapunov function candidate and let S be any level surface of V, that is,*

$$S(c_0) \; = \; \{x \in \mathbb{R}^n | V(x) = c_0\} \tag{D.14}$$

for some constant $c_0 > 0$. Then a solution $x(t)$ of Equation (D.1) is uniformly ultimately bounded with respect to S if

$$\dot{V} \; = \; \frac{\partial V}{\partial x} f(x) < 0 \tag{D.15}$$

for x outside of S.

If $\dot{V}$ is negative outside of S then the solution trajectory outside of S must be pointing toward S. Once the trajectory reaches S we may or may not be able to draw further conclusions about the system, except that the trajectory is trapped inside S.

D.3 GLOBAL AND EXPONENTIAL STABILITY

The condition $\dot{V} < 0$ along solution trajectories of a given system guarantees only local asymptotic stability even if the condition holds globally. In order to show global (asymptotic) stability the Lyapunov function V must satisfy an additional condition, known as **radial unboundedness.**

Definition D.7 *Suppose $V : \mathbb{R}^n \to \mathbb{R}$ be a continuously differentiable function. $V(x)$ is said to be* **radially unbounded** *if*

$$V(x) \to \infty \quad \text{as} \quad \|x\| \to \infty$$

With this additional property for V we can state

Theorem 11 *Let $V : \mathbb{R}^n \to \mathbb{R}$ be Lyapunov function candidate for the system given by Equation (D.1) and suppose that V is radially unbounded. Then $\dot{V} < 0$ implies that $x = 0$ is globally asymptotically stable.*

A stronger notion than asymptotic stability is that of **exponential stability.**

Definition D.8 *The equilibrium $x = 0$ of the system given by Equation (D.1) is* **exponentially stable** *if there are positive, real constants α and γ such that*

$$\|x(t)\| \leq \alpha \|x(0)\| e^{-\lambda t} \qquad \text{for all } t > 0 \tag{D.16}$$

The exponential stability is local or global depending on whether or not the inequality (D.16) holds for all initial conditions $x(0) \in \mathbb{R}^n$.

A sufficient condition for exponential stability is the following

Theorem 12 *Suppose that V is a Lyapunov function candidate for the system given by Equation (D.1) such that*

$$\begin{aligned} K_1\|x\|^p \leq V(x) \leq K_2\|x\|^p \\ \dot{V} \leq -K_3\|x\|^p \end{aligned} \tag{D.17}$$

where K_1, K_2, K_3, and p are positive constants. Then the origin $x = 0$ is exponentially stable. Moreover, if the inequalities (D.17) hold globally, then $x = 0$ is globally exponentially stable.

D.4 LYAPUNOV STABILITY FOR LINEAR SYSTEMS

Consider the linear system given by Equation (D.4) and let

$$V(x) = x^T P x \tag{D.18}$$

be a Lyapunov function candidate, where P is symmetric and positive definite. Computing $\dot{V}$ along solutions of Equation (D.4) yields

$$\begin{aligned} \dot{V} &= \dot{x}^T P x + x^T P \dot{x} \\ &= x^T (A^T P + PA) x \\ &= -x^T Q x \end{aligned} \tag{D.19}$$

where we have defined Q as

$$A^T P + PA = -Q \tag{D.20}$$

Theorem 9 says that if Q given by Equation (D.20) is positive definite (it is automatically symmetric since P is), then the linear system given by Equation (D.4) is stable. One approach that we can take is to first fix Q to be symmetric, positive definite and solve Equation (D.20), which is called the **matrix Lyapunov equation**, for P. If a symmetric positive definite solution P can be found to this equation, then Equation (D.4) is stable and $x^T P x$ is a Lyapunov function for the linear system (D.4). The converse to this statement also holds. In fact, we can summarize these statements as

Theorem 13 *Given an $n \times n$ matrix A, then all eigenvalues of A have negative real part if and only if for every symmetric positive definite $n \times n$ matrix Q, Equation (D.20) has a unique positive definite solution P.*

Thus, we can reduce the determination of stability of a linear system to the solution of a system of linear equations.

D.5 LASALLE'S THEOREM

The strict inequality in (D.13) showing asymptotic stability may be difficult to obtain for a given system and Lyapunov function candidate. We therefore discuss **LaSalle's invariance principle** in this section, which can be used to prove asymptotic stability even when V is only negative semi-definite.

The main difficulty in the use of Lyapunov stability theory is finding suitable Lyapunov functions satisfying $\dot{V} < 0$ in order to prove asymptotic stability. LaSalle's invariance principle), or LaSalle's theorem gives us a tool to determine the asymptotic properties of a system in the weaker case that V is only negative semidefinite, that is, when $\dot{V} \leq 0$.

The version of LaSalle's Theorem here follows closely the development in [63]. Consider the nonlinear system

$$\dot{x} = f(x) \quad x \in \mathbb{R}^n \tag{D.21}$$

where f is a smooth vector field on $\mathbb{R}^n$ with $f(0) = 0$.

Definition D.9 Positively Invariant Set

A set M is **positively invariant** *with respect to the system (D.21) if*

$$x(0) \in M \implies x(t) \in M, \text{ for all } t > 0$$

Theorem 14 LaSalle's Theorem

Let D be a region in $\mathbb{R}^n$ and let $\Omega \subset D$ be a compact set that is positively invariant with respect to the nonlinear system (D.21). Let $V :\longrightarrow \mathbb{R}$ be a continuously differentiable function such that $\dot{V} \leq 0$ in Ω. Let E be the set of all points in Ω where $\dot{V} = 0$. Let M be the largest invariant set in E. Then every solution starting in Ω approaches M as $t \to \infty$.

As a corollary to LaSalle's Theorem it follows that the equilibrium solution $x = 0$ of Equation (D.21) is asymptotically stable if V does not vanish identically along any solution of Equation (D.21) other than the null solution, that is, if the only solution of Equation (D.21) satisfying

$$\dot{V} \equiv 0 \tag{D.22}$$

is the null solution.

Bibliography

[1] C. Abdallah, D.M. Dawson, P. Dorato, and M. Jamshidi. A survey of robust control of rigid robots. *IEEE Control Systems Magazine*, 11(2):24–30, February 1991.

[2] R. Abraham and J. E. Marsden. *Foundations of Mechanics.* The Benjamin/Cummings Pub. Co., Inc., London, 1978.

[3] R. Anderson and M.W. Spong. Hybrid impedance control of robot manipulators. *IEEE Trans. on Robotics and Automation*, 1988.

[4] Russell L. Anderson. Dynamic sensing in a ping-pong playing robot. *IEEE Transaction on Robotics and Automation*, 5(6):723–739, 1989.

[5] R. L. Andersson. *A Robot Ping-Pong Player. Experiment in Real-Time Intelligent Control.* MIT Press, Cambridge, MA, 1988.

[6] H. Asada and J.A. Cro-Granito. Kinematic and static characterization of wrist joints and their optimal design. In *Proc. IEEE Conf. on Robotics and Automation,*, St. Louis, MO, 1985.

[7] K.J. Åstrom and Tore Hagglund. *PID Controllers: Theory, Design, and Tuning.* Instrument Society of America, 1995.

[8] S. Barnett. *Matrix Methods for Engineers and Scientists.* McGraw-Hill, London, 1979.

[9] Jerome Barraquand and Jean-Claude Latombe. Robot motion planning: A distributed representation approach. *International Journal of Robotics Research*, 10(6):628–649, December 1991.

[10] H. Berghuis and H. Nijmeijer. A passivity approach to controller-observer design for robots. *IEEE Trans. on Robotics and Automation*, 9:740–754, 1993.

[11] D. Bertsekas. *Nonlinear Programming.* Athena Scientific, Belmont, MA, second edition, 1999.

[12] William M. Boothby. *An Introduction to Differentiable Manifolds and Riemannian Geometry.* Academic Press, 1986.

[13] O. Botema and B. Roth. *Theoretical Kinematics.* North Holland, Amsterdam, 1979.

[14] B. Brogliato, I. D. Landau, and R. Lozano. Adaptive motion control of robot manipulators: A unified approach based on passivity. *Int. J. of Robust and Nonlinear Control*, 1:187–202, 1991.

[15] R. Brooks and T. Lozano-Perez. A subdivision algorithm in configuration space for findpath with rotation. In *Proc. Int. Joint Conf. on Art. Intell.*, pages 799–806, 1983.

[16] J. F. Canny. *The Complexity of Robot Motion Planning.* MIT Press, Cambridge, MA, 1988.

[17] F. Chaumette. Potential problems of stability and convergence in image-based and position-based visual servoing. In D. Kriegman, G. Hager, and S. Morse, editors, *The confluence of vision and control*, volume 237 of *Lecture Notes in Control and Information Sciences*, pages 66–78. Springer-Verlag, 1998.

[18] H. Choset, K. M. Lynch, S. Hutchinson, G. Kantor, W. Burgard, L. E. Kavraki, and S. Thrun. *Principles of Robot Motion: Theory, Algorithms, and Implementations.* MIT Press, Cambridge, MA, 2005.

[19] J. C. Colson and N. D. Perreira. Kinematic arrangements used in industrial robots. In *Proc. 13th International Symposium on Industrial Robots*, 1983.

[20] P. I. Corke and S. A. Hutchinson. A new partitioned approach to image-based visual servo control. *IEEE Trans. on Robotics and Automation*, 17(4):507–515, August 2001.

[21] M. Corless and G. Leitmann. Continuous state feedback guaranteeing uniform ultimate boundedness for uncertain dynamic systems. *IEEE Transactions on Automatic Control*, 26:1139–1144, 1981.

[22] J.J. Craig. *Adaptive Control of Mechanical Manipulators.* Addison-Wesley, Reading, MA, 1988.

[23] M. L. Curtiss. *Matrix Groups*. Springer-Verlag, New York, NY, second edition, 1984.

[24] V. S. Cvetković and M. Vukobratović. One robust, dynamic control algorithm for manipulation systems. *International Journal of Robotics Research*, 1(4):15–28, winter 1982.

[25] C. Canudas de Wit et.al. *Theory of Robot Control.* Springer Verlag, Berlin, 1996.

[26] K. Deguchi. Optimal motion control for image-based visual servoing by decoupling translation and rotation. In *Proc. Int. Conf. Intelligent Robots and Systems*, pages 705–711, October 1998.

[27] J. Denavit and R. S. Hartenberg. A kinematic notation for lower pair mechanisms. *Applied Mechanics*, 22:215–221, 1955.

[28] J. C. Doyle, B. A. Francis, and A. R. Tannenbaum. *Feedback Control Theory.* Macmillan Publishing Company, New York, NY, 1992.

[29] J. Duffy. *Analysis of Mechanisms and Robot Manipulators.* John Wiley and Sons, Inc., New York, NY, 1980.

[30] J. Duffy. The fallacy of modern hybrid control theory that is based on orthogonal complements of twist and wrench spaces. *J. Robot Syst.*, 7:139–144, 1990.

[31] B. Espiau, F. Chaumette, and P. Rives. A New Approach to Visual Servoing in Robotics. *IEEE Transactions on Robotics and Automation*, 8:313–326, 1992.

[32] S. E. Fahlman. A planning system for robot construction tasks. *Artificial Intelligence*, 5:1–49, 1974.

[33] O.D. Faugeras. *Three-Dimensional Computer Vision.* MIT Press, Cambridge, MA, 1993.

[34] J. T. Feddema, C. S. George Lee, and O. R. Mitchell. Weighted Selection of Image Features for Resolved Rate Visual Feedback Control. *IEEE Transactions on Robotics and Automation*, 7:31–47, 1991.

[35] J.T. Feddema and O.R. Mitchell. Vision-guided servoing with feature-based trajectory generation. *IEEE Trans. on Robotics and Automation*, 5(5):691–700, October 1989.

[36] R. Fikes and N. Nilsson. STRIPS: A new approach to the eapplication of theorem proving to problem solving. *Artificial Intelligence*, 2:189–208, 1971.

[37] A.F. Filippov. *Differential Equations with Discontinuous Right Hand Sides.* Kluwer, Dordrecht, 1988.

[38] D. Forsyth and J. Ponce. *Computer Vision: A Modern Approach.* Prentice Hall, Upper Saddle River, NJ, 2003.

[39] G. F. Franklin, J. D. Powell, and M. L. Workman. *Digital Control of Dynamic Systems.* Addison-Wesley, 2nd edition, 1990.

[40] S. H. Friedberg, A. J. Insel, and L. E. Spence. *Linear Algebra.* Prentice-Hall, Englewood Cliffs, NJ, 1979.

[41] K.S. Fu, R. C. Gonzalez, and C.S.G. Lee. *Robotics: Control Sensing, Vision, and Intelligence.* McGraw-Hill, St Louis, MO, 1987.

[42] M. Gautier and W. Khalil. On the identification of the inertial parameters of robots. In *IEEE Conf. on Decision and Control*, pages 2264–2269, 1988.

[43] M. Gautier and W. Khalil. Direct calculation of minimum set of inertial parameters of serial robots. *IEEE Trans. on Robotics and Automation*, 6:368–373, 1990.

[44] S.S. Ge, T.H. Lee, and C.J. Harris. *Adaptive Neural Network Control of Robotic Manipulators.* World Scientific, Singapore, 1998.

[45] A. A. Goldenberg, B. Benhabib, and R. G. Fenton. A complete generalized solution to the inverse kinematics of robots. *IEEE J. Robotics and Automation*, RA-1(1):14–20, 1985.

[46] H. Goldstein. *Classical Mechanics.* Addison-Wesley, Reading, MA, 1974.

[47] G.H. Golub and C.F. Van Loan. *Matrix Computations.* The Johns Hopkins University Press, 1983.

[48] Irving M. Gottlieb. *Electric Motors and Control Techniques.* TAB Books, New York, NY, 1994.

[49] W. M. Grimm. Robustness analysis of nonlinear decoupling for elastic-joint robots. *IEEE Trans. on Robotics and Automation*, 6:373–377, 1990.

[50] R. M. Haralick and L. G. Shapiro. *Computer and Robot Vision, Vols. I and II.* Addison-Wesley Pub. Co., Inc., Reading, MA, 1993.

[51] N. Hogan. Impedance control: An approach to manipulation: Parts I–III. *ASME J. of Dynamic Systems, Measurement, and Control*, 107:1–24, 1985.

[52] J.M. Hollerbach. A recursive formulation of Lagrangian manipulator dynamics and a comparative study of dynamics formulation complexity. *IEEE Trans. on Systems, Man, and Cybernetics*, SMC-10(11):730–736, Nov 1980.

[53] J.M. Hollerbach and S. Gideon. Wrist-partitioned inverse kinematic accelerations and manipulator dynamics. *International Journal of Robotics Research*, 4:61–76, 1983.

[54] B. K. P. Horn. *Robot Vision.* MIT Press, Cambridge, 1986.

[55] R. Horowitz and M. Tomizuka. An adaptive control scheme for mechanical manipulators - compensation of nonlinearities and decoupling control. *ASME Journal of Dynamic Systems, Meas. and Control*, 108:127–135, 1986.

[56] A. Isidori. *Nonlinear Control Systems.* Springer-Verlag, New York, NY, 1999.

[57] J.J. Uicker Jr., J. Denavit, and R. S. Hartenberg. An iterative method for the displacement analysis of spatial mechanisms. *Trans. Applied Mechanics*, 31 Series E:309–314, 1964.

[58] T. Kailath. *Linear Systems.* Prentice-Hall, Englewood Cliffs, NJ, 1980.

[59] R. Kalman. Contributions to the theory of optimal control. *Boletin de la Sociedad Matemática Mexicana*, 5:102–119, 1960.

[60] Subbarao Kambhampati and Larry S. Davis. Multiresolution path planning for mobile robots. *IEEE Journal of Robotics and Automation*, 2(3):135–145, September 1986.

[61] Lydia E. Kavraki. *Random Networks in Configuration Space for Fast Path Planning.* PhD thesis, Stanford University, Stanford, CA, 1994.

[62] Lydia E. Kavraki, Petr Švestka, Jean-Claude Latombe, and Mark H. Overmars. Probabilistic roadmaps for path planning in high-dimensional configuration spaces. *IEEE Trans. on Robotics and Automation*, 12(4):566–580, August 1996.

[63] H.K. Khalil. *Nonlinear Systems, 2nd Ed.* Prentic-Hall, Englewood Cliffs, NJ, 1996.

[64] O. Khatib. Real-time obstacle avoidance for manipulators and mobile robots. *International Journal of Robotics Research*, 5(1):90–98, 1986.

[65] D.E. Koditschek. The application of total energy as a Lyapunov function for mechanical control systems. In et.al. J.E. Marsden, editor, *Dynamics and Control of Multibody Systems*, volume 97, pages 131–157. AMS, 1989.

[66] D.E. Koditschek. Robot planning and control via potential functions. In *The Robotics Review 1*, pages 349–367. MIT Press, 1989.

[67] A.J. Krener and I. Isidori. Linearization by output injection and nonlinear observers. *Syst. Control Lett.*, 3:47–52, 1983.

[68] A.J. Krener and W. Respondek. Nonlinear observers with linearizable error dynamics. *SIAM J. Control and Optimization*, 23(2):197–216, 1985.

[69] K. Kreutz. On manipulator control by exact linearization. *IEEE Transactions on Automatic Control*, 34(7):763–767, July 1989.

[70] B.C. Kuo. *Control Systems.* Prentice Hall, 1982.

[71] C. Lanczos. *The Variational Principles of Mechanics.* University of Toronto Press, Toronto, CA, 4th edition, 1970.

[72] I.D. Landau and R. Horowitz. Synthesis of adaptive controllers for robotic manipulators using a passive feedback systems approach. *Int. J. of Adaptive Control and Signal Processing*, 3:23–38, 1989.

[73] J. C. Latombe. *Robot Motion Planning.* Kluwer Academic Publishers, Boston, MA, 1991.

[74] C.S.G. Lee. Robot arm kinematics, dynamics, and control. *Computer*, 15(12):62–80, 1982.

[75] C.S.G. Lee, R. C. Gonzales, and K. S. Fu. *Tutorial on Robotics.* IEEE Computer Society Press, Silver Spring, MD, 1983.

[76] C.S.G. Lee and M. Ziegler. A geometric approach in solving the inverse kinematics of PUMA robots. *IEEE Trans. Aero. and Elect. Sys.*, AES-20(6):695–706, 1984.

[77] F.L. Lewis, S. Jagannathan, and A. Yesildirek. *Neural Network Control of Robot Manipulators and Nonlinear Systems.* Taylor and Francis, London, 1999.

[78] Ming Lin and Dinesh Manocha. Efficient contact determination in dynamic environments. *International Journal of Computational Geometry and Applications*, 7(1):123–151, 1997.

[79] T. Lozano-Perez. Spatial planning: A configuration space approach. *IEEE Transactions on Computers*, February 1983.

[80] A. De Luca. Dynamic control of robots with joint elasticity. In *IEEE ROBOT'88*, pages 152–158, Philadelphia, PA, April 1988.

[81] D.G. Luenberger. Observing the state of a linear system. *IEEE Trans. Mil. Electronics*, MIL-8:74–80, 1964.

[82] J.Y.S. Luh, M. W. Walker, and R.P.C. Paul. On-line computational scheme for mechanical manipulators. *ASME J. of Dynamic Systems*, 102:69–76, 1980.

[83] Yi Ma, Stefano Soatto, Jana Kosecka, and Shankar Sastry. *An Invitation to 3-D Vision: From Images to Geometric Models.* Springer-Verlag, New York, NY, 2003.

[84] E. Malis, F. Chaumette, and S. Boudet. 2-1/2-D visual servoing. *IEEE Trans. on Robotics and Automation*, 15(2):238–250, April 1999.

[85] R. Marino and M.W. Spong. Nonlinear control techniques for flexible joint manipulators: A single link case study. In *Proceedings of IEEE Conference on Robotics and Automation*, pages 1030–1026, San Francisco, CA, 1986.

[86] B.R. Markiewicz. Analysis of the computed torque drive method and comparison with conventional position servo for a computer-controlled manipulator. Technical Report TM 33-601, Jet Propulsion Laboratory, Pasadena, CA, March 1973.

[87] D. Marr. *Vision.* Freeman, San Francisco, CA, 1982.

[88] J. E. Marsden and T. S. Ratiu. *Introduction to Mechanics and Symmetry.* Springer Verlag, New York, NY, second edition, 1999.

[89] M.T. Mason. Compliance and force control for computer controlled manipulators. *IEEE Trans. on Systems, Man, and Cybernetics*, 14:418–432, 1981.

[90] R.H. Middleton and G. Goodwin. Adaptive computed torque control of robot manipulators. *Systems and Control Letters*, 1987.

[91] Brian Mirtich. V-Clip: Fast and robust polyhedral collision detection. Technical Report TR97-05, Mitsubishi Electric Research Laboratory, 201 Broadway, Cambridge, MA 02139, June 1997.

[92] G. Morel, T. Liebezeit, J. Szewczyk, S. Boudet, and J. Pot. Explicit incorporation of 2D constraints in vision based control of robot manipulators. In Peter Corke and James Trevelyan, editors, *Experimental Robotics VI*, volume 250 of *Lecture Notes in Control and Information Sciences*, pages 99–108. Springer-Verlag, 2000. ISBN: 1 85233 210 7.

[93] R.M. Murray, Z. Li, and S.S. Sastry. *A Mathematical Introduction to Robotics.* CRC Press, Boca Raton, FL, 1994.

[94] B. Nelson and P. K. Khosla. Integrating sensor placement and visual tracking strategies. In *Proceedings of IEEE Conference on Robotics and Automation*, pages 1351–1356, 1994.

[95] B. J. Nelson and P. K. Khosla. The resolvability ellipsoid for visual servoing. In *Proc. IEEE Computer Society Conference on Computer Vision and Pattern Recognition*, pages 829–832, 1994.

[96] J. Nethery and M.W. Spong. Robotica. *IEEE Robotics Magazine*, 1(1), 1995.

[97] S. Nicosia, F. Nicolo, and D. Lentini. Dynamical control of industrial robots with elastic and dissipative joints. In *Proceedings of the IFAC 8th Triennial World Congress*, pages 1933–1939, Kyoto, Japan, 1981.

[98] N. Nilsson. A mobile automaton: An application of artificial intellgience techniques. In *Proc. Int. Joint Conf. on Art. Intell.*, 1969.

[99] R. Ortega and M.W. Spong. Adaptive control of robot manipulators: A tutorial. *Automatica*, 25(6):877–888, 1989.

[100] Mark H. Overmars and Petr Švestka. A probabilistic learning approach to motion planning. In *Proceedings of Workshop on Algorithmic Foundations of Robotics*, pages 19–37, 1994.

[101] K. Passino and S. Yurkovich. *Fuzzy Control.* Addison Wesley, Menlo Park, CA, 1998.

[102] R. Paul. *Robot Manipulators: Mathematics, Programming and Control.* MIT Press, Cambridge, MA, 1982.

[103] R. P. Paul, B. E. Shimano, and G. Mayer. Kinematic control equations for simple manipulators. *IEEE Trans. Systems, Man., and Cybernetics*, SMC-ll(6):339–455, 1981.

[104] R.P. Paul. Modeling, trajectory calculation, and servoing of a computer controlled arm. Technical Report AIM 177, Stanford University Artificial Intelligence Laboratory, Palo Alto, CA, Nov 1972.

[105] D. L. Pieper. *The Kinematics of Manipulators under Computer Control.* PhD thesis, Stanford University, 1968.

[106] W. Press, B.Flannery, S. Teukolsky, and W. Vetterling. *Numerical Recipes in C.* Cambridge University Press, 1988.

[107] M.H. Raibert and J.J. Craig. Hybrid position/force control of manipulators. *ASME*, 102:126–133, 1981.

[108] J. N. Reddy and M.L. Rasmussen. *Advanced Engineering Analysis.* John Wiley and Sons, Inc., New York, NY, 1982.

[109] Azriel Rosenfeld and Avi Kak. *Digital Picture Processing.* Academic Press, New York, NY, 1982.

[110] C. Samson, M. Le Borgne, and B. Espiau. *Robot Control: The Task Function Approach.* Clarendon Press, Oxford, England, 1992.

[111] A. C. Sanderson, L. E. Weiss, and C. P. Neuman. Dynamic sensor-based control of robots with visual feedback. *IEEE Trans. on Robotics and Automation*, RA-3(5):404–417, October 1987.

[112] J. T. Schwartz, M. Sharir, and J. Hopcroft, editors. *Planning, Geometry, and Complexity of Robot Motion.* Ablex, Norwood, NJ, 1987.

[113] M. Shahinpoor. The exact inverse kinematics solutions for the Rhino XR-2 robot. *Robotics Age*, 7(8):6–14, 1985.

[114] R. Sharma and S. Hutchinson. Motion perceptibility and its application to active vision-based servo control. *IEEE Trans. on Robotics and Automation*, 13(4):607–617, August 1997.

[115] R. Sharma and S. A. Hutchinson. On the observability of robot motion under active camera control. In *Proceedings of IEEE Conference on Robotics and Automation*, pages 162–167, May 1994.

[116] D.B. Silver. On the equivalence of lagrangian and Newton-Euler dynamics for manipulators. *International Journal of Robotics Research*, 1(2), 1982.

[117] J.-J. E. Slotine and W. Li. Adaptive manipulator control: A case study. *IEEE Trans. on Automatic Control*, 33:995–1003, 1988.

[118] J.J.-E. Slotine. The robust control of robot manipulators. *Int. J. Robotics Research*, 4(2):49–64, 1985.

[119] I. S. Sokolnikoff and R. M. Redheffer. *Mathematical Methods of Physics and Modern Engineering*. McGraw-Hill, New York, NY, 1958.

[120] M. Spivak. *A comprehensive introduction to differential geometry*. Publish or Perish, Inc., Berkeley, CA, second edition, 1979.

[121] M. W. Spong, R. Ortega, and R. Kelly. Comments on 'adaptive manipulator control: A case study'. *IEEE Trans. on Automatic Control*, 35:761–762, 1990.

[122] M.W. Spong. Modeling and control of elastic joint robots. *Transactions of the ASME, J. Dynamic Systems, Measurement and Control*, 109:310–319, December 1987.

[123] M.W. Spong. On the robust control of robot manipulators. *IEEE Transactions on Automatic Control*, AC-37(11):1782–1786, November 1992.

[124] M.W. Spong, F.L. Lewis, and C.T. Abdallah. *Robot Control: Dynamics, Motion Planning, and Analysis*. IEEE Press, 1992.

[125] M.W. Spong, J.S. Thorp, and J. Kleinwaks. The control of robot manipulators with bounded input part II: robustness and disturbance rejection. In *IEEE Conf. on Decision and Control*, Las Vegas, December 1984.

[126] M.W. Spong, J.S. Thorp, and J. Kleinwaks. The control of robot manipulators with bounded input. *IEEE Transactions on Automatic Control*, 31(3):483–490, March 1986.

[127] M.W. Spong and M. Vidyasagar. Robust linear compensator design for nonlinear robotic control. *IEEE Transactions on Robotics and Automation*, RA-3(4):345–351, August 1987.

[128] C.H. Su and C. W. Radcliffe. *Kinematics and Mechanisms Design.* John Wiley and Sons, Inc., New York, NY, 1978.

[129] R.J. Su. On the linear equivalents of nonlinear systems. *Syst. Control Lett.*, 2, 1981.

[130] L. Sweet and M.C. Good. Re-definition of the robot motion control problem: Effects of plant dynamics, drive system constraints, and user requirements. In *Proc. 23rd IEEE Conf. on Decision and Control*, pages 724–731, Las Vegas, December 1984.

[131] L. M. Sweet and M. C. Good. Redefinition of the robot motion control problem. *IEEE Control Systems Magazine*, 5(3):18–24, 1985.

[132] M. Takegaki and S. Arimoto. A new feedback method for dynamic control of manipulators. *Journal of Dynamic Systems, Measurement, and Control*, 102:119–125, 1981.

[133] E. Trucco and A. Verri. *Introductory Techniques for 3-D Computer Vision.* Prentice Hall, Upper Saddle River, NJ, 1998.

[134] L. Tsai and A. Morgan. Solving the kinematics of the most general six- and five-degree-of-freedom manipulators by continuation methods. In *Proc. ASME Mechanisms Conference*, Boston, October 1984.

[135] Gino van den Bergen. A fast and robust GJK implementation for collision detection of convex objects. *Journal of Graphics Tools*, 4(2):7–25, 1999.

[136] Gino van den Bergen. *User's Guide to the SOLID Interference Detection Library.* Eindhoven University of Technology, Eindhoven, The Netherlands, 1999.

[137] M. Vidyasagar. *Nonlinear Systems Analysis, 2nd Ed.* Prentice Hall, Englewood Cliffs, NJ, 1993.

[138] D. E. Whitney. The mathematics of coordinated control of prosthetic arms and manipulators. *J. Dyn. Sys., Meas. Cont.*, December 1972.

[139] E. T. Whittaker. *Dynamics of Particles and Rigid Bodies.* Cambridge University Press, London, 1904.

[140] W. Wolovich. *Robotics: Basic Analysis and Design.* Holt, Rinehart, and Winston, New York, NY, 1985.

[141] W. Wonham. On pole assignment in multi-input controllable linear systems. *IEEE Trans. Aut. Cont*, 12(6):660 – 665, December 1967.

[142] T. Yoshikawa. Manipulability of robotic mechanisms. *International Journal of Robotics Research*, 4(2):3–9, 1985.

Index